THE GEOENGINEERED TRANSHUMAN

"When we are struck by the evil of the Wetiko mind-virus, we lose our humanity and become like robots and automatons. In *The Geoengineered Transhuman*, Elana Freeland lays bare for all to see how this process is invisibly playing out all around—and inside—of us. I am filled with gratitude for Freeland, for she has written an amazing book, one of the most important books of our time."

Paul Levy, author of *Wetiko* and *The Quantum Revelation*

"Elana Freeland's *The Geoengineered Transhuman* is a must-read for every human being alive. She explains like no other the interconnectedness of technocratic transhumanism, geoengineering, synthetic biology, 5G, and secret government programs. She is a pioneer who has been sounding the alarm for decades. This fourth book of hers is just as needed as her other masterpieces in these challenging times of the self-assembly nanotechnology attack on the human species and all life."

Ana Maria Mihalcea, M.D., Ph.D., author of *Light Medicine*

"Arguably one of the most important issues of our time is the encroaching transhumanist agenda to remake the world and humanity through strategies such as geoengineering and synthetic engineering. I know of no other book as deeply researched on this topic as *The Geoengineered Transhuman*. Freeland not only explores the Deep State's attempt to achieve full-spectrum dominance over planet Earth, she also reveals the insidious attempt to hijack the trajectory of human biological evolution. Luckily for us, Freeland comprehends the spiritual dimension of the forces that humanity is up against and offers the reader ways to meaningfully participate through the soul of consciousness. Read this book—and be enlightened!"

Kingsley L. Dennis, author of
The Struggle for Your Mind

"*The Geoengineered Transhuman* totally exposes the transhumanist agenda, a nefarious plan to subjugate the human race that is now becoming visible. The stronghearted among us must spread the word before this transformation takes over the human soul. Elana Freeland's book is truthful and deeply painful to read. For example, weather modification and the militarization of space are heating our planet and damaging our health. Our taxes fund this secret plan, yet who ever agreed to this? We need to consider the hard truths in this excellent and comprehensive work. She believes we must see the whole spectrum to save our planet, and I agree! To avoid becoming an AI robot afflicted with synthetic biology, read this book!"

Barbara Hand Clow, author of
Awakening the Planetary Mind

"Elana Freeland is a dedicated and relentless researcher who serves us all. I have worked with her for several decades, and together we have witnessed the battlefield against humanity unfold. There is a time to consider and ponder and debate, but there is also a time to understand and comprehend. I recommend that we return Elana's effort, read this book, and understand the reality of our times without delay. Elana continues to help us tirelessly, and now it is time for your gauntlet."

Clifford E. Carnicom, president of
the Carnicom Institute

"Freeland bravely exposes the hidden dangers of our love affair with technology."

Susan B. Martinez, Ph.D., author of
Delusions in Science and Spirituality

THE GEOENGINEERED TRANSHUMAN

The Hidden Technologies of HAARP, Chemtrails, 5G/6G, Nanotechnology, Synthetic Biology, and the Scientific Effort to Transform Humanity

A Sacred Planet Book

ELANA FREELAND

Bear & Company
Rochester, Vermont

Bear & Company
One Park Street
Rochester, Vermont 05767
www.BearandCompanyBooks.com

Text stock is SFI certified

Bear & Company is a division of Inner Traditions International

Sacred Planet Books are curated by Richard Grossinger, Inner Traditions editorial board member and cofounder and former publisher of North Atlantic Books. The Sacred Planet collection, published under the umbrella of the Inner Traditions family of imprints, includes works on the themes of consciousness, cosmology, alternative medicine, dreams, climate, permaculture, alchemy, shamanic studies, oracles, astrology, crystals, hyperobjects, locutions, and subtle bodies.

An earlier version of this book was self-published in 2021 under the title *Geoengineered Transhumanism: How the Environment Has Been Weaponized by Chemicals, Electromagnetics, & Nanotechnology for Synthetic Biology.*

Cataloging-in-Publication Data for this title is available from the Library of Congress

ISBN 978-1-59143-512-9 (print)
ISBN 978-1-59143-513-6 (ebook)

Printed and bound in the United States by Lake Book Manufacturing, LLC
The text stock is SFI certified. The Sustainable Forestry Initiative® program promotes sustainable forest management.

10 9 8 7 6 5 4 3 2 1

Text design and layout by Debbie Glogover
This book was typeset in Garamond Premier Pro with Gill Sans MT Pro and Nobel used as display fonts

To send correspondence to the author of this book, mail a first-class letter to the author c/o Inner Traditions • Bear & Company, One Park Street, Rochester, VT 05767, and we will forward the communication, or contact the author directly at **https://www.elanafreeland.com**.

To Billy "The HAARP Man" Hayes (1949–2022), who endured the CIA and its MK-Ultra mind-control program, dedicating his life to exposing the secret space program's geoengineering of our world, which encompasses chemistry, electromagnetism, nanotechnology, and synthetic biology, to covertly move us into Transhumanism. Without Billy, I would not have kept going to uncover the extent of the HAARP/Space Fence lockdown in its totality.

The forging together of human nature with the nature of the machine will be a significant problem for the rest of the Earth's evolution.

RUDOLF STEINER

Contents

Introduction

> *It is not rational to assume that wars are always fought only the way we know them . . . We must permit ourselves to think of war as a silent, unnoticed, slow working but deliberate destruction of life on a planet, a satellite, or even a star. We must be ready to change every single view we have held sacred if it contradicts new facts and new experiences. War may be going on right now with no one being aware of it, with men dying, with trees bending like rubber hoses, green pastures turning into dust bowls, and with academic and civil institutions explaining it all away as with "just this" or "just that." In short, it may turn out correct what one would otherwise feel inclined to ascribe to a schizophrenic mind—namely, that instead of shooting at the victims of war with bullets, one could very well sap life energy out of the war victims with machines which operate according to the orgonomic potential of the Cosmic Energy.*
>
> Wilhelm Reich, MD (1897–1957)

Why devote years of research and writing about geoengineering and related subjects—chemtrails, ionospheric heaters, weather modification, weather warfare, climate change—topics that the mainstream consistently lies to us about? Why persist in getting information about this cloaked, classified technology out to people when elected officials, scientific experts, and mainstream media seem bent on keeping us ignorant of these covert programs?

The above quote by Wilhelm Reich, MD, an outspoken critic of all forms of institutionalized authority and a vociferous advocate of individual freedom, reflects my own thinking on this matter. Yes, the Cold War that Reich was referencing supposedly ended in 1991, but the National Security Act of 1947, which among other things established the CIA, certainly didn't end, with Russia still required to play the part of the villain in cover stories, and "national security" still demanding that our government keep things hidden from us "for our own good," including the historical truth about covert programs like Operation Paperclip, the secret CIA transfer of Nazi Third Reich scientists, doctors, and technicians—many accused of war crimes, mass murder, and slavery—to the United States in one of the most complex, nefarious, and jealously guarded government secrets of the twentieth century.

Getting to the bottom of such a "plausible deniability" script has taken a veritable army of citizen researchers like myself decades of studying assassinations and deciphering the U.S. military's secret space program and geopolitics behind it in order to expose truths about what the atmospheric wireless "smart grid" now in place around and on the Earth—the Space Fence—is really about. It took me nearly a decade to decode the *epigenetics* behind the all-but-invisible chemical, electromagnetic, and nano-sized "implements of control" being implanted during these thirty years of nano-loaded aerosol spraying, GMO agrobusiness, and now mRNA inoculations designed to *lock in* brain computer interface (BCI) à la artificial intelligence in the newly minted transhuman society that Human 1.0 is now unwittingly being subjected to.

Whether the term *transhuman* points to slave diminishment or superhuman enhancement—the latter promised by futurist Ray Kurzweil, now Google's lead innovator of AI "machine learning and language processing" and cofounder of the Singularity University—we are being stealthily forced to enter a cyborg metaverse of Human 2.0.[1] Thus, the delivery system called chemtrails* is loaded with nanoscale heavy metals, red blood cells, polymer fibers, natural and synthetic fungi, oxides, proteins, unbelievably

*A *nanometer* measures one-billionth of a meter. Anything in nanos easily slides through the blood-brain barrier. The term *chemtrails* is directly from the cover of a U.S. Air Force Academy chemistry manual for future pilots. As well, the term *conspiracy theory* was invented by the CIA to silence doubts about the murder of President John F. Kennedy and has been a useful trigger term for public dismissal ever since.

tiny microprocessors, microrouters, micro-circuit boards, and more, none of which is jet engine debris, all of it for breathing and ingesting for the varying levels of Transhumanism.

As we consider all the preparation that has gone into the twenty-first-century epigenetic approach to changing the planet and "human nature"—from the onset of electricity in the nineteenth century and the airplane in the early twentieth century to the simultaneous ground-breaking scientific discoveries of Nikola Tesla, we begin to question our naïve assumption that the intent behind science and technology is always about service to humanity as a whole. First, we note that the theme of war is never far from the trajectory that technology is forced by powerful cabals to take. Tesla sought to release free energy to the masses, and yet his Wardenclyffe Tower on Long Island led not to free wireless communication and power generation but to the explosion in 1908 in Tunguska, Siberia, that sparked a half century of U.S.-Soviet experimentation for weapons potentials of extremely low frequency (ELF) waves and psychotronic scalar grids that led to the High-frequency Active Auroral Research Program (HAARP) in Gakona, Alaska, and its ionospheric heater beam technology (phased array antennas) that by 2013 gave the U.S. Air Force control over the ionosphere itself.*

The public has been deeply misled about "climate change," "global warming," and carbons, the three primary cover stories that purposefully hide the truth that chemtrails / geoengineering is neither a "conspiracy theory" nor "solar radiation management" requiring "carbon dioxide removal." In reality, geoengineering is a weapon system whose wireless, ionized atmosphere is maintained to serve the creation and control of an AI-driven planet and its transhuman Human 2.0 energy source. The United Nations Committee on Disarmament figured out a way around legal scrutiny by giving the go-ahead to "research projects" like HAARP while promoting a "climate change" narrative to cover for the intentional heating of our atmosphere and the "persistent contrails" that in fact are loaded with chemicals and nanotechnology.

Under Project Cloverleaf, the joint U.S.-Canada covert operation begun in the 1990s for which commercial and military jets seeded nanoparticles

*The ionosphere surrounds the Earth 50–600 miles / 80–1,000 km above the troposphere we breathe from. It is naturally ionized by solar radiation but has been weaponized with electromagnetic technologies like ionospheric heaters and lasers.

into the atmosphere to support an ionospheric heater technology that now controls both the ionosphere and the jet stream, our atmosphere has been chemically and electromagnetically transformed as has the soil, the vegetation, the oceans, the climate, and Human 1.0 bodies and brains.

With the birth of the airplane in the early twentieth century, *the Earth's atmosphere became the primary war theater*—from two world wars dependent upon air assaults to nuclear fission over White Sands, New Mexico, Hiroshima and Nagasaki, Japan, in 1945, to atmospheric "nuclear tests" like Operation Argus in the Van Allen radiation belt in the 1950s. At the same time, chemical warfare "experiments" were continued over domestic civilian populations—for example, aerosols loaded with zinc cadmium sulfide over future "smart cities" like St. Louis—and during the "Vietnam conflict" in the 1960s and early 1970s, Agent Orange, sarin, and VX nerve gas. The 1978 Environmental Modification Convention, an international treaty prohibiting the military or other hostile use of environmental modification techniques, arrived too little, too late to prohibit treating the atmosphere as a war theater, much less to prohibit manipulating planetary weather as a weapon.

Since the U.S. Air Force took control over the ionosphere in 2013, our atmosphere has been kept *ionized*, which basically means battle-ready as an active electrified antenna.* HAARP and other ionospheric heaters (including those mobile and smaller) are fully operational; strategically placed, these "sky heaters" and other Space Fence technologies sustain literally thousands of miles of artificially sustained plasma cloud cover consisting of jet-delivered nanochemicals shepherded by pulsed high-frequency beams. To give you an idea of where these mechanisms are placed and just how many there are, I've included figure I.1.

Because the electromagnetic waves are virtually invisible, this weapon system has been able to remain hush-hush for three decades, its true purpose concealed by one cover story or another. A host of questions arise, like where does the money come from to finance these operations? Black budget? Private foundations? Hidden system of finance? Blackmail of U.S. government agencies operating as front organizations? Certainly, secret budgets

*David Walker, deputy assistant secretary of the U.S. Air Force for Science, Technology and Engineering, admits that HAARP was created to control the ionosphere. The effective radiated power (ERP) of HAARP alone can be measured in terms of billions of watts of electrical power.[2]

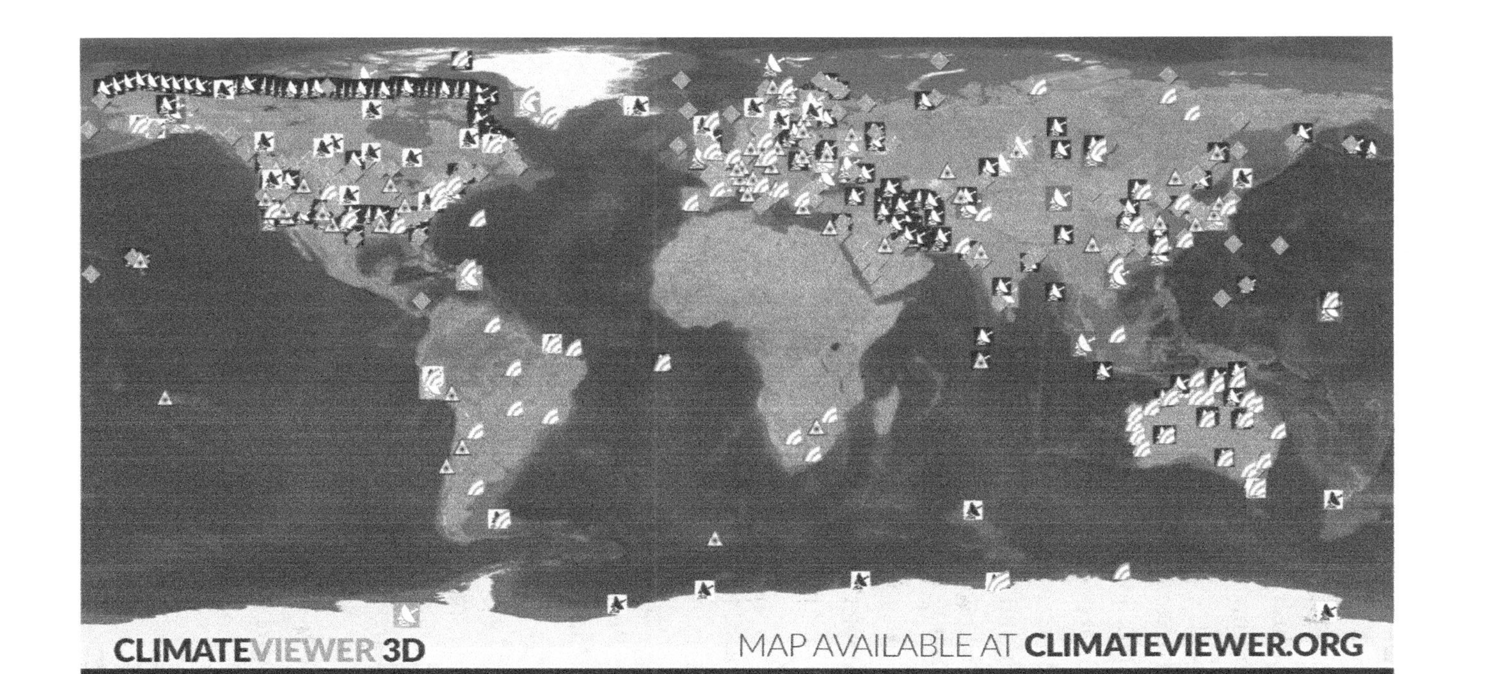

Fig. I.1. ClimateViewer-3D, ionospheric heater, space weather modification map included to give you an idea of the widespread scale and reach of these devices. It is difficult to discern one type of device from another in black and white, so visit the climateviewer.org website to see the image in color and to start your own research.

James F. Lee Jr., climateviewer.com

have been redirected from a host of public funds, like the $21 trillion missing from the U.S. budget, including a huge transfer of $800 billion from the U.S. Treasury to the U.S. Army to "reconcile past years."[3] Catherine Austin Fitts, former assistant secretary of Housing and Urban Development under George H. W. Bush and now president of Solari, Inc., publisher of the Solari Report, and managing member of Solari Investment Advisory Services, calls the secret space program a space-based economy. Fitts has presented at the 2014 Secret Space Program Conference in San Mateo, California,[4] at the Dark Journalist Secret Space Program Conference in Austin, Texas, in November 2015, and at GlobalBEM, the Breakthrough Energy Movement conference held in the Netherlands in 2019.

The civilian and military entities behind the secret space program adopted the Pentagon *doctrine of full-spectrum dominance* after September 11, 2001: military control over air, land, maritime, subterranean, and space war theaters as well as electromagnetic and information *fifth generation warfare,** replete with layers of propaganda like UN Agenda 2030, promoted to a gullible public as a program for "sustainable development."

All around the planet and embedded in forests and GMO agrobusiness land, from one 15-minute "smart city" to another, we have been plugged into the smart grid that Lockheed Martin calls the *Space Fence.*† It began back in the Reagan–George H. W. Bush–Dick Cheney era of the 1980s under the "Star Wars" Strategic Defense Initiative (SDI) with the Space Fence, at that time a belt of radar installations across the southern continental United States from sea to shining sea. The U.S. Air Force claimed to be tracking space junk and missiles, but the Space Fence has always been about developing a global network-centric warfare system to eventually be run by artificial intelligence so as to *master the Earth and the human domain.*

The Space Fence is a comprehensive weapon system encompassing a massive infrastructure of low-earth orbit (LEO) and near-earth orbit (NEO), mobile and stationary ionospheric heaters, radar installations,

---

*Fifth generation warfare is an irregular warfare primarily conducted *nonkinetically* through emerging technologies like the Internet of Things, Internet of Nano Things, Internet of Bodies, Internet of Medicine, etc.

†Lockheed Martin and Raytheon are the two top corporations of the "military-industrial complex" that outgoing President Eisenhower warned Americans about so long ago.

CERN's Large Hadron Collider, cloud-seeding and ionization pluviculture (rainmaking) technology, microwave towers and power lines, fracking wells, "green" solar panels, wind turbines—all connected to GMO labs and fields, synthetic biology, inoculations, Wi-Fi, 5G/6G, the Internet of Things (IoTs) and carbon-based nanotechnology delivered by jets, rockets, drones, and syringes, directed energy weapons (DEWs) targeting individuals and populations for mind control and Transhumanism. *The key is that it is all designed and calibrated to work together.*

Part 1 of this book is written to orient you as to how this multilevel system works as a basic geoengineering infrastructure. Part 2 focuses on how geoengineering is employed as a weapon, from magnetism to nuclear releases of ionized and nonionized energies, to what it means that life (*bios)* is under assault and increasingly subject to synthetic biology (*synbio*) and Transhumanism controlled by 5G/6G millimeter and terahertz waves. Unless we understand how Nature and the Divine are being usurped *and the intent behind it*, and how we are all plugged into the Space Fence, we will be unable to defend ourselves and chart a new course for those of us who manage to remain human in the coming metaverse run by artificial intelligence.

Many who have not read my previous geoengineering books may find what I posit to be overwhelming and feel the despair of "what can *I* do?" But true human beings are endowed with an innate connection to the Divine and the ability to exercise free will—something that machine intelligence, no matter how technologically advanced, lacks, just as it lacks creativity, which is an expression of our connection to our spiritual selves. As quantum physics points out, consciousness precedes materiality and in fact shapes all material forms. Yes, we must first be aware of what has been going on in secret for decades, as burying our head in the sand and denying the truth of what is happening is to empower the shadow elements within the elite brotherhoods to continue as they have. The solution begins with lifting our consciousness into ways of thinking that are different from the ways of thinking that created this problem in the first place. It begins on an individual level, but again, as quantum physics points out, each individual is a ripple that either assists or interrupts the transhuman juggernaut hoodwinking us into a machine metaverse. The choice is ours. If we so choose, we will be empowered far beyond what we have been programmed to not recognize.

There are clear signs that more and more of humanity *is* waking up . . .

As I indicate in the final chapter on 5G/6G, our solution to geoengineered Transhumanism begins with electromagnetics and reconfiguring our assumptions about "progress" and technology. As I have reconfigured my own relationship with electromagnetics, so I have continued my acid-alkaline balance in the organic food that I have prepared and consumed for a half century. We must take back our health from profit institutions and "experts," most of whom are ignorant of what geoengineering has epigenetically done to our environment and bodies in preparation for a "stealth" Transhumanism.

Is the Divine model of *Homo sapiens* worth saving? Let's rethink what it means to be human in a technological era, including how far the scope of "evil" now extends. If television and blockbuster movies have been your source of knowledge about what being human is all about, I doubt that you will care all that much whether *Homo sapiens* continues or not, given that you have already been technologically separated from a large and deep swath of the humanity you as a child shared with the billions who have come to the Earth before and plan to come still. You may even be looking forward to being an "enhanced" transhuman. That said, I still suspect that those reading a book like this hear a *call to life* that may be impossible to silence even now, when Nature "is not now as it hath been of yore," as recorded by nineteenth-century English poet William Wordsworth in, "Ode: Intimations of Immortality from Recollections of Early Childhood":

> *There was a time when meadow, grove, and stream,*
> *The earth, and every common sight,*
> *To me did seem*
> *Apparelled in celestial light,*
> *The glory and the freshness of a dream . . .*

Fig. I.2. Artist Anne Gibbons, "Weather Warfare."

PART I

Nanotechnology in the Sky with Diamonds

1

Atmospheric Nanoscience

Aside from nanotech's potential as a weapon of mass destruction, it could also make possible totally novel forms of violence and oppression. Nanotechnology could theoretically be used to make mind-control systems, invisible and mobile eavesdropping devices, or unimaginably horrific tools of torture.

ADAM KEIPER,
"THE NANOTECHNOLOGY REVOLUTION,"
THE NEW ATLANTIS (SUMMER 2003)

When we look up at the sky, it looks enough like what we consider "normal" that people scurrying about on planet Earth, distracted by all their "doings," will be the last to know about the geoengineering that has utterly changed not only our atmosphere but humanity and planet Earth herself. That the sky is not the deep blue it used to be is chalked up to industrial pollution; that the weather is less cyclic and more extreme is blamed on "climate change."

The term *geoengineering* may be relatively new to some people, but terms like *nanoparticles* and *nanotechnology* are now part of our everyday vocabulary thanks to the Big Pharma mRNA serums that have been injected into people's bodies over the last few years.

I look forward to the day—hopefully not far off, as people everywhere are waking up—when we can look up at the sky and wonder how is it that we were fooled for so long.

THE WEAPONIZATON OF WEATHER

Our weather is now weaponized. The U.S. military has claimed for almost thirty years that "we will own the weather" by 2025 (and apparently sooner, as we shall see in the following chapters).[1] The ability to make plasma clouds, electrostatically charged snow, extreme winds, "wildfires," and various other atmospheric events is part of the manipulation of our natural weather to create synthetic weather. The aerosols sprayed overhead, widely known as "chemtrails," have been distributed for the last twenty-five years. They have been loaded with conductive nanometals as well as genetically engineered fungi, red blood cells, nanobots, nanosensors, and nano-synbio creations piggybacked onto polymer fibers (see chapter 2, "Those White Lines in the Sky," as well as appendix 5, "Substances Used in Chemical Spraying Operations"). These Frankenstein creations now permeate the soil, tree bark and roots, aquifers, and all food growing in the open air. We are breathing and ingesting these substances. And altogether they are weakening the human immune system, fogging our brains, and making us ill with endless autoimmune conditions. All the while we are conditioned by "experts" and heavily indoctrinated so that we can no longer see the difference between what is real and natural and artifice. Carbon is being blamed to cover for the true culprit behind these weather extremes: the clandestine atmospheric nanoscience going on in our now fully ionized atmosphere. At present, all planetary changes for the worse are filed under the "climate change" narrative (the new term having replaced the previous term, *global warming*).

The chemicalizing of Earth's atmosphere and the use of electromagnetics for weather engineering skews all other planetary data and leaves scientists, PhDs, and lay researchers alike in a perpetual state of cognitive dissonance, the state of discomfort felt when two or more modes of thought contradict each other, usually resulting in an aversion to facts that may challenge one's reality. Nuclear chemist J. Marvin Herndon"[2] noticed what was unfolding in the sky over his hometown two decades after Project Cloverleaf, the joint U.S.-Canadian program of aerial chemical spraying, had been in full swing. In a study published in a scientific journal in 2015, Dr. Herndon concluded that the "ultra-fine airborne particulate matter of undisclosed composition"—i.e., nanotechnology—is dangerous to our health:

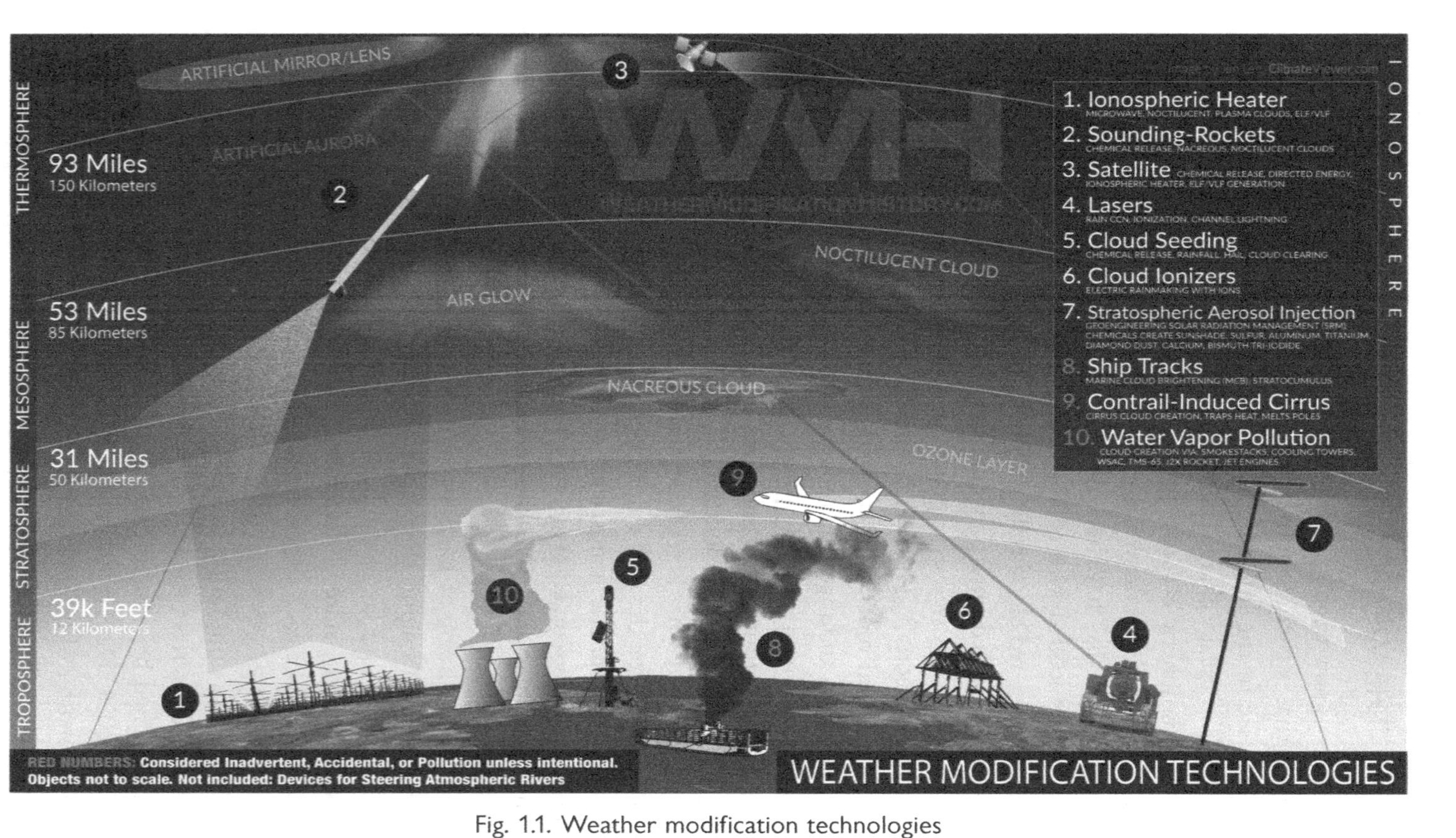

Fig. 1.1. Weather modification technologies

Image courtesy of James F. Lee Jr., climateviewer.com

> The dry warm air above San Diego is not conducive to the formation of jet contrails, which are ice condensate. By November 2014 the tanker-jets were busy every day crisscrossing the sky spraying their aerial graffiti. In a matter of minutes, the aerosol trails would start to diffuse, eventually forming cirrus-like clouds that further diffuse to form a white haze that scattered sunlight, often occluding or dimming the sun. Aerosol spraying was occasionally so intense as to make the otherwise cloudless blue sky overcast, some areas of sky turning brownish. Sometimes the navigation lights of the tanker-jets were visible as they worked at night, their trails obscuring the stars overhead; by dawn, the normally clear-blue morning sky already had a milky white haze. Regardless, aerosol spraying often continued throughout the day.[3]

That same year, CBS-affiliated KSLA in Louisiana investigated the possible source of a substance that fell to Earth in the wake of high-altitude chemtrailing. High levels of barium (6.8 parts per million, or ppm), lead (8.2 ppm), and traces of arsenic, chromium, cadmium, selenium, and silver were found, with barium being more than six times the EPA toxic level. Symptoms of barium poisoning begin with stomach and chest pains and blood pressure problems.[4]

DUSTY PLASMAS, SMART DUST, NEURAL DUST

What are we really seeing when we look up and *think* we're looking at the same blue sky and puffy clouds as those of yesteryear, before the onslaught of atmospheric nanoscience, before the program of artificial atmospheric air ionization was officially launched in 1994 as Project Cloverleaf?[5] What we are seeing is electromagnetically charged chemical aerosols in cloud form—*dusty plasmas* full of charged nanoparticles, all of it cooked up by nanoparticle factories and delivered into the stratosphere (the layer just above the troposphere that we breathe from) via military and commercial jets. These nanoparticles are incredibly small carbon molecules often engineered with tinier-than-you-can-imagine, gigaflop, radio-controlled microprocessors that are then renamed "smart dust."

Smart dust falls like bad fairy dust in chemical trails; it looks like iridescent glitter, otherwise known as MEMS (microelectromechanical sensors) and GEMS (global environmental MEMS sensors).[6] This smart dust can record everything, but most seriously it is being inhaled by people everywhere. Not surprisingly, smart dust research is funded by the U.S. Department of Defense (DOD). Fitted with computing power, sensing equipment, wireless radios, and long battery life, this smart dust makes possible the observation and relay of mountains of real-time data about people, cities, and the natural environment not to mention what it does to our blood and brain once inhaled.[7]

Engineered nanoparticles are alchemically turned into smart dust by vaporizing materials via plasma heat, pressure, and volume, plasma being superheated matter so hot that the electrons are ripped away from the atoms, forming an ionized gas. Once the temperature decreases, the resulting vapor becomes supersaturated with trillions of nanoparticles via nucleation, condensation, and coagulation.[8]

What is the purpose of creating artificial plasma plumes of charged microdust and nanodust particles of metal oxides? We are told it is to study the effects of this dusty plasma in our atmosphere while observing terrestrial weather, but it is actually to "own the weather" *and transform our atmosphere for transhuman 2.0* by creating artificial plasma plumes of charged microdust and nanodust particles of metal oxides—what we are deceived into still thinking of as "clouds"—to make the atmosphere more electromagnetically conductive.

All plasma manifestations "have in common a kind of 'non-atomic' state of matter where the electrons of an atom and its nucleus exist 'apart' from each other, and in a kind of 'roaming free'—though, as we shall see, not 'amorphous'—state. A plasma consisting of uranium or of helium, for example, would have the *electrons* of uranium or helium atoms, and the *nuclei* of uranium or helium atoms all present in a particular region, but *not* bound together in their familiar state *as* atoms. Or to put it differently, since the matter of the elements comprising a plasma still exist with all their protons, neutrons, and electrons 'intact and present' but just not arranged *in atoms*, one may also say that in plasmas, matter exists in a *sub-atomic* or 'incomplete atomic' condition. —Quoted from Oxford scholar Joseph P. Farrell's 2023 book *The Demon in the Ekur: Angels, Demons, Plasmas, Patristics, and Pyramids* (Adventures Unlimited Press); in reference to Robert Temple, *A New Science of Heaven: How the Science of Plasma Changes Our Understanding of Physical and Spiritual Reality* (London: Coronet, 2021).

Plasma physics—the study of the subatomic particles that make up plasma—has only existed since the 1930s. The term *plasma* ("to mold") was coined by Nobel Prize–winning American chemist and physicist Irving Langmuir (1881–1957) to describe how electrons entering plasma behave as a well-organized collective (hive or swarm), much like a biological organism, and similar to the colorless fluid found in living blood cells, which is also known as a plasma.

We are accustomed to hearing about plasma in fluorescent lamps, neon lights, plasma TVs, plasma lighters, the aurora borealis, the Sun and the stars. All have to do with *light* because the motion of electrons and ions inside plasma generates electrical and magnetic fields, the electrical fields accelerating charged particles to high speeds while magnetic fields (the real architects of the universe) guide and confine them. Magnetism determines the character, motion, and shape of ionized matter, otherwise known as plasma. The auroras at the North and South Poles are magnetized plasmas accelerating solar electrons to 10^3 electron volts (eV) (solar flares being 10^9-10^{10} eV). In deep space, X-rays and gamma rays are produced by magnetized plasmas whose accelerating electrons stream from the Sun.

In 1983 the Ariel 3 and 4 weather satellites discovered *a permanent plasma duct ciphoning ions from the magnetosphere into the stratosphere of the entire North American continent.*[9] The cause appeared to be an enormous amount of PLHR (power line harmonic resonance) percolating endless electromagnetism. With the advent of the HAARP "web" of ionospheric heaters in the Northern Hemisphere, Birkeland currents—the magnetic lines of force that connect the Earth to the plasmasphere up through the magnetosphere and into the ionosphere—were inducing more plasma ducts to form. In 2015, an Australian graduate student at University of Sydney named Cleo Loi photographed three huge plasma tubes in 3D above the Earth over the Western Australian desert with the Murchison Widefield Array radio telescope.[10] The following year during a sunset over Pacifica, California, a woman watched the Sun split into multiple layers followed by green flashes.[11] She sent the video to astronomer

Fig. 1.2. Stationary ionospheric heaters from around the world. There are also smaller mobile ionospheric heaters in all shapes, sizes, and contrivances.
From James F. Lee Jr.'s site Weather Modification History

Andrew T. Young at San Diego State University who confirmed at least three ducts and that she was looking up through one of the ducts.

Less than a year later, NASA went public with the massive VLF (very low frequency)* "CD-Rom ring" ("impenetrable barrier") around the equator and going all the way to the inner edge of the Van Allen radiation belt 400+ miles

*Very low frequency (VLF) is, according to Billy "The HAARP Man" Hayes, "the mainstay in many engineering, scientific, and military operations."

above the Earth. NASA's narrative is that it has been "accidentally" caused by military radio communications signals, the "strange" solar-charged particles "caught within Earth's magnetic field" simply "anthropogenic impacts" from "chemical release experiments, high-frequency wave heating of the ionosphere, and the interaction of VLF waves with the radiation belts."* Satellite debris? Magnetic nanoparticles? Outright nanotechnology?

HAARP phased array antennas have been used to generate radio-reflective plasma by steering radiation into the ionosphere. Now, keeping the upper atmosphere flush with ionized plasma requires Alfvén wave magnetohydrodynamics (MHD)† generators to "bomb" the skies with plasma, plus nanometals inside tiny CubeSats, jets, and drones to create metal vapor that reactions with atmospheric oxygen to produce more and more dusty plasma.[13] Are the U.S. Air Force "plasma bombs" and nanometal vapors about increasing ions to boost wireless communications and dry out the air for transmissions?[14] Is that what the *plasma orbs* are that are being created in space and then delivered into the troposphere by radio control?[15]

As ionized matter, plasma is less dense than solids or liquids, with no fixed shape or volume. Unlike normal gases, plasma's charged ions roam freely, the atoms having been stripped of their electrons. Plasmas can thus conduct electricity, make magnetic fields, and be held in place by magnetic fields (fusion power). An example is the ISS mentioned above, located 254 miles above us. The ISS runs on 120 volts DC with a plasma cloud or cocoon of complex currents of electrons and ions in constant flux, extending hundreds of feet around it.‡ Upon returning to Earth, ISS crew and guests exhibit the faintly acrid smell of ozone generated from the stress on the metallic mass of the craft.

Back in 1990, NASA launched its Combined Release and Radiation Effects Satellite (CRRES) in an elliptical geosynchronous transfer orbit

*The 2016 U.S. Patent 9491911-B2, "Method for modifying environmental conditions with ring comprised of magnetic material," describes the man-made metallic ring around the equator whose "magnetic climate control material deflects and absorbs rays of the sun . . . deploying a magnetic climate control material to a local area in the thermosphere and the exosphere. Reflective, absorptive nanopowders (0.1–100 microns) are fed into the ring by stratospheric jets, sounding rockets, and satellites "equipped with electromagnets."[12]

†MHD determines the dynamics of fluids that conduct electricity (like plasma and liquid metals). Fluid motion generates magnetic fields that confine plasma along the magnetic lines of force.

‡Traveling at 17,500 miles per hour in a consistent plasma-microgravity environment. The ISS is pursuing many biological experiments that cannot be performed on Earth.

22,236 miles up. Its payload consisted of twenty-four canisters of chemicals to be released and ionized by the Sun's UV light to create elongated plasma clouds along the magnetic field lines known as Birkeland currents that connect our Earth and the Sun. These chemical releases into "Earth space"—the magnetosphere and ionosphere space environment, described in a CRRES press release as a "dynamic ocean of invisible magnetic and electrical fields and particles"—were augmented by massive chemical releases from ten sounding rockets (instrument-containing rockets designed to take measurements and perform scientific experiments) launched from the Arecibo Observatory in Puerto Rico (until it collapsed on December 1, 2020), and Space Fence Radar on the Kwajalein Atoll in the Marshall Islands. Needless to say, we all breathed in those chemicals.

Mixed in with the dusty nano-charged plasma raining down on us from the stratosphere are *smart dust* and *neural dust* consisting of magneto-aerotactic bacteria nanobots, algal-based nanobots, and DNA-based nanobots—all synthetic. "Smart" technologies such as these nanobots (essentially nano-sized robots) are used in remote networks feeding machine-learning computers—artificial intelligence, or AI—dedicated to collecting and implanting biometrics, communications, tracking, tracing, targeting, and so forth.

In 2002, Michael Sailor and colleagues at UC San Diego developed *smart dust*, the ultratiny, wireless, microelectromechanical sensors, or MEMS, that detect everything and everyone emitting a frequency. Smart dust is composed of modified silicon particles ultrasonically fragmented.[16] Smart dust is essentially an Internet of Things (IoT) device that wirelessly monitors light, frequencies, temperature, humidity, magnetism, and chemical signatures, "acting as nerve-endings in an ad hoc distributed network that provides full spectrum intelligence."[17] *Magnetic smart dust* is composed of magnetic, amphiphilic (polar and nonpolar) particles encasing droplets of organic and aqueous (water-based) solvents, which in the presence of an external magnetic field will move, falling from the sky like glitter, tracking movement, money, biometrics, temperature, chemical signatures, etc.[18] Once heated, this chemical concoction, called M-spray, magically transforms minuscule inanimate objects into bots:

> capable of adapting to a wide variety of explorative purposes . . . By putting on this [gluelike] magnetic coat, we can turn any objects into a

robot and control their locomotion," explained [Dr. Shen Yajing from City University of Hong Kong]. "The M-spray we developed can stick on the targeted object and activate it when driven by a magnetic field . . . reprogramming the millibot's movement mode is something that can be enacted on demand."[19]

Is electrospray magnetic dust involved in the quorum sensing* phalanx swarms of nanobots being utilized in geoengineered fires?† (More about geoengineered fires below.)[20]

Fitbits for the nervous system is a good description for *neural dust*, which is composed of quorum sensory nanobot swarms made with electronics. Neural dust replaces the old-fashioned wire electrodes of the past with a *brain-computer interface (BCI)* that can be remotely directed by ultrasound to monitor nerves, muscles, and organs in real time. Once inside the body after being breathed in, ingested, or inoculated, neural and smart dust travel through the blood to the cortex while being directed by ultrasound to monitor nerves, muscles, and organs in real time, energized by the *mesh electronics nanotechnology* undergoing development inside the bodies living in a forest of beam-steering 5G phased array antennas.

> [Quorum] sensors are about the size of a large grain of sand [and] contain a piezoelectric crystal that converts ultrasound vibrations from outside the body into electricity to power a tiny, on-board transistor that is in contact with a nerve or muscle fiber. A voltage spike in the fiber alters the circuit and the vibration of the crystal, which changes the echo detected by the ultrasound receiver, typically the same device that generates the vibrations. The slight change, called backscatter, allows them to determine the voltage.[21]

Mesh electronics created in the brain by nanoelectronics produce neural interfaces with brain cells so as to affect thinking, emotions, memory, behavior, judgment, and decision-making. This is *precision electronic medicine*.[22]

---

*Quorum sensing is a type of cell signaling by gene regulation.

†In drone footage of 2017–2018 California "wildfires," phalanxes of "embers" supposedly being blown by nonexistent wind are actually armies of nanobots responding to transmissions detailing which homes to burn and where to go next.

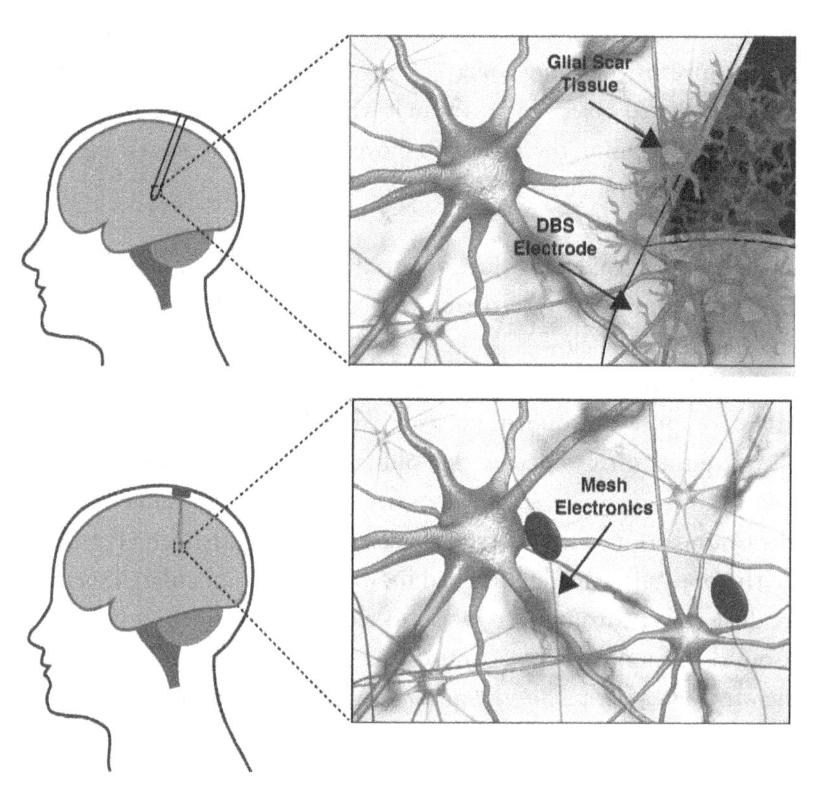

Fig. 1.3. Brain-computer interface (BCI) "mesh electronics." Charles M. Lieber, PhD, of Harvard University worked on this with Shaun R. Patel of Harvard Medical School. Lieber, former Chair of Harvard University's Chemistry and Chemical Biology Department, and two Chinese nationals were charged in 2020 under the China Initiative, a program established by the Department of Justice in 2018 to investigate academic espionage at American universities. Lieber was charged with hiding his ties to Chinese sources of funding, a talent recruitment program, and universities, including the Wuhan Institute of Technology.

Image from Thomas, Liji, "Mesh Electronics Could Make Brain Stimulation the New Therapeutic Norm," NewsMedical.net, September 6, 2019. Read more about the concept in Patel, Shaun R., and Charles M. Lieber, "Precision Electronic Medicine in the Brain," *Nature Biology* 37, no. 9 (September 2019):1007–12.

Neural dust can be wirelessly remote controlled to monitor the human brain from the inside. In fact, the network of tiny implantable sensors that comprise neural dust functions like an MRI (magnetic resonance imaging) inside the brain, recording data on nearby neurons and transmitting it back out. Neural dust nanoparticles are coated with polymer and contain a sensor capable of measuring electrical activity in neurons. Neural dust can thus be monitored by components powered by ultrasound outside the body, which is much more efficient than radio waves for targeting the tiny subatomic world embedded deep in the body and brain, as ultrasound transmits ten million times more power than electromagnetic waves at the same scale.[23]

In fact, electromagnetic waves generate a damaging amount of heat because of the amount of energy the body absorbs and the signal-to-noise ratios at the subatomic scale. Given that the neural dust has nanofibers as well as nanoparticles in it, the insoluble crystalline nanoparticles of titanium dioxide, silicon dioxide, sodium nitrate, and iron oxide vibrate / spin / rotate in the body *as if in a blender* and end up shredding the epithelial lining of blood vessels and the digestive tract, as well as cutting holes in cell walls, damaging DNA and mitochondria.[24]

This "blender" theory was derived by the British independent researcher Robin D. P. Watson who spent years working with government doctors and scientists. The symptoms Watson describes—shredding blood vessels, cutting holes in cell walls, thrombosis, etc.—sound exceedingly like the debilitating symptoms following the mRNA / spike "clot shot" inoculations.

> When H. G. Wells wrote *The War of the Worlds*, he imagined creatures from outer space attacking humanity using advanced technology. Humanity is finally saved by our most basic of life forms, the bacteria which co-inhabit this planet. But what if, instead of humanity, the tiniest of life forms—DNA, mitochondria and all the numerous cells which inhabit this planet—came under attack. This alarming scenario is actually in fact exactly what is happening with the use of nanotechnology. There is profound loss of biodiversity on this planet, and most of us [have] been too busy to notice that this is absolutely deliberate.[25]

Since 2012, Watson has maintained that "almost continuous thrombogenesis occurs with sustained exposure to insoluble nanoparticles,"[26] not to mention the body's attempt to repair blood vessels by forming plaque.

Stimulating neurons with nanoparticles via light or sound or magnetic fields is all the rage, especially now that experiments are not limited to labs, but are out in the open, thanks to the ubiquitous chemical aerosols being sprayed above us. Imagine inhaling gold nanoparticles used to absorb and convert optogenetic pulsed light to heat in order to switch brain cells on and off; spherical iron oxide nanoparticles giving off heat when exposed to an alternating magnetic field; and controlling intrinsic fields deep within the brain with magnetoelectric nanoparticles (MENs). As one study demonstrates,

> 20 billion nanoparticles [were inserted] into the brains of mice. They then switched on a magnetic field, aiming it at the clump of nanoparticles to induce an electric field. An electroencephalogram [EEG] showed that the region surrounded by nanoparticles lit up . . . "When MENs are exposed to even an extremely low frequency magnetic field, they generate their own local electric field at the same frequency," says [Sakhrat Khizroev of Florida International University in Miami]. "In turn, the electric field can directly couple to the electric circuitry of the neural network" . . . Running it in reverse . . . our brain states would then become input parameters for computers.[27]

IONOSPHERIC HEATERS: HAARP

The High-frequency Active Auroral Research Program, best known as HAARP, advertised as "the world's most capable high-power, high-frequency transmitter for study of the ionosphere,"[28] is located in the Arctic Circle, at Gakona, Alaska, with additional instruments at Poker Flat Research Range. HAARP is integral to the U.S. Department of Defense's "we will own the weather" manifesto. The installation keeps the atmosphere ionized and antenna-primed for the doctrine of full-spectrum dominance defined by the DOD as "the cumulative effect of dominance in the air, land, maritime, and space domains and information environment, which includes cyberspace, that permits the conduct[ing] of joint operations without effective opposition or prohibitive interference."[29]

Based on futurist inventor Nikola Tesla's discoveries from bouncing radio waves off the ionosphere since 1924—discoveries that he hoped would end in free energy for everyone on the planet—HAARP was designed as a weapon that could create and utilize scalar standing waves*:

> This [HAARP] invention provides the ability to put unprecedented amounts of power in the Earth's atmosphere at strategic locations and

*See appendix 2, "Quantum, Scalar, and Hyperspace."

> to maintain the power injection level, particularly if random pulsing is employed, in a manner far more precise and better controlled than heretofore accomplished . . . HAARP [is] the largest ionospheric heater in the world, located at a latitude most conductive to putting Eastlund's invention into practice.[30]

By the mid-1970s, ionospheric heating experiments were being conducted out of Platteville, Colorado; Armidale, New South Wales, Australia; and Arecibo, Puerto Rico. By the late 1970s, the Max Planck Institute had built the European Incoherent Scatter Radar (EISCAT), a HAARP-like 100-megawatt ionospheric heater located in Tromsø, Norway. In 2018, China's version of HAARP, known as the Sanya Incoherent Scatter Radar (SYISR), was installed in the South China Sea area.[31] Then there is KAIRA, the ionospheric heater in Finland, completed by 2012; and LOFAR, the Netherlands' version of HAARP, completed by 2010 and utilized by seven European nations.

> Near the villages of Jiefang ("Liberation") and Kan'erjing ("Underground Caverns"), Google Earth gives us an aerial shot of what might be China's next generation of HAARP arrays. Nathan Cohen, inventor and patent-holder of fractal arrays and CEO of Fractal Antenna Systems, Inc. [. . .], describes the symmetrical fractal shapes in the satellite photograph . . . two banks of three arrays for two separate bands, and one bank of two arrays for another. You can't tell the operational frequencies from the spacings. The panels are many wavelengths across, but we don't know how many. It is a multiband array antenna farm with flat array panels.[32]

The HAARP ionospheric sky heater is a phased array antenna farm consisting of 180 units organized in fifteen columns by twelve rows. Phased array antennas such as HAARP and similar installations in other countries (4G and 5G cell phones also use phased array antennas) act synchronously as one large antenna to form exceedingly powerful beams to transmit directed energy (DE). Originally jointly managed by the U.S. Air Force, Navy, and DARPA (Defense Advanced Research Projects Agency), since 2013 HAARP has been able to successfully control the planetary ionosphere 49 to 428 miles above the Earth by means of phased array beams that cause ions (charged electrons) to spiral down along Earth's magnetic field lines, thus binding Earth's upper and lower atmospheres to one another, the ionosphere being positively charged and the Earth negatively charged. This ionospheric binding, coupled with endless infusion into the atmosphere of conductive heavy metal nanoparticles ("chaff" of aluminum, barium, chromium, strontium, lithium, etc.), keeps our atmosphere ionized—antenna and battery-ready—for trillions of wireless transmissions.

By forcing radio waves into a narrow beam that vibrates the ionosphere and produces a repetitive excitation that heats our troposphere, the lowest layer of our atmosphere (hence the "global warming" cover story), HAARP and other ionospheric heaters basically hold the planet in thrall—what I call in my second book on geoengineering *Space Fence lockdown*. In July 2010, the thermosphere, located fifty miles above Earth, collapsed, then rebounded as a result of HAARP activity. NASA and the Naval Research Lab claimed it "stumped researchers," calling it a "Space Age record"[33] while the truth was that it was caused by tampering with the upper atmosphere. Harvard University geoengineer David Keith was trotted out to blame carbon and greenhouse gases for the warming that in reality is caused by ionospheric tampering. In a 2010 article, Keith equates dangerous "side effects" with "photophoretic levitation."[34] Some of those side effects definitely involve our health.

American physicist Bernard Eastlund's 1987 patent, "Method and apparatus for altering a region in the earth's atmosphere, ionosphere, and/or magnetosphere," gave birth to HAARP.[35] The patent language reveals that the weird weather phenomena being blamed on "global warming"/"climate change" is not due to carbons, but to electromagnetically manipulated electron cyclotron frequencies from the ionosphere to increase ion density in Earth's atmosphere and heat it up. High gain sources of EM radiation from directed energy weapons like HAARP increase power density without the attenuation (loss of power) the inverse square law says antennas should have.[36] Again and again, it becomes apparent that the technology and science we are confronting is neither Newton's nor Einstein's, but Tesla's, and not at all in the way he envisioned.

As a result of HAARP's atmospheric heating, along with some thirty years of nanoparticle aerosals being injected in the air via chemtrails, conductivity between the lower and upper atmospheres has basically become contiguous, making our atmosphere a thoroughly charged sky antenna of fully ionized plasma. In 2003, independent scientist Clifford Carnicom,* founder of the Carnicom Institute (carnicominstitute.org), measured the conductivity of the lower atmosphere with a 200,000-volt Van de Graaf electrostatic generator. After determining the spark length, he realized that the fundamental electrical nature of the atmosphere had been radically *ionized* as a result of the ongoing aerosol operations.[38]

Right from the beginning, HAARP and other classified chemical trail "research projects" have smacked of intelligence operations.[39] Though the U.S. Air Force Academy used the term *chemtrails* in its Chemistry 101 manual, that term has been turned into a pejorative by the CIA, much as the term *conspiracy theorist* was used to discredit anyone who doubted the official narrative about the Kennedy assassination. Clifford Carnicom initially thought it strange when he received an August 23, 2000, "chemtrails are contrails" letter from Lieutenant Colonel Michael K. Gibson, a "master intelligence officer" with the U.S. Air Force.[40] Also see appendix 7 "Visitors to www.Carnicom.com, Aug. 26, 1999" for a long list of agencies discretely checking on Clifford Carnicom's discoveries from his chemical and microscopic examinations of the condensation and HEPA filter detritus he collected. This constituted surveillance, as he received no official responses from agencies like the Environmental Protection Agency (EPA) and Centers for Disease Control and Prevention (CDC) that he had contacted with requests to duplicate his findings, due to his increasing alarm.

Strategically placed "sky heaters" like HAARP are now smaller and mobile and stretch around the globe, operating under plasma cirrus cloud cover originating from commercial and military jet chemtrails—the brew that we have been breathing for going on three decades. And yet what are we told? That it is all due to excess carbon "climate change." Meanwhile, people in Riga, Latvia, and Gomel, Belarus, wake up at 4 a.m. robbed of their Schumann resonance "breathing" with the planet, forced into another brain entrainment entirely.

---

*Carnicom served as "a technical research scientist acting in a professional capacity supporting analysis and development of major Department of Defense physical and weapons modeling systems, with extensive computer programming and system application development experience. He has held a Top Secret/SCI clearance. He was appointed for and completed two years of intensive graduate level studies in mathematics, statistics, computer science, and geodesy under the auspices of the Department of Defense."[37]

THE SPACE FENCE

Similar to the way the term *chemtrails* began with the U.S. Air Force Academy, the term *Space Fence* originated with the military through its number one global defense contractor, Lockheed Martin. Planet Earth is now trapped beneath an ionized plasma cloud cover that runs interference with our God-given cosmic frequencies (see pages 16–17 for some discussion of plasma). Basically, we are inside a spider's web that people call the smart grid and the military calls the Space Fence—from the conductive CD-ROM-like ring around the equator to satellites, lasers, and an exactly placed and calibrated infrastructure of ionospheric heaters, radar installations, microwave towers, NexRad and SBX "golf ball" radar, "green energy" wind farms, fracking wells, gas and water lines, electric power lines and fiber optic cables.[41]

Once HAARP had achieved control over the ionosphere and Project Cloverleaf was in place, the Space Fence too was up and running, the explanation to the public being "global warming,"[42] as per the 2002 paper, "Earth Rings for Planetary Environmental Control," presented by government contractor Star Technology and Research at the 53rd International Astronautical Congress: increased warming, melting polar ice caps, frequency of cold waves, intense droughts and heat waves, frequent and severe floods, increased fires and wildfires, dangerous thunderstorms, more intense and destructive storms, tsunamis, volcanic activity, loss of biodiversity and increased animal extinction, strained ocean life, diminished food and water, spread of disease, and economic consequences of all the aforementioned. The old Strategic Defense Initiative (SDI) narrative during the Reagan administration of the 1980s—when the Space Fence was a mere band of radar installations across the southern United States and control over the ionosphere lay over a decade in the future—was that the Space Fence was for tracking missiles and space debris.

Now, the 260 Strategic Defense Initiative (SDI) Space Fence installations around the world are engaged in ultralow frequency (ULF) and extremely low frequency (ELF) manipulation that encompasses the ULF nanometallic ring around the equator and relay system of sixteen Starfire optical lasers are used to manipulate the jet stream for weather deliveries *and* profitable weather derivatives.

Control over the weather makes money. From *Chemtrails, HAARP* (2014):

> Remember Enron and the energy-deregulating scandal? Enron Weather gave birth to weather derivatives back when it basically meant offering utilities a hedge against spiking temperatures in the summer and tanking temperatures in the winter. Enron internationalized weather derivatives by taking it to the UK, Norway, Australia, Hong Kong, Toyko, and Osaka. When Enron finally went bankrupt, UBS Warburg bought their trading desk *and* Enron Weather.[43]
>
> By privatizing and deregulating the energy sector, Enron and other big energy and finance players like Willis Group Holdings (terrorism risk), Koch Industries (commodities trading, ventures, investments, etc.), and Pxre Reinsurance Company were able to manipulate and arbitrage—i.e., capitalize on an imbalance between markets—remarkably well. For a while.
>
> In those same early Enron days, HAARP was coming online and the aerial aerosol assault revving up. Coincidence, huh?[44]

Defense Department contractor Lockheed Martin's description says, "Space Fence, now the world's most advanced radar, provides uncued detection, tracking and accurate measurement of space objects, including satellites and orbital debris, primarily in low-earth orbit (LEO). The new radar permits the detection of much smaller microsatellites and debris than current systems. It also significantly improves the timeliness with which operators can detect space events. The flexibility and sensitivity of the system also provides coverage of objects in geosynchronous orbit while maintaining the surveillance fence."[45]

THE CONTROLLED JET STREAMS

Back in 1993, an early version of the HAARP ionospheric heater located in Russia, dubbed "the Russian Woodpecker,"[46] altered the jet stream to create epic flooding across the U.S. Midwest by relaying pulses along the Ground Wave Emergency Network system (GWEN), whose stated purpose was to protect U.S. communications systems during a high-altitude nuclear explosion. The epic flooding resulted in five atmospheric rivers of vapor crisscrossing the continental U.S., some of them 420 to 480 miles wide and capable of moving 340 pounds of water per second 2 miles above the Earth. This pump-and-dump operation involved directing the flow of electricity to steer the jet stream so as to dump masses of rain on predesignated geographic areas.

Simply put, Earth's jet streams—fast-flowing air currents in the upper atmosphere—are caused by a combination of planetary rotation and atmospheric heating, if allowed to proceed naturally. Control (full-spectrum dominance) over the jet streams, on the other hand, can be achieved by transmission of a HAARP high-frequency (HF) modulation into the ionosphere so as to generate Alfvén waves, ionized gas loaded with charged particles that respond collectively to electromagnetic forces. It is Alfvén waves that create extremely low frequency (ELF) harmonics in the Earth's atmosphere, thus setting the stage for *cyclotronic resonance*, the HAARP mechanism by which very low-strength electromagnetic fields, in concert with the Earth's geomagnetic field, can produce major biological effects by concentrating field energy upon specific particles like the crucial ions of sodium, calcium, potassium, and lithium.

As early as 2005, independent scientist Clifford Carnicom observed how the potassium ion (K^+) was being atmospherically targeted for "biological interference."[47] Martin Pall, PhD, a professor emeritus of biochemistry at Washington State University,[48] has done extensive research into 4G and 5G technologies and verified that our neurons need high densities of voltage-gated calcium channels (VGCCs) for the calcium-signaling necessary to release neurotransmitters and maintain synapsis between nerve cells. Chronic electromagnetic radiation exposure—ionized or nonionized—can cause excessive Ca^{2+} levels in the mitochondria* and produce cell death (apoptosis), similar to what follows double-strand breaks in cellular DNA and increases and releases in hormone levels that exhaust the body and lead to autoimmune conditions.

Greater by far than the outmoded big bang model of the origin of the universe is the *electric plasma theory* of Hannes Alfvén (1908–1995), which states that the universe had no beginning (and has no foreseeable end), and that the electric and magnetic forces that comprise plasma are responsible for the organization of all matter in our universe. Comprising 99.999 percent of all matter, magnetized plasmas, or Alfvén waves, detected

*Mitochondria are the powerhouses of our cells, converting nutrients and oxygen into energy. They have their own DNA. Damaged mitochondria = altered DNA.[49]

by telescope in deep space produce X-rays and gamma rays whose accelerating electrons stream from the Sun.* As magnetically charged plasma, Alfvén waves exhibit behavior similar to fluids and gases, but with the added complexity of containing magnetic (and sometimes electric) fields.

Before control over the ionosphere was achieved by the HAARP system, the polar jet stream in the Northern Hemisphere measured 6–7.5 miles high and 5.5 to 7.4 miles above sea level, and the subtropical jet stream measured 6–10 miles high and 6.25 to 10 miles above sea level. Both jet streams travel from west to east. Now, the polar jet stream is reported to be 80 miles high and 80 miles above sea level. The speed of the polar jet stream was once more than 450 miles per hour, but now with 1,000 mile-per-hour winds, who knows? Another lower, fragmented jet stream with strong electrical currents measures 50 miles high and 50 miles above Earth in the ionosphere. Seemingly, the altitude and speed of the polar and subtropical jet streams have been artificially increased (i.e., weaponized).[50]

According to *Webster's*, scientism (as opposed to science) is "an exaggerated trust in the efficacy of the methods of natural science applied to all areas of investigation." It is scientism that has kept the truth about the changed electrical dynamics of Earth—from weather, to Earth-Sun interconnectivity via Birkeland currents and Alfvén waves—confused in the public mind. This has been purposeful, given that geoengineering is weaponized and thereby classified, predicated on secrecy and deception as the natural planet is being morphed into an artificial planet for transhumans. The polar jet stream is an excellent example of this. Now erratic, it is manipulated to flow from Siberia to Texas, and pull warm air north to melt Arctic ice so that 15,600 km of fiber-optic cable can be laid for a 24-terabit connection between Tokyo and London.[51]

"They are flooding all of California. Farmers will be destroyed—California is the main food basket of the country, and this Santa Cruz flooding is just the start . . . Does this jet stream seem normal to you at all? What happened to a wide level of rain coming in and spreading over a broad area without such concentrated heavy downpours?" Lucretia Smith, Ashland, Oregon, January 1, 2023

---

*Telescopes can now span 73 octaves of the electromagnetic spectrum.

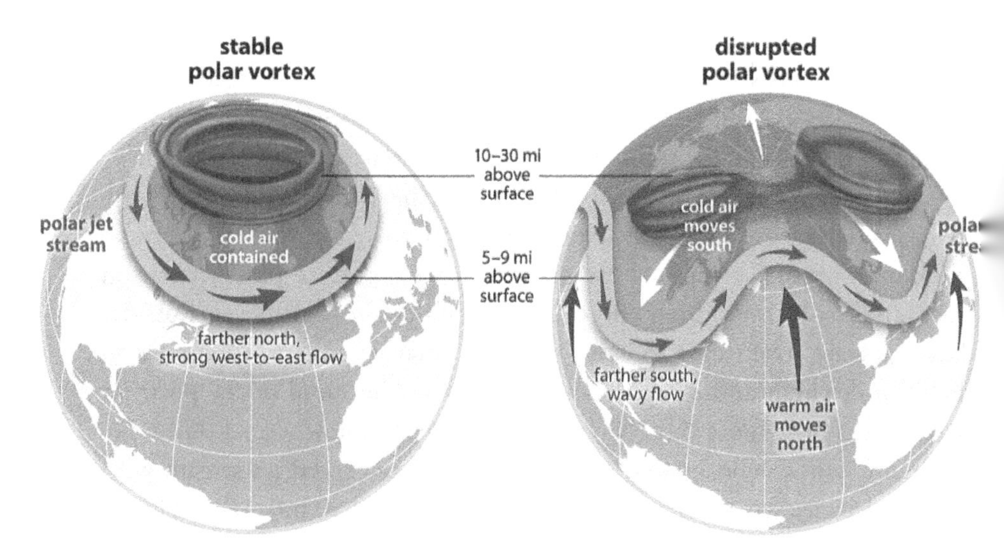

Fig. 1.4. The jet stream is now controlled and manipulated. It can even be dragged south as far as Texas.
National Oceanic and Atmospheric Administration

Alfvén wave generation also induces solar activity, perhaps because the Sun is basically plasma.* Meanwhile on Earth, to induce earthquakes and tsunamis, the ground-based HAARP or its sea-based HAARP equivalent, the U.S. Navy's orbital test vehicle the SDX-1, beam earthquake frequencies (2.5 Hz) up into the ionosphere, which then bounces the signal beam back to a preselected target area. Besides the sea-based

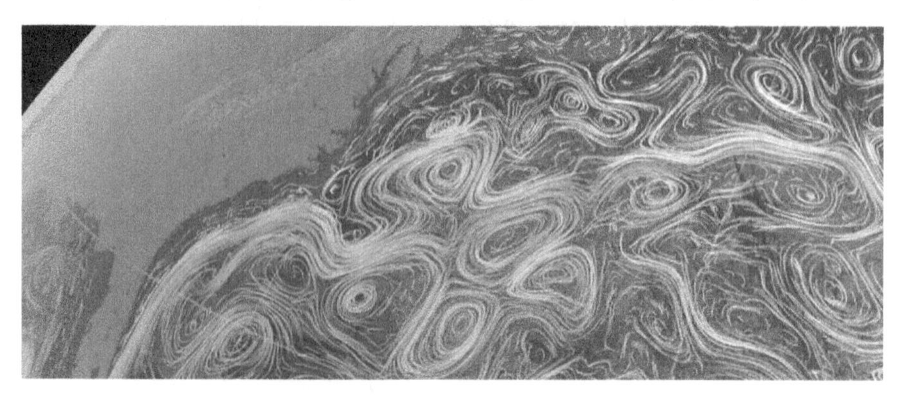

Fig. 1.5. Gulf Stream NASA, Florida-Carolinas coast, U.S.A.
NASA/Goddard Space Flight Center Scientific Visualization Studio

SDX-1, the air-based X-37B, the U.S. Space Force's unmanned, resusable "spaceplane," is in constant orbit, ever ready to steer the jet stream and beam earthquake- or tornado-inducing frequencies from directly over a target area.

> The US Air Force is preparing to execute a "climate chaos" event to facilitate the Vatican UN "Climate Change" New World Order agenda. The USAF Space Command redeployed their weather modifying and earthquake inducing Orbital HAARP called the X-37B *aboard* an Atlas 5 rocket on May 20, 2015. . . . Before the X-37B was built, they induced earthquakes and tsunamis by beaming the earthquake frequency (2.5 Hz) into the ionosphere from ground-based HAARP Alaska or the U.S. Navy sea-based HAARP, SBX-1, and the ionosphere deflected the very narrow beam to a preselected target on the earth's surface or ocean floor. With the X-37B now in orbit, they can beam the narrow-beam, earthquake-inducing radio frequencies from directly over the target (Haiti, Japan, Iran, waters off Puerto Rico or California, Yellowstone National Park, New Madrid fault line).[53]

MANIPULATING EARTH'S MAGNETIC FIELDS

Did you know that the universe is alive with natural electric plasma currents organized into sheets of filaments and cells with magnetic field–aligned boundaries (Birkeland currents, Alfvén waves, etc.) that charge the aurora borealis and form systems of galaxies linked like pearls on a string?

As above, so below. The mechanisms by which the *secret space program* pursues its agenda of full-spectrum dominance over planet Earth and the Sun run corollary to the mechanisms used for the Space Fence lockdown. Natural and manmade plasma, Alfvén waves, and Birkeland currents play a big role in these nefarious activities. Under cover of "climate change" remediation, multiple operations in both the troposphere and

---

*"Scientists using the joint European Space Agency (ESA)/NASA Solar and Heliospheric Observatory (SOHO) spacecraft have discovered 'jet streams' or 'rivers' of hot, electrically charged gas called plasma flowing beneath the surface of the Sun."[52]

stratosphere are being bankrolled by carbon credits and stolen Federal Accounting Standards Advisory Board (FASAB) trillions.*

On July 1, 1960, President Eisenhower (1953-1961) signed the executive order that moved U.S. space operations—the secret space program deeply buried in "national security" since the covert entry of Operation Paperclip Nazi scientists dedicated to space and mind control—from the U.S. Army to NASA, a civilian agency not answerable to Congress. Equipment valued at $100 million† and 4,700 civilian employees, plus Operation Paperclip aerospace engineer and space architect SS officer Wernher von Braun and his Nazi rocket team, were transferred from the U.S. Army's Redstone Arsenal in Huntsville, Alabama (and Cape Canaveral, Florida) to NASA's Marshall Space Flight Center.

Since then, the revolving door of the military-industrial-intelligence complex, now called *public-private partnerships*, has remained silent about the secret space program. The entire topic of space is like a Gorgon's head sprouting hypnotic, devouring snakes, which, if one goes up against them looking for facts instead of fiction, ends in being turned to stone: professionally ostracized, destitute, or in danger. The only ray of hope that the space program might be extricated from layers of disinformation has been the recent deep research of Mark Skidmore, professor of economics at Michigan State University, and investment analyst Catherine Austin Fitts, both of whom worked hard to unearth the fact that the *$21 trillion* in defense transactions missing between 1998 and 2015 led to the secret space program and "breakaway civilization" behind it.[54]

Because of the artificial atmospheric air ionization program implemented since HAARP achieved control over the ionosphere in 2013, our atmosphere and its climate and weather have been transformed, which leads to the fact that what we are breathing is chemically and electromagnetically transforming us. We live inside an ionized, highly conductive atmospheric dome loaded with trillions of nanosensors and nanobots monitor us and charged dry plasma clouds (ionized gas full of positive ions and free electrons). Electromagnetic radiation is everywhere, including in our lungs, blood, and brain.

By electromagnetically tweaking and adjusting the plasma density

*See appendix 8, "Movers and Shakers."

†One dollar in 1960 was worth eight dollars today.

all around us, kinetic energy can be built up or dispersed. Clouds are seeded with chemical agents as well as perfluorocarbon tracers (or tracers made of aluminized glass chaff fibers and sulfur hexafluoride) to measure atmospheric motion and provide information on plasma cloud structure. Rain composition can be adjusted chemically and electromagnetically to sustain the electrolytic water that supports military C5ISR operations (command, control, communications, computers, combat, intelligence, surveillance, and reconnaissance). Thunder and lightning can be provoked in our atmospheric "chamber" by using pulsed high-frequency sound waves (acoustic shockwaves or "booms in the air") to electrically adjust the atmosphere as well as to disable electronic systems and even shatter satellites.[55] Photons—particles representing quantums of light or other electromagnetic radiation—can be converted to electrons in the upper atmosphere, and metallic nanoparticles delivered by jets and industry (including nuclear) can be increased or decreased in the lower atmosphere.

Operations like those laid out in the following two patents depend on maintaining this calibrated plasma "antenna in the sky":

> US-20070238252-A1 (2005), "Cosmic particle ignition of artificially ionized plasma patterns in the atmosphere," describes how "the applications are useful for telecommunications, weather control, lightning protection and defense applications."*

> US-8981261-B1 (2012), "Method and system for shockwave attenuation via electromagnetic arc," describes how "denser-than-air" plasma shields over smart cities "will work by superheating the air between the intended target and the blast, using an arc generator to create a plasma shield, using methods which might include high-intensity laser pulses, conductive iron pellets, sacrificial conductors, projectiles trailing electric wires or magnetic induction."[57]

The advent of artificial atmospheric air ionization that actually began 150 years ago with industrial chemical pollution and the advent

*Plasma-based weapons are rife, the effects being presented as space weather, like the 620-mile-wide swirling mass of plasma called a "space hurricane."[56]

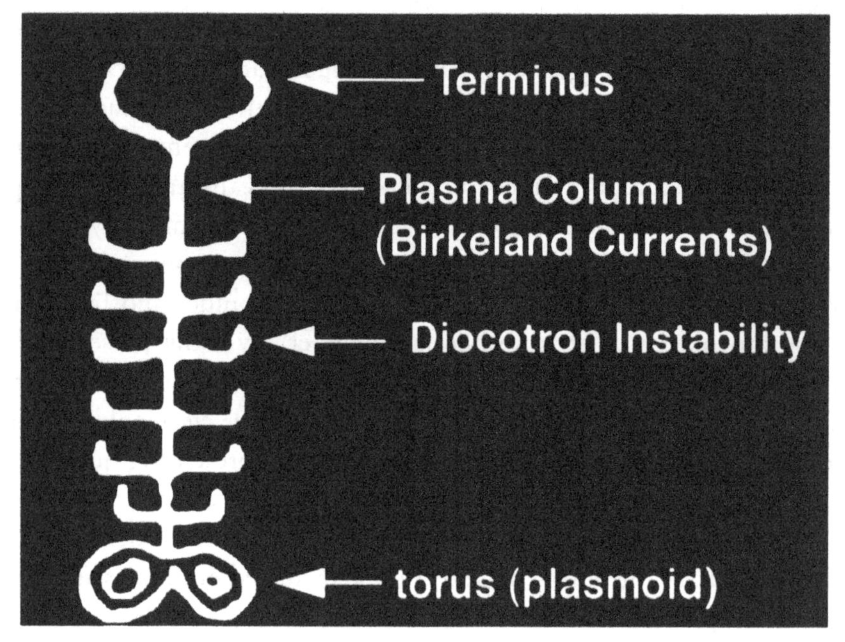

Fig. 1.6. Ladder of heaven, backbone of the sky. Did this rock pictograph in Kayenta, Arizona, depict a cosmic plasma event?
S. Schirott, "Plasma Scientist Anthony Peratt Meets the Electric Universe," *The Thunderbolts Project*, April 8, 2016

of electrification has certainly modified our lower atmosphere. None of us alive today has ever experienced an atmosphere that still serves biological life's O_2–CO_2 cycle. Way back when, therapeutic ionized oxygen followed from (1) decay of radiation from the Sun, cosmic radiation, and radioactive materials on Earth; (2) the 400,000-volt gradient electric field between the ionosphere and the Earth's surface; (3) natural wind (air masses) and weather fronts; and (4) falling (precipitating) water. Now, we are constantly penetrated by beams of electrons in plasma-radiating microwaves, an effect not dissimilar to the beams of billions of iPhones. The lower atmosphere (troposphere) accumulates more energy than it dissipates due to its now increased electrical field strength. Free electric charges make plasma electrically conductive (+), but even more importantly for all of life on Earth, plasma-radiating microwaves couple strongly with the electromagnetic fields around us, given that our bodies are organic and bioelectromagnetic.

CREATING WEATHER PHENOMENA

While averring they are maintaining the Earth's "radiation budget," i.e., the energy entering, reflected, absorbed, and emitted by Earth, geoengineers consistently leave out or lie about such manipulations as accelerating, eliminating, or retroengineering nature's electromagnetic and plasma phenomena, whether it's particle accelerators and Alfvén wave electrojets at the Poles, or juggling Birkeland currents to manipulate the ionosphere and solar events, or turning high-frequency ionospheric perturbations into weaponized high-frequency energy forms. As NASA (Not A Space Agency) glibly notes, "A [radiation] budget that's out of balance can cause the temperature of the atmosphere to increase or decrease and eventually affect our climate."[58] It is important to remember that short-term weather and long-term climate are viewed—as is the "radiation budget," in military terms—as *force multipliers*. In other words, weapons.

Both natural and man-made Alfvén waves and Birkeland currents are now manipulated to generate earthquakes and hurricanes, and to increase the speed of electrical propagation and conductivity between the ionosphere and atmosphere, the planets, and the Sun. This is accomplished by the high-powered phased array antennas wielded by ionospheric heaters like HAARP as well as air- and sea-based counterparts like the navy spacecraft referenced earlier.

> There's so much more to the [HAARP] story: ionospheric charging, Alfvén waves, solar radiation influx, DEW [directed energy weapons] technology, etc. They say that hurricanes are a product of evaporation and water temperature, with no reference to atmospheric charge potential. They say that ion channeling is the mechanism for propagating energy, but don't mention Alfvén wave downward cascades (Townsend avalanches). HAARP, which [Bernard] Eastlund designed, didn't simply heat an area of the ionosphere; it created Alfvén waves that energize the entire planet and induce solar activity via out-of-phase Birkeland current filaments between the Earth and Sun . . . Their next move is already in place to blame the Sun for climate change.[59]

An electrostatic field is a high-voltage field in which no current flows. A lot can be done with such a field in a plasma atmosphere, rather like rubbing an inflated balloon and sticking it to a wall, or being shocked by a doorknob after walking across a carpet. The static charge is held by electron transfer after the two surfaces separate. This is, in fact, exactly how electrostatic charge is retained in an atmosphere loaded with charged nanoparticles of metal and used to create the chemically nucleated snow that does not melt because it is composed of polymer/Mylar particulates storing static energy.* In Amarillo, Texas, on May 1, 2013, the temperature was 100 degrees Fahrenheit, but snow fell all the same, thanks to the static electricity that forced ice to nucleate high above the Earth. This "thundersnow" is the result of chemistry and the stored static electricity potential between the ground and upper cloud masses.

Electrostatic phenomena such as the circular holes blasted through cloud cover and artificial fixed rainbows are often due to transmitter-generated computer algorithms, like these "rainbows" seen in New Zealand by Rose and Greg:

> Greg and I witnessed over fifty rainbows through our beautiful Westhaven Inlet and on the way home. Every second corner had one or two, and double rainbows were over the waterways. The difference between these rainbows and what we would call "normal" was that they were static, not the ever-elusive "pot of gold" rainbows. They stayed in one spot; we could drive through them and then look back on them.

Similar effects can be produced by bombarding light with ions via laser-induced breakdown spectroscopy (LIBS), such that it is absorbed and reemitted in spectra to produce artificial rainbows. NASA's Orbiting Rainbows project manipulates and controls engineered dusty plasma aerosols via radio frequencies, optics, and microwaves:

> Our objective is to investigate the conditions to manipulate and maintain the shape of an orbiting cloud of dust-like matter so that it can function as an ultra-lightweight surface with useful and adaptable electromagnetic characteristics, for instance, in the optical, RF, or micro-

*Ice nucleation is chemically created with carbon nanomaterials: carboxylated graphene nanoflakes, graphene oxide, oxidized single-walled carbon nanotubes, and oxidized multi-walled carbon nanotubes.[60]

> wave bands . . . A cloud of highly reflective particles of micron size acting coherently in a specific electromagnetic band, just like an aerosol in suspension in the atmosphere, would reflect the Sun's light much like a rainbow.[61]

Making rainbows, albeit synthetic ones, sounds innocuous enough, even friendly, but the true purpose of this technology is far more insidious: the constant production of plasma cloud cover—"spatially disordered dust-like objects that can be optically manipulated"—cyber-morphed into static, sculpted, picture-perfect cumulus clouds that may be *watching us as we watch them.* As NASA notes, "The engineering of distributed ensembles of spacecraft swarms [of nanotechnology] to shape an orbiting cloud of micron-sized objects . . . can operate as an adaptive optical imaging sensor."[62]

MANIPULATING BIOLOGICAL FORMS

A whole other way of looking at electrostatic fields has to do with biology, namely what is called "the Ebner Effect." In the 1980s, at Ciba Pharmaceuticals,* biologists Guido Ebner and Heinz Schürch conducted a series of electrostatic field experiments with grain seeds and fish eggs. From their 1991 U.S. Patent 5048458-A, "Method of breeding fish":

> Unexpectedly, primeval organisms grew out of these seeds and eggs: a fern that no botanist was able to identify; primeval corn with up to twelve ears per stalk; wheat that was ready to be harvested in just four to six weeks; and giant trout extinct in Europe for 130 years, with so-called salmon hooks. It was as if the electrostatic field made these organisms access their genetic memories.

Activating the "primeval code" in living organisms fits with biology *and* synthetic biology in creating genetically modified "food" (GMOs) without the kind of genetic engineering that Monsanto is infamous for, but with electrobiology. As Ebner and Schürch put it, "Our experiments do not involve a mutation of the organism in question, which in the case

*In 1996, Ciba and Sandoz merged and created Novartis; the agricultural division was outsourced as Syngenta.

of genetic engineering involves channeling an additional gene into the organism. No entirely new organism is created. In the electrostatic field, only the gene expression is altered—the retrieval of the existing gene. . . . The Third World in particular could benefit from this method of cultivation that is both environmentally friendly and inexpensive."[63]

Notably, once the Ebner Effect was patented, it went black. A decade later in 2001, Ebner and Schürch both died "unexpectedly"; their untimely and *simultaneous* deaths curtailed publication of their discoveries in scientific journals. That the Ebner Effect is about altering gene expression and *not* about genetic engineering—shades of the recent mRNA "gene drive" debate—may be why a shroud of secrecy seems to hang over the Ebner Effect (just as it has hung over the recent mandated inoculations).

Ebner's two sons have done their best to keep alive their father's ecological alternative to the controversial genetically engineered seeds of the international agro-multinationals.

In a lecture delivered at the World Mysteries Forum in Basel, Switzerland, in 2008, Daniel Ebner explained his and his father's experiments: "It was as if these organisms accessed their own genetic memories on command in the electric field, a phenomenon which the English biochemist Rupert Sheldrake, for instance, believes is possible."[64]

The dirty politics of the agrobiz "Frankenfood" / Big Pharma industries run by Monsanto, Bayer, BASF, DuPont, Dow Chemical, Syngenta, and all the rest continue to do their best to hide their abuse of the Ebner Effect in the gene drive, but quantum physics and epigenetic inquiries continue to pursue the epigenetics working in tandem between geoengineering and biology, like how lightning during thunderstorms can be used to influence the electrostatic field generated between the Earth and sky to influence organisms' biology. In previous Earth ages, there were more electric thunderstorms. Is climate being manipulated to produce "extreme weather," more and stronger storms, more lightning, including ground-to-cloud lightning from the Earth as well as the heavens?[65] Has the natural environment become a lab for an entirely manmade evolution to replace the old Darwinian model of evolution?

When I look up at our ionized atmosphere and realize how much lightning is now manmade from positive-negative ion dances being generated from the Earth *and* atmosphere, I wonder how far the Ebner Effect has really been taken.

Fig. 1.7. The 6-megavolt Tesla Tower (a Marx generator, not a Tesla coil) at the High Voltage Research Center outside Moscow produces massive 200-meter volts of lightning and plasma arcs.

From "Soviet-era 'Tesla Tower' restarted with spectacular lightning bolts," RT, August 20, 2014

MANIPULATING THE SCHUMANN RESONANCE

Life on Earth is not just *affected* by the Schumann resonance; it is *shaped* by it. The vibrational forces heard for eons as the sacred sound *Om* synchronize our metabolic functions, including our circadian rhythms, with Earth's rhythmic cycles. Our brains are literally locked on to Earth's harmonic, which points to the *depth* of humanity's relationship with this planet in particular. Schumann frequency pulses connect and influence all living systems on Earth.

Inside the massive system of large phased array antennas of ionospheric heaters, LiDAR (laser imaging, detection, and ranging), and photographic space surveillance sensors linked together by satellite and terrestrial communications systems of the Space Fence ("smart grid" lockdown), our brains are monitored as they fluctuate from 8.5 Hz to 16.5 Hz in the Earth-ionosphere cavity in which we live: theta (4–7.5 Hz, which underlies various aspects of cognition, including learning, memory, and spatial navigation); alpha (7.5–14 Hz, which governs the state of relaxation and meditation), and beta (14–40 Hz, which characterizes a strongly engaged mind). According to the Global Coherence Monitoring System (GCMS), a worldwide network of sensitive magnetic field detectors that monitors fluctuations in Earth's geomagnetic fields and resonances in the ionosphere, "Resonances can be observed at around 7.8, 14, 20, 26, 33, 39, and 45 Hz, with a daily variation of about ± 0.5 Hz, which is caused by the daily increase and decrease in the ionization of the ionosphere due to UV radiation from the sun."[66]

Just due to UV from the Sun? *Really?*

Normally the Schumann standing scalar wave "rings" the Earth-ionosphere cavity like a bell thirty-four miles up, but it is equally true that when ionospheric heaters play their ULF frequencies in the Schumann resonance range of 7–8 Hz, the ionosphere "rings," as ions tumble into our lower atmosphere, where they are intensified by the conductive metal nanoparticles being zapped by radio frequency and microwaves. In other words, the Schumann resonance is being intentionally manipulated to *lower* our consciousness, not raise it.

THE POLES, THE SUN, AND BIRKELAND CURRENTS

In the 1990s, the military underwent a radical revision called the RMA (Revolution in Military Affairs). The majority of Americans never even noticed. If they did notice, they thought it just had to do with the military and not domestic affairs, knowing nothing of *dual use technologies* nor *the doctrine of full-spectrum dominance* over planet Earth and our solar system.

In 1996, NASA published *The Future is Now! Future Strategic Issues / Future Warfare [Circa 2025]*, a PowerPoint slideshow put together by Dennis M. Bushnell, chief scientist at NASA Langley Research Center.[67] Though only a PowerPoint, study it and you will see that much foreseen and planned for in 1996 has come to pass.

The extreme activity caused by HAARP-driven agendas in the Arctic Circle, including shifts in magnetic north,* and in Antarctica, points to how pivotal Birkeland currents are when it comes to full-spectrum dominance over our star, the Sun. Full-spectrum dominance thus begins with the Poles.

*"One magnetized patch is beneath Canada while the other is under Siberia," and it's not necessarily due to "the flow of materials in our planet's core."[68]

Birkeland currents, so-named after their discoverer, Norwegian physicist Kristian Birkeland (1867–1917), are a critical ingredient in a variety of plasma processes on Earth, such as the aurora borealis and radio emissions, as well as found throughout our galaxy. They connect our Poles with the Sun,* ionize atoms, create "funnels" of energy for the creation of electrojets, and perturb Alfvén waves in the ionosphere. Birkeland currents can be naturally twisted by zapping gas plasma with electricity. Increase the ion flow along the magnetic lines of force and it will be mirrored back toward Earth; increase the charge potential in the ionosphere and "fountain" up along the coherent inner core of Birkeland currents into the Sun's electromagnetic circuit. *Increase the charge potential of the Sun and voltage can be induced to increase solar activity*, as per Tesla in 1901.

With Nikola Tesla's visionary words ringing in our ears—"Man could tap the Breast of Mother Sun and release her energy toward Earth as needed, magnetic as well as light"—it is essential to grasp the fact that Alfvén waves also provide the ELF harmonics necessary for Earth-Sun communications that extend to all the planets in our solar system by inducing solar activity via sympathetic resonance between the Sun and Earth. In other words, we're seeing the future unfold.† With the help of ionospheric heaters in the Northern Hemisphere, the bipolar maser outflow from the two poles of our star can be increased.

Since the start of Operation Deep Freeze, the code name for the U.S. military assault on Antarctica beginning in 1955, and going back even further to Admiral Richard E. Byrd's expeditions to Antarctica in the 1920s, the Earth's Poles have been hotbeds of activity due to the Birkeland current boundary flows that connect Earth to the Sun via the Poles. For optimal Space Fence operations, Deep Freeze engineers are tasked with maintaining, calibrating, and experimenting with the balance between magnetic south and magnetic north. The truth is that the struggle over who will control magnetic north and the plasma energy pouring from the Poles is far from resolved. Control over magnetic north has everything to do with global power because the Earth is basically a magnet.

---

*The same appears to be true of other planets.[69]

†Much of this is from an August 1, 2016, email from Christopher Fontenot who has extensive training in nuclear and electrical engineering. Now, his primary interest is in the Electric Universe theory and its implications.

Before this electromagnetic age, extremely low frequency (ELF) fields emanated from natural solar and cosmic events—lightning, geomagnetic storms, volcanic activity, earthquakes, and the Schumann resonance. Now, however, cycles like solar minimums and events like eclipses, sunspots, and CMEs are all grist for the mill for those bent on turning Earth into a machine.

2

Those White Lines in the Sky

"Sixty billion gallons of kerosene-based jet fuel are burned worldwide every year by military and commercial aircraft, with 26 billion gallons burned in the continental U.S. alone, representing the single largest atmospheric chemical exposure to both civilians and military personnel. Additional exposures occur through contact with groundwater or soil contaminated by raw fuel constituents."[1] Yet regulation of aircraft exhaust is nonexistent, and major NATO and NASA fuel formulae remain classified.

The CIA-linked contractor Evergreen International Aviation's[2] 2010 patent for an "Enhanced aerial delivery system"[3] describes:

> an enhanced aerial delivery system [that] addresses issues raised when large quantities of fluids, powders, and other agent materials are to be transported in and aerially dispersed by aircraft. Some aspects include positioning and securing of tanks aboard the aircraft . . . Other aspects address coupling of the tanks and associated piping to lessen structural effects upon the aircraft. Further aspects deal with channeling, containing, and dumping stray agent materials that have escaped from the agent tanks on board the aircraft.

The challenges facing such a task as chemtrailing the entire planet are not inconsiderable, as the patent language explains:

> Aerial delivery systems receive, transport, and disperse fluids, powders, or other substances from aircraft to terrain below for various reasons. In certain cases including fire fighting, weather control, decontamination

> exercises, and geotechnical applications, it is desirable for large quantities of materials to be dispersed with each trip of the aircraft since areas for dispersion of the materials can be vast, travel distances between receiving and dispersion points can be great, and response time to complete a job can be demanding. These and other applications where large quantities of materials are to be aerially dispersed present particular issues regarding aircraft control, safety and other issues that unfortunately conventional approaches have not addressed.

While carbon is now taking the rap for the covert chemtrail program that has been going on for thirty years now,[4] military and commercial jets and sounding rockets continue to load chemical trails with particulate matter and conductive nanometals like aluminum, titanium, boron, barium, strontium, lithium, europium, calcium, and other substances (see appendix 5, "Substances Used in Chemical Spraying Operations"), all to keep our atmosphere fully ionized and battery-ready for a variety of "experimental agendas."

Ulrike Lohmann of ETH Zurich, a public research university in Zurich, Switzerland, is known for her research into aerosal particles in clouds. In the 2016 documentary film *Overcast: An Investigation into Climate Engineering,* Professor Lohmann discusses her team's study of aircraft engine exhaust and jet fuel from various turbines at Zurich Airport—not just the soot and particulate matter, but the "very rare" chemical composition measurements. The team confirmed the presence of sixteen different metals, particularly aluminum and barium.

The above 2010 patent language references "tanks" and "piping," "dumping," "weather control," and "geotechnical applications." Various ducts and bleed-air valves probably serve as "dual use" nozzles in line with the exhaust output and fuel "evac" nozzles off the wings, which exude pyrolytic compounds of carbon, chlorine, calcium, copper, aluminum, silicon, sulfur, phosphorus, iron, potassium, titanium, chromium, and magnesium from the combustion chamber, which contributes to the *aerotoxic syndrome* suffered by aircraft personnel.[5] Added to this chemical cocktail, cabin air includes lubricants, "stray agent materials" in the dual-use duct system.

An Italian documentary called "Prohibited Interview"[6] offers excellent, thorough testimony by Enrico Giannini, then an aircraft loading operator at Malpensa Airport in Milan, Italy. As a loading operator, Giannini observed the underbellies of commercial jets, the ducts along the wings for condensation drainage, special ducts said to be "sensors" along the blades of the turbine and predrilled holes under the fuselage. Clearing the ducts of water and oil often included having to *push out* materials that nebulized (converted to a fine spray) as they were discharged. Giannini concluded that the core components of geoengineering might be in the fuel, but other elements, including catalysts, are ducted from tanks and canisters in the extreme aft of the jet.

Regarding the aerotoxic syndrome, now they have installed 5G throughout the aircraft so passengers can download their own movies and games during the flight. Are they aerosoling "5G Fixer in a Can"? What kind of synergistic effect does 5G have on the chemical burden already in place? Those inventive engineers at The Mister, Inc., have "developed easy-to-manufacture nanoparticles in the shape of the 5G designation, internally code-named 5G-hype nanoparticles packed into a standard can pressurized by harmless nitrogen for convenient use and no safety issues."[7] No safety issues, *really*? As one 2020 investigation states:

> The fifth generation (5G) of radiofrequency (RF) radiation is [being] implemented globally without investigating the risks to human health and the environment . . . The evaluation of RF radiation health risks from 5G technology is ignored . . . There seems to be a cartel of individuals monopolizing evaluation committees, thus reinforcing the no-risk paradigm. We believe that this activity should qualify as scientific misconduct.[8]

FOGGING TECHNOLOGY

Then there's the disinfectant spray "fogging" of passenger jets and schools, nanoparticle nasal sprays, and chemically saturated masks. Under the guise of disinfecting an aircraft, fogging nanotechnology is sprayed in jet cabins between flights, producing symptoms like nausea; hearing and equilibrium loss; seizures; and cognitive impairment, including an inability to process information or even to speak, read, or write. Mayte Abad and her travel partner were badly nanofog-poisoned on two humanitarian flights from Central America to Canada via George Bush Intercontinental/Houston

Airport. They claim that rigorous detox can help, but their subsequent extreme electromagnetic sensitivity has never gone away—in fact, Mayte Abad is now a walking Geiger counter: "The nanobots in my brain make sure I can sense with inhuman accuracy every single deviation in microwave frequency and feel every 5G tower, meter, streetlight, and receiver long before I see it."*

Fogging nanotech consists of the nanochemicals Bacoban and Viraclean, both of which include the World War I chemical warfare agents benzalkonium chloride, benzyl chloride, and chloroacetophenone, as well as their derivatives: ammonia, radioactive aluminum, and radioactive mercury. These are the very same nanochemicals used in schools and all public spaces, subways, buses, and rental cars. After the electrostatic chemical blast, a GermFalcon UVC (ultraviolet C radiation) "terminal disinfection" sweep delivers a potent UV mercury vapor into the cabin air, the same mercury neurotoxin in ion propulsion rocket engines.† Travelers board the plane immediately after application, the nanochemicals in their bodies self-replicating for up to ten days, the positively charged atoms attracted to neutral or negatively charged atoms in the blood and brain.

AEROSOL-DISPENSING AIRCRAFT

The airline industry uses less fuel than you think.

A jet engine sits forward of the wings on pylons so as to create a magnetic field for vortex liftoff. Once the jet goes into cruise mode, you can hear the engine change pitch. The video "Jet Fuel Hoax Airbus A380 Exposed free energy truth"[10] begins by questioning how 255 tons of fuel could be loaded into the wings of the Airbus A380. Once aloft, however, an aircraft carrying such a load, along with 853 passengers, is possible, but not when the plane is on the ground during liftoff. So how does liftoff occur?

Consider the bumblebee's tiny wings, which resonate the hollow cavity next to its larynx while lifting its relatively massive weight. Once its resonance equals that of the magnetic field around its body, a magnetic

*Thanks to Mayte Abad (Facebook, September 14, 2020) for alerting me to this practice.
†Ion propulsion uses powerful magnets to push through charged nanoparticles at high speed to generate thrust.[9]

bubble forms, and the bee levitates into the air, where it can fly at a speed of up to thirty miles per hour, its wings steering but not adding to the thrust. Similarly, 90 percent of a jet's thrust comes from compressed bypass air, not from the fuel combustion chamber.

Austrian scientist and naturalist Viktor Schauberger (1885–1958),[11] the "father of bioenergetics," together with his contemporary, Russian scientist Viktor Grebennikov, studied levitation and propulsion, implosion, and vortices not with air, but with natural flowing water. The principle that governs rivers and the natural spiral / vortical patterns of Nature also governs aircraft in flight contending with an air flow similar to water. *Jet engineering uses the vortex to "suck" a jet aloft and into the sky.* Treated in this way, conversion of matter (air, particulates, gases) creates a "biological axis" along the axis in front of a jet's turbofans* that then produces what Schauberger called "synthesis electricity"—natural electricity produced by the action of a toroid in an electrostatic field. This *strake effect* occurs when engine strakes (i.e., aerodynamic surfaces on the nose, wings, and rear of the aircraft) create a vortex that sucks the jet upward as it moves forward. Since childhood, I have loved that particular moment of sudden levitation during takeoff! For the hovering or stalling of heavy jets, small vortex generators energize airflow over the wings and tail surface of the craft.

In virtually all commercial aircraft today, turbofans generate thrust by drawing in air with a fan. While most of the air goes around the engine, some passes through it, drawn in by a compressor composed of many blades attached to a shaft. The compressor pressurizes the air, greatly increasing its temperature. The hot air is then forced into a combustion chamber, where it's sprayed with fuel and ignited, resulting in hot, high-pressure gas that, as it expands, spins the engine turbine before blasting out of a nozzle *at the rear of the engine*, thrusting the aircraft forward. Because all the engine's components are connected through a central shaft, a rotating turbine not only drives the low-pressure compressor, it also spins the fan, providing additional thrust.[12] (A compelling theory is that compressed air moves the jet forward, with fuel being used only for takeoff, landing, and taxiing.)

---

*The *turbofan* (*fanjet*) is the air-breathing jet engine widely used in aircraft propulsion. A *turbine* is a steam-powered machine whose shaft produces electricity through movement.

Fig. 2.1. Mega-spraying operation in northeast Ohio.
Nick Rogers, www.nickrogersphotography.com

If this is how jets achieve thrust, then the chemical effluvium from the combustion chamber only occurs in quantity during takeoff and landing. Thus the "contrails" that form from what is spewing out of jets—often along the wings, not from the rear of the aircraft beneath the tail—must be from chemicals ejected from a supplementary duct system, as Mr. Giannini indicated in the Italian video interview "Prohibited Interview" discussed above.

Aerosol-dispensing aircraft fly under a cloak of secrecy. Air traffic control uses passive radar, not active, which means that aircraft transmit their call numbers with a transponder that can be turned off, or they can use fake identifiers. Not only jet aircraft, but ships too can be made to "go dark," meaning they no longer register on automatic identification systems so they can vanish from radar.[13] These phantom jets ("phantom" referring to the radar-absorbent material that turns on and off) lay long aerosol trails, then disappear on radar *and even visually, due to cloaking.**

Aerosol-dispensing aircraft have a perennial green-light whether under air traffic control or not. And space surveillance of aircraft is made possible by the PROBA-V satellite (less than a cubic meter in size), which picks up some 25 million positions of more than 15,000 jets with its Automatic Dependent Surveillance Broadcast.[15] Thanks to remote telemetry, computers run all aspects of jets in flight, even to filing flight plans.

If the jet is using a Rolls Royce engine, then Rolls Royce supercomputers monitor the engine and receive telemetry reports. According to pilot hearsay, dissemination of chemicals is often conducted by on-the-ground supercomputers that instruct the flight guidance system as to where and when to release the chemical payloads per their frequency signatures, similar to how the flight termination system takes full control over drones as well as military or passenger jets.†

---

*Radar-absorbent material, or RAM, is a coating containing carbonyl iron ferrite ("iron ball" paint or tiles). When radar encounters RAM, a magnetic field forms inside the metallic elements and an alternating polarity dissipates the signal.[14]

†The PFN-TRAC system "has been put in place already at many levels and is being used to aggressively go in, take over, control, track and manage any system, whether it be an airplane, car, motorcycle, human being or animal."[16]

Get the picture? Pilots may be allowed to control their instruments, but who's directing their brainwaves from the ground?

> Mind control has made its way into airplanes. . . . The [brain-computer interface, i.e., mind control] system monitors and interprets signals in brain electrical activity that translate to movement commands. Once a pilot is connected to the system using a set of 32 electrodes, he can monitor his brain activity using a brain-computer interface. The pilot focuses on arrows on the interface that will tell the computer what he wants the plane to do.[17]

SUPPLEMENTARY ATMOSPHERIC DELIVERY SYSTEMS

Beside the jet deliveries of chemicals loaded with nanoparticles (and therefore nanotechnology) are three other supplementary delivery systems beyond the normal pollution of carbons, etc. Two are generated at sea, one on land.

Marine cloud brightening sprays *sea spray aerosols* up into the atmosphere to make clouds that reflect sunlight back into space while at the same time feeding and directing geoengineered extreme weather events like hurricanes (cyclones).

Ship tracks—ship plumes from ship smokestacks of U.S. Navy, Merchant Marine, and commercial ships head skyward to make, feed, and steer weather systems along with what the jets are providing while keeping ships hidden from satellite eyes and influencing communications. The sulfur nanoparticles in marine diesel and CO_2 provide nuclei for cloud condensation, which was highly useful during the June 1994 Monterey Area Ship Track (MAST) experiments off the California coast to concoct sea spray aerosols so as to nucleate nanoparticles for producing artificial clouds.[18] Engines of aircraft, motor vehicles, and ships all emit *chemiions* that play an important role in particle formation.[19]

Like the chemtrail chemicals from fuel and supplementary chemical additives, ship tracks distribute nanometals like lithium oxide (Li202). Interestingly, lithium is highly water-reactive and psychoactive; combined with superheated saltwater (or blood?), an electrostatic / conductive state can be created.

In 2016, Oregon activist Ann Fillmore, PhD, wrote the essay "Aerosol Experiments Using Lithium and Psychoactive Drugs Over Oregon" (*PositiveHealthOnline*, Issue 228, February 2016) about what an anonymous government whistleblower had informed her: that in 2015, lithium aerosols had been dropped on Oregon Coast towns after which Sociological Research Division operatives had been sent out to collect hearsay health complaints: lethargy, thirst, stomach distress, sudden weight gain, muscle and joint pain, twitching, loss of appetite, slurred speech, blurred vision, confusion / hallucinations, ersatz goiter, impotence, endocrine disruptions causing severe menses, kidney pain, skin rashes, hair loss, etc.—"the test population's behavioral traits like consumer habits, political engagement levels, and awareness of geoengineering programs."[20]

The whistleblower called himself "Locke" (no doubt after the Enlightenment philosopher John Locke, 1632–1704), describing himself as "an employee of a weather data collection company and, by proxy, a subcontractor for the National Weather Service office in [central Oregon] . . . I collected data, e.g., soil samples, that were used to direct spraying operations for the last three years." He discovered quantities of lithium used to manufacture air stagnation in the Rogue and Umpqua Valleys as well as much of the Oregon Coast south of Florence. As Fillmore observed, "I noticed the new method of holding the lithium haze cloud in place first over the area in northern California where the massive fires have hit . . ."

As I wrote in *Under an Ionized Sky*:

"Holding the lithium haze cloud in place" sounds distinctly like ship tracks. The most active part of ship tracks lies *above*, where a plasma-induced iCloud "computer" can be pinned and wedged between the frequencies of a higher altitude chemclouds and low-altitude chemclouds whose ice particle count has been increased by laser- and radio-frequency-induced plasma. Introduce anhydrous ammonia (NH_3) at specific points during the lithium/water

> reaction and the mid-atmosphere can be stimulated. Supercharge the two layers of chemclouds and run a laser beam between them to create a super-antenna that gathers and stores messages. Set up networks, communicate with satellites and other processors, etc. Strike the first beam with another beam (scalar interferometry) and you can store hard data on a virtual "CD."*

Direct steam condensation for generating power—steam releases from nuclear reactors, power plants, etc.—often depends upon evaporative *wet surface air cooler (WSAC)* technology used to fuel geoengineered weather systems. The 7,000+ power plants in the U.S. have large cooling towers (WSACs) that produce *water vapor generation fueling.* The cover story is open-loop water to cool closed-loop fluid from the power plant, but like the microwave (cell) towers built to produce far more power than cell phone connectivity requires, WSACs are built for "dual use." The power plant stack contents mix with cooling tower water vapor as both ascend to mix with descending chemtrails loaded with nanoparticles.†

WSAC steam particles ascend, the moisture feeding what is descending from jets and rockets, thus multiplying the chemical effect. Note that nuclear reactor stacks mean radioactive nanoparticles are in the chemical brew of chemtrail synthetic cloud cover, all synergizing with the descending conductive nanometals and Mylar nanoparticles that we then breathe in.

Extreme weather events are not due to thousand-year cycles or "inland sheared tropical depressions"; they are geoengineered using commercial and military aircraft, ship tracks, and in-place water vapor generation (WSAC). Spray nano-laden aerosols from jets, ships, and factories into the atmosphere to capture moisture, then release billions of tons of rainwater with electromagnetic pulsing, and *voilà,* an atmospheric river on demand.

---

*Review *Under an Ionized Sky: From Chemtrails to Space Fence Lockdown* (Feral House, 2018). Strike the first beam with two other beams and a 3D hologram is created, like the hologram jet that hit the Trade Tower on September 11, 2001.

†Thanks to WeatherWar101's August 17, 2016, email and NexRad YouTubes.

OZONE DEPLETION

Ozone (O_3) is an upper atmosphere (10–15 miles up) bluish gas that absorbs the bulk of UVB (ultraviolet B radiation). Ozone depletion in the stratosphere (the middle layer of our atmosphere) and its increase in the troposphere (the lowest layer of our atmosphere) is conventionally attributed to burning fossil fuels (cars, jets) that produce nitrogen oxides. But Christopher Fontenot, the former U.S. Navy nuclear propulsion engineer mentioned earlier,* has a whole other take on ozone depletion: namely, the relativistic electron acceleration in our ionosphere (30 to 600 miles above Earth) due to the geoengineered high-frequency modulation of Alfvén waves and the ionospheric heaters that ionize Earth's electrojets with endless nanoparticles delivered by jets and rockets.[21]

Solar radiation management—"a climate engineering strategy to reduce temperature increases due to global climate change . . . [by] increasing the concentration of aerosol particles in the stratosphere"[22]—is generally what people think of when they hear the word *geoengineering* because this is how the mainstream media have conditioned people to think of the entirety of the classified geoengineering program. Independent scientist Clifford Carnicom's 2005 "Seven Geoengineering Operations" gives a more honest portrayal.

For example, the *Stratospheric Aerosol Injection (SAI)* solar geoengineering program spraying large quantities of sulfur dioxide into the stratosphere is most definitely a chemical operation not so much in service to "climate change" as other operations. The geoengineered fires called "wildfires" contribute 10 to 20 percent of the sulfur that then becomes carbonyl sulfide (COS), a long-lived form of sulfur in the troposphere.

Besides creating acid rain, these sulfur injections lead to more and more ozone depletion. So it goes when such "save-the-planet" programs are backed by ExxonMobil and Shell Oil.

---

*Fontenot is a proponent of Electric Universe theory, which says that electricity is the engine behind a long list of natural and astrophysical spectacles. For decades he devoted himself to educating the public about the dangers of smart grids, smart meters, and radiofrequencies. His site A Microwaved Planet no longer exists as he has withdrawn to a private life he richly deserves.

SEVEN GEOENGINEERING OPERATIONS IN SERVICE TO THE SECRET SPACE PROGRAM

Thanks to independent scientist Clifford Carnicom for the basic paradigm.

Weather / Climate
Chemical / Electromagnetic
Planetary / Geophysical
Directed Energy Weapons (DEWs)
Surveillance / Neural manipulation
Nanotechnology / Digital *synbio* / Transhumanism
Cloaking / detection of exotic propulsion craft and plasma life forms

WEATHER / CLIMATE

Chemical / electromagnetic creation of plasma cirrus cloud cover, a conductive C4 atmospheric matrix for building and steering extreme weather events, manipulating the ionosphere to charge, and millions of other wireless operations. Weather is the *sine qua non* of every Space Age operation in our atmosphere.

CHEMICAL / ELECTROMAGNETIC

Electrical weather events, volcanic eruptions, earthquakes, etc., all of which serve to keep the atmosphere charged. Plasma and antimatter experiments and "farming." Creation and manipulation of Birkeland currents, Alfven "whistler" waves, rotating electrical fields, etc. Plasma rotation under EM field stress to convert transverse waves into longitudinal (scalar) waves and vice versa.

PLANETARY / GEOPHYSICAL

Utilize engineered droughts, fires, floods, hurricanes, tornados, earthquakes, storms, polar vortices, solar cycles for environmental modification and disaster capitalism. Earth harvesting for REITs (real estate investment trusts). Solar experiments like the Sun simulator; as Tesla said, "Man could tap the breast of Mother Sun and release her energy toward Earth as needed, magnetic as well as light."

DIRECTED ENERGY WEAPONS (DEWS)

The 1990s Revolution in Military Affairs (RMA) has changed warfare to the quest for full-spectrum dominance of both the electromagnetic spectrum and the cultural civilian spectrum. Instabilities of systems are identified and weaknesses exploited by means of remote applications of electromagnetic signatures, pulses, and frequencies. Scalar interferometry of ionospheric heaters, lasers / masers, particle beams, HPMs, etc.

SURVEILLANCE / NEURAL MANIPULATION

Artificial intelligence (AI), mind control / remote neural monitoring (RNM), EM frequency targeting of populations, 5G / 6G remote access to DNA, etc.

NANOTECHNOLOGY / DIGITAL *SYNBIO* / TRANSHUMANISM

"Hive mind" aerosol / GMO / vaccination delivery of nanotechnology (sensors, microprocessors, electro-optics, Morgellons, etc.) to be inhaled and ingested. Remote genetic engineering (epigenetics, optogenetics). Pandemics delivered via 5G and disease frequency. Replace Nature with virtual reality and cyborgs.

OBSCURATION / DETECTION OF EXOTIC PROPULSION CRAFT AND PLASMA LIFE FORMS

Hiding and demonstrating manmade exotic crafts, generally triangular and in plasma cirrus cloud cover, seemingly capable of producing their own plasma cloud cover; plasma life forms visible with infrared, some ancient inhabitants of planet Earth, some manmade as self-replicating, evolving inorganic life forms.

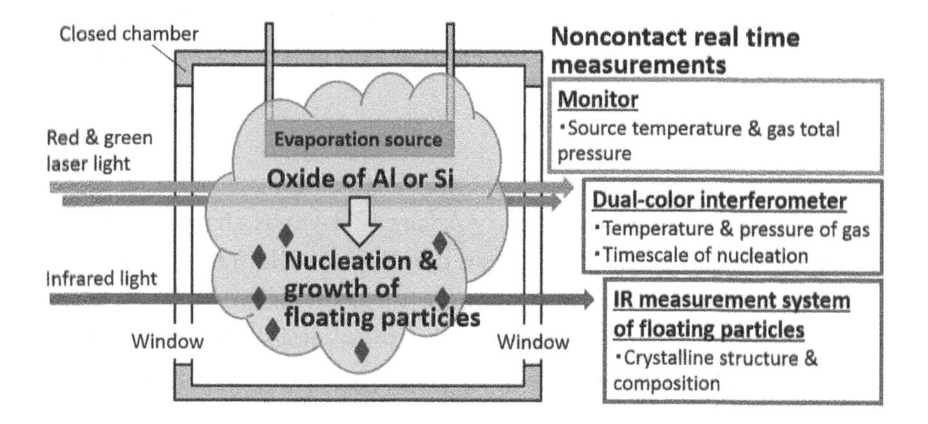

Fig. 2.2. Nucleation and growth of oxide nanoparticles, aka star "crumbs." Hokkaido University: "We identified the initial conditions required to form nanoparticles, the basic building blocks of earth-like planets. In particular, we searched for conditions where nucleation of oxidized aluminum and silica could easily occur, and we identified the first nanoparticles which govern the evolution of cosmic dust."
Hokkaido University, Japan, Sept. 28, 2015

NANOMETALS

Metals and minerals go with planet Earth, don't they? Even sunlight has spectroscopic lines of iron, gold, and other metal "lights" in it. Ore deposits and their veins lie deep in the Earth. Without magnesium, plants would not form chlorophyll or build their bodies out of air and water. We could not breathe without iron. Anemia follows from a lack of cobalt in the soil. Throughout our bodies, bits of all kinds of minerals are distributed, like calcium carbonate and phosphate of lime in our bones, all with important functions, all needed for the physical body to retain its health integrity and some for consciousness itself.

STAR "CRUMBS"

"We identified the initial conditions required to form nanoparticles, the basic building blocks of Earth-like planets. In particular, we searched for conditions where nucleation of oxidized aluminum and silica could easily occur, and we identified the first nanoparticles which govern the evolution of cosmic dust."—Press release, "Successful Launch of the Sounding Rocket S-520-30 Experiment using a microgravity environment to reproduce star 'crumbs,'" Hokkaido University, Japan, Sept. 28, 2015

Truthfully, the metals in our bodies and in all life on Earth come from the cosmos. Meteoric iron contains iron, cobalt, nickel, copper, and other minerals. Plants take these cosmic forces and make from them their chlorophyll life blood. All animals and humans depend on the cosmic rhythm of their inner organs for the quality of their blood, and it's minerals from the cosmos that make it happen: iron from Mars, copper from Venus, gold from the Sun, lead from Saturn, tin from Jupiter, silver from the Moon,

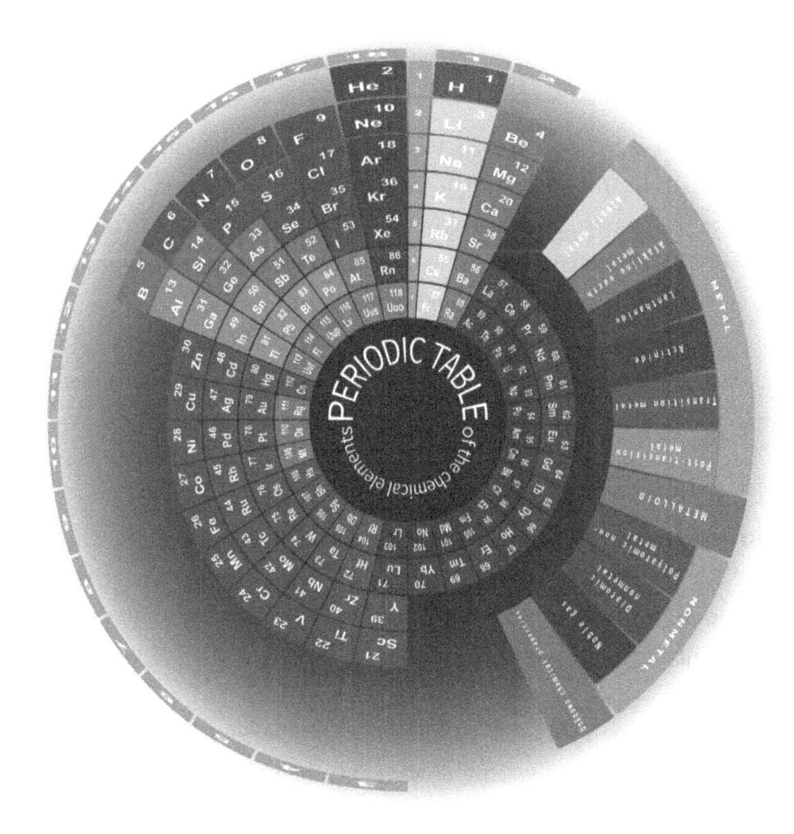

Fig. 2.3. The periodic table of elements.
Image ID 47259048 by Danielz1 | Dreamstime.com

mercury from the planet Mercury. Copper, silver, and mercury are alkaline and excellent conductors, primarily affecting the metabolic system, while iron, lead, and tin are acidic and poor conductors, affecting the nervous-sensory system.

As the ancient Tao Te Ching warns, "The world is a sacred vessel which must not be tampered with or grabbed after. To tamper with it is to spoil it, and to grasp it is to lose it." The problem with manufactured "dusts" consisting of conductive nanometals raining down every day on all of planetary life is twofold: the sheer quantity of the spraying for nearly three decades now, and the alchemical forcing of metal synergies that affect changes in all organic life forms on Earth, including the soil in which we grow our food. Such acts are not only misguided and negligent: they are intentional and therefore fall into the category of pure evil.

Barium, strontium, calcium, and magnesium are released together in the upper atmosphere for their ability to ionize gases as low pressure.

"This particular group of metals together—barium (Ba), strontium (Sr), calcium (Ca), and magnesium (Mg)—are cold cathode emitters used industrially for their unusual special property of having a low electron work function. The important thing to understand is that this special group of unique elements was chosen to aid in the ionization of gases at low pressures, just like a neon tube does, just like the air in the atmosphere does. That makes them really useful for HAARP devices." —Geoengineering researcher and targeted individual CAROLYN WILLIAMS PALIT in 2007

In my early days of searching the Internet for facts about the classified HAARP project and "chemtrails" being dumped into the atmosphere, targeted geoengineering researcher Carolyn Williams Palit was extremely valuable; eventually, she was eliminated:

> We are dealing with Star Wars [the Strategic Defense Initiative (SDI) program]. It involves the combination of chemtrails for creating an atmosphere that will support electromagnetic waves, ground-based electromagnetic field oscillators called gyrotrons, and ionospheric heaters. Particulates make directed energy weapons work better. It has to do with

> "steady state" and particle density for plasma beam propagation. They spray barium powders and let it photo-ionize from the ultraviolet light of the sun. Then they make an aluminum plasma generated by "zapping" the metal cations [+ charged ions or atoms] that are in the spray with either electromagnetics from HAARP, the gyrotron system on the ground [the Ground Wave Emergency Network (GWEN)], or space-based lasers. The barium makes the aluminum plasma more particulate-dense. This means they can make a denser plasma than they normally could from just ionizing the atmosphere or the air. More density means that these particles which are colliding into each other will become more charged because there are more of them present to collide. What they are ultimately trying to do up there is create charged-particle plasma beam weapons. Chemtrails are the medium—GWEN pulse radars, the various HAARPs, and space-based lasers are the method, or more simply, *Chemtrails are the medium, directed energy is the method.* Spray and zap.[23]

Nanoparticles of barium, strontium, calcium, and magnesium are released in the upper atmosphere to aid in the ionization of gases (i.e., plasma) at low pressure. Once they fall earthward, we inhale and ingest their nanoparticles and grow food in soil contaminated with them.

Calcium

Barium and strontium fool the body into thinking barium is calcium and strontium is magnesium; cell phones, through a microwave network, cause our neurons to release calcium ions, which makes us tired, irritable, and emotional. *Stress* is a key word when it comes to what electromagnetism can do to calcium.

"Barium and strontium bind together inside the body. Because they are very similar to calcium and magnesium, the body is fooled into taking them up. They then get into the soft body parts and cause untold medical conditions." —Email, Paul Stephen Cox, researcher, March 21, 2018

It is child's play to transmit an ELF-modulated signal to be broadcast by the entire cell phone network, if need be. By this means, all cell phone users can be behaviorally modified or forced to develop cancer from low-level microwave exposure from the phones, which stresses the neural network through constant calcium ion efflux and interference with bioelectric fields.*

Mentioned earlier, Martin Pall, PhD, the professor emeritus of biochemistry at Washington State University who has done extensive research into 4G and 5G technologies and the constant electrosmog caused by the Wi-Fi fields we live in,[25] points out that our negatively affected stem cells produce oxidative stress on the voltage-gated calcium channels (VGCCs) in the plasma membranes around muscle, neural, and glial cells† by loading a million ions per second into our cells via the voltage sensors in the plasma membrane surrounding each cell. The increase in cellular calcium, nitric oxide, and excessive signaling confuses the immune system and produces a host of chronic autoimmune *symptoms* named to sound like new *diseases*: lupus, rheumatoid arthritis, Crohn's disease, irritable bowel syndrome, Type 1 diabetes, chronic fatigue, fibromyalgia, and others.‡ Complex calcium signaling overload is the domino effect behind many autoimmune conditions, from cataracts and the breakdown of the blood-brain barrier, to lowered nocturnal melatonin (and increased nocturnal norepinephrine) and metabolic weakening.

Barium

In 2005, Clifford Carnicom, whose scientific background includes technical work for the Department of Defense, used spectral analysis to determine that the combination of nanometals he was collecting from precipitation and a HEPA filter had seemingly been designed to interfere with the resonant frequency of our potassium ions when hit by ELF radia-

*"Mounting evidences suggest possible non-thermal biological effects of radiofrequency electromagnetic radiation (RF-EMR) on brain and behavior. Behavioral studies have particularly concentrated on the effects of RF-EMR on learning, memory, anxiety, and locomotion."[24]

†This plasma is not the same plasma of plasma physics. Blood plasma makes up 55 percent of our blood and acts as a gatekeeper between the blood and circulatory system. It is a light yellow liquid that carries water, salts, enzymes, hormones, the proteins albumin and fibrinogen, immunoglobulins (antibodies), and clotting factors to parts of the body that need it, plus removes waste from the body's cells.

‡The trend of misdiagnosing symptoms as diseases is often based on ignorance of the role of electromagnetic radiation, as supported by numerous studies, one of which is Kıvrak et al., "Effects of Electromagnetic Fields Exposure on the Antioxidant Defense System."[26]

tion. Barium blocks the passive efflux of intracellular potassium, *the* essential alkali. Potassium depletion can lead to heart fibrillation, arrythmia, and heart attacks. Without potassium, we are vulnerable to Big Pharma broadcasts of gain-of-function disease frequencies such as what has been called the Covid "virus" (i.e., not a virus—see chapter 9, "*Synbio*"). A lack of potassium chlorate affects the lungs; a lack of potassium phosphate, the cerebrum; a lack of potassium sulfate, the solar plexus and colon. It may be the same for the sodium and chloride ions, which act as neuron action potentials, varieties of action impulses sent by neurons to parts of the body that want to move.

Given the power of carrier frequencies in a chemically ionized atmosphere, it appears that the central nervous system of human beings is being intentionally targeted for transhumanist modification. If interference signals are superimposed on the natural signals generated by the body (e.g., by using artificially created centimeter waves as a carrier), the brain could be presented with simulated states that we consciously perceive but which do not exist in reality . . .

> In a "psychotronic war" using microwaves modulated by using ELF waves, it would no longer be necessary to kill whole armies by inducing cardiac or respiratory irregular signals. The enemy can simply be incapacitated by disturbing their states of balance or confusing the ability to think logically.[27]

Microwaves make me think of Xiuhcoatl, the fire serpent of the Aztecs. Legend says that the success of the Aztec sun god Huitzilopochtli in battle was due to his skill in wielding the fire serpent. Was the fire serpent a high-tech weapon emitting high-power microwaves (HPMs), lasers, or plasmas?

Released into the atmosphere and ionized by light and UV radiation, hygroscopic barium removes water from the air and sets up a conductive layer that acts as a filter to short out Earth's electromagnetic current (800–1,000 volts per meter potential gradient)* in order to keep the

---

*The Earth's charge potential is no doubt increasing, due to the increasing power density of EM radiation. It used to be much lower (e.g., 100–300 volts per meter). This artificially induced electromagnetic current impacts solar activity via the Sun-Earth connective Birkeland currents and, as you can imagine, has untold impact on human health.

Fig. 2.4. The Aztec sun god Huitzilopochtli.
From the *Codex Telleriano-Remensis*, sixteenth century

trillions of nanoparticles, nanobots, and nanosensors discharged in chemtrails suspended. Unlike short-lived isotopes that induce conduction and gather and direct moisture, ionized barium uses the Sun's radiation to form a chemical rain that causes radiolysis (molecular damage caused by ionizing radiation).

Metamaterials—materials engineered to have properties that are rarely observed in naturally occurring materials—are basically artificial and

Fig. 2.5. Aztec sculpture of the fire serpent Xiuhcoatl.
Photo by Simon Burchell

include engineered nanoparticles (ENPs). A 2013 paper in *Environmental Science & Technology* points to metal oxide nanoparticles, fullerenes (molecules of carbon in the form of hollow spheres, ellipsoids, tubes, etc.) and carbon nanotubes (CNTs)—all of which are implicated in chemical aerosols.

> . . . it is currently postulated that ENPs cause nonspecific oxidative damage and that the resulting stress may be the predominant cause of DNA damage and subsequent genotoxicity.[28]

The barium ENPs we are breathing in are much more refractive (light-bending, as opposed to light-reflective)* than barium in Nature. After absorbing high levels of UV emissions because of its low electron volt work function, the barium ENP ionizes the emissions and re-emits them into our visible spectrum.

Barium has complex synergistic capabilities. Barium-strontium-titanate has the ability to share calcium electrons (valence) that are highly fluoride-reactive. Barium's synergy with hydrogen fluoride is said to remove fluoride and other acids from atmospheric suspension. In the mid-1980s, electrical engineer James E. Phelps at Oak Ridge National Laboratory measured high-frequency (HF) emissions and tied them to national health damage and changing local weather, after which Oak Ridge DOE (Department of Energy) and TVA (Tennessee Valley Authority) studies led to major DOE chemtrail operations following Phelps's prescription for *air pharmacology* to treat the air with chemicals that mitigate and offset the effects of toxic chemicals in the air such as hydrogen fluoride and chlorine. By 1999, Phelps was blowing the whistle about worker illnesses at the Oak Ridge National Laboratory, referencing how burning soluble uranium fluorides releases hydrogen fluoride and other toxic fluorides into the air, along with chemical catalysts like methyl cyanide.[29] Lung damage and high blood calcium mean HF exposures that lead to symptoms we have heard about for years: arthritis, sore joints, thinking impairment, rashes, and fatigue. HF calcium impacts nerve myelin and kills off mitochondria in cells, leading to heart spasms and attacks.

Magnesium

Introduction to Atmospheric Chemistry, by Peter V. Hobbs, published in the year 2000, is considered a foundational textbook on atmospheric chemistry, and yet it strangely does not discuss magnesium, titanium, aluminum, barium, or calcium.† The year following publication of Hobbs' book, Carnicom noted that substantial amounts of elemental

**Atmospheric refraction*: when direction of the propagation of electromagnetic radiation or sound waves in the atmosphere is changed due to density gradients in the air.

†In fact, the publisher demanded that Clifford Carnicom remove Hobbs's chart of atmospheric elements from his website.[30]

magnesium kept showing up as crystals in his environmental samples. Two-thirds the weight of aluminum, magnesium is as conductive as copper and aluminum and can be ionized by the Sun's UV. Over twenty years ago, Carnicom reported that "evidence continues to accumulate that certain metals, i.e., magnesium and barium, as well as certain biologicals and fibrous components, are established as the core elements of the aerosol operations in progress."[31]

Depletion of real (as opposed to synthetic) magnesium as a result of the chemical soup raining down on us has robbed us of the very element that enhances the binding of oxygen to heme proteins so as to produce healthy red blood cells (hemoglobin and iron), and therefore oxygen transport and utilization. A lack of adequate amounts of natural magnesium is why Morgellons sufferers (see chapter 9, "*Synbio*," for a discussion of the Morgellons pathogen being delivered by geoengineering) swear by Epsom salt baths (MgSO, magnesium sulfate) as well as magnesium supplements taken with zinc, liquid magnesium spray, and magnesium-rich foods in a balanced acid-alkaline diet. ENP magnesium depletes us of natural magnesium.

Strontium and Fluoride

Strontium, an alkaline earth element that is highly chemically reactive, is essential to geoengineers, as it allows them to create charged metallic aerosols for artificial dusty plasmas[32] (see chapter 1) in *noctilucent* ("night-shining") *clouds* that form in the upper atmosphere some fifty miles above the equator where it is bone-dry (no water vapor). Noctilucent clouds, normally only visible at twilight, are now made by bombarding ice at the Poles with radio frequencies to create charged, diamondlike structures in dusty plasma. This is primarily the cause of the release of methane at the Poles.

A discussion of strontium leads us to a deeper consideration of ionized radiation in the air and fluoride in public water, and the synergy between these two elements in the human body. The atmospheric nuclear tests that took place between July 16, 1945, and November 4, 1962, poisoned the atmosphere with one billion grams of radium from strontium-90, whose half-life is 28.8 years. Add high-voltage electricity, radar microwaves, nuclear energy, and heat released by spent fuel rods from nuclear reactors, and it's easy to see how the "deadened orgone" (the state of no life force, or chi) that psychiatrist Wilhelm Reich (1897–1957) talked about, when the atmospheric energy seems dead and even toxic, has been depleting the life force of our living planet and its inhabitants for almost a century. In 1950, Reich's Oranur experiment tested the effects of concentrated orgone energy on nuclear energy. Among other discoveries, Reich proved that deadened orgone clouds cause drought and desertification.*

Throughout the Cold War, the affinity between strontium and fluoride to form highly toxic, extremely reactive strontium-90 plus fluoride (Sr90F2) was consistently lied about in the name of "national security." The practice of fluoridating water, which began in the United States in 1945 shortly after the atomic bombing of Hiroshima and Nagasaki, has since then dramatically increased the presence of highly insoluble and radioactive strontium-90 in the entire world population. Strontium-90 plus fluoride settles in bones and teeth and leads to genetic mutations.

Were these environmental releases intended to further weaken Human 1.0 for transhumanist modification? Fluoride rarely occurs naturally in biological molecules,† so what was the reasoning behind accustoming Americans to drinking and bathing in fluoridated water? In 1960, it was 16.6 percent of 180.7 million Americans; it's now 62.4 percent of 323 million Americans.[33]

Fluoride compounds like hydrogen fluoride as well as fluoride nanometals descending from the upper and lower atmospheres are pulled out of the atmosphere and neutralized to some degree by barium, but not nearly enough to counter all the ways fluoride compounds employed by the military-industrial-intelligence complex synergistically‡ combine to be breathed in or ingested with the fluoride water—like PERTRAS (the perfluorocarbon tracer system) that tags air masses with a PFC (perfluorocarbon) compound by loading aircraft so they

*The chemical plasma clouds of today are often darkened by deadened orgone, not by withheld rain.

†The pineal gland sequesters fluoride from the bloodstream, but for how long when under constant exposure? Fluoride either lays the groundwork for or produces effects like hearing voices (schizophrenia), and if it impairs the left occipital lobe, one's willpower is impaired.

‡*Synergy:* the interaction or cooperation of two or more substances to produce a combined effect greater than the sum of their separate effects.

disperse PFCs as atmospheric tracers (30 kg PFCs per flight) through their hot exhaust (dusty plasma) or in HALO flights (high altitude and long range research) through the spray nozzles on the fuselage exterior. PFCs then find their way into our lungs, and are in blood plasma processing, blood substitutes, drug deliveries, and liquid ventilation for deep diving and space travel. *PFCs do not metabolize.* Flu-like symptoms may arise, from light fever and myalgia to a 24-hour intense fever, arterial hypertension, tachycardia, high white blood cell count, and thrombocytopenia.

Fuel will have either hydrazine (N_2H_4) or liquid ammonia (NH_3) while the oxidizer employed is selected from *the group consisting of liquid fluorine (F_2), chlorine trifluoride (ClF_3) and oxygen difluoride (OF_2).* Hydrazine may dissolve as barium chloride ($BaCl_2$) or barium nitrate ($Ba(NO_3)_2$) or a combination of the two; the same goes for using liquid ammonia. Then there is the oxidizer, the cryogenic liquid fluorine F_2.[34]

Then there's the decades of SCoPEx (stratospheric controlled perturbation experiment) whose latest balloon launch in Sweden has been refused and tabled,[35] and the irony that HALO-type aircraft release *sulfur hexafluoride* (SF_6) in order to trace pollutants.* Are the 45 kg of SF_6 gases ejected from fuel-oil nozzles and the one-sixteenth-inch jet in the baggage compartment just more of the chemical trails delivery system?† The release of SF_6 and perfluoromethylcyclohexane (C_7F_{14}) "into urban air" describes the path of entry of the world's most powerful greenhouse gas (22,200X more heat-trapping than CO_2). Tracing pollutants with pollutants . . . Where is the common sense if the intention is not the demise of Human 1.0?

Since November 22, 2016, under the 1970 Toxic Substances Control Act, the Fluoride Action Network, which calls water fluoridation "the most damaging environmental pollutant of the Cold War," has pursued a case against the U.S. Environmental Protection Agency to ban fluoride in public water supplies due to the risks posed to the brain as well as for its inhibiting of enzyme action, including the enzymes necessary

*Typically, another HALO acronym exists for the sake of confusion and cover: High Altitude Lidar Observatory.

†"Liquid-feed flame spray pyrolysis (LF-FSP) is a general aerosol combustion route to unagglomerated and often single crystal mixed-metal oxide nanopowders with exact control of composition."[36]

for cell oxidation.* The fluoride lawsuit trial began the week of June 8, 2020.[39] Expert witnesses for the plaintiff confirmed that sodium fluoride is a pesticide, and that fluoride parallels lead neurotoxicity in terms of an IQ drop of 3 to 4 points. Danish environmental epidemiologist Phillip Grandjean, an expert on mercury neurotoxicity, testified that he was threatened by the Harvard Dental School regarding publication of his neurotoxicology studies on fluoride.† Finally on August 24, 2024 the court ruled on behalf of the Fluoride Action Network, deeming fluoridation an "unreasonable risk" to the health of children, and the EPA will be forced to regulate it as such.[41]

The most electronegative and reactive element of the entire periodic table, fluoride is difficult to work with and yet commonly used in pharmaceuticals (comprising 20 to 30 percent), possibly because of the tight bonds it forms with carbon, thus making it difficult for enzymes to break down drug deliveries too quickly. Antibiotics, anti-inflammatories, and antidepressants all have fluoride in them.

Nanoparticle (NP) Titanium and Titanium Dioxide

Despite being a Group 2B carcinogen (cancer) and as far from being a nutrient as you can imagine, nanoparticle *titanium* (NP Ti) is in everything from toothpaste to soaps to cake frosting ("food additives" like E171) to sunscreen. For our purposes, we will examine its jet fuel air pharmacology properties that make titanium dioxide nanoparticles (NP TiO_2) prone to becoming electromagnetic when activated by the Sun's UV radiation. NP TiO_2 is excellent for cloud-seeding and rain enhancement, due to its absorbent nature, but in the lungs and through the skin, NP TiO_2 can load the body with oxidative stress, as well. In fact, titanium oxide (TiO) is an excellent semiconductor and photo catalyst.

How much of the influence on our health is due to the chemical titanium itself, and how much is due to the nano-size of the oxidated version of the chemical? NP gold, NP zinc oxide, NP silica oxide—all are found in processed foods. The nano-size makes it easy to deliver these agents to

*Fluoroacetate is used to introduce fluoride into organic molecules, similar to how soil bacteria are made to convert molecules into polyketides, the molecules that incorporate acetate.[37] Sodium fluoroacetate is used to poison rodents and coyotes by impairing their oxidative metabolism.[38]

†"Due to its high toxicity, fluoride has long been used as a pesticide."[40]

human cells, including brain cells. Gold NPs have an innate ability to bind with biomolecules and possess excellent optical properties for LiDAR (light detection and ranging) biotagging with quantum dot color coding for threat level security clearance, vaccination status, etc. The NP TiO_2 food additive E171 shreds the surface of the vascular system but is insoluble and so bioaccumulates in organs as a semiconductor ready to be activated by 5G / 6G.

Certainly, NP TiO_2 is a dual-use technology,* not merely a neutral "additive" or harmless "nanoscale material," as the U.S. Food and Drug Administration (FDA) would have it.

Everything is about electromagnetics and conductivity.

Triboelectric nanogenerators are composed of titanium and aluminum oxide or titanium and silicone dioxide and are used to harvest energy from just about anything with a surface—wind, ocean currents, sound, etc. The *triboelectric effect* is how electrons flow across barriers, with temperature playing a major role.[42]

Earlier, I quoted Robin D. P. Watson who connects titanium dioxide nanoparticles with the Covid-19 penchant for blood clots. Beginning in May 2020, alarm was raised that post-Covid-19 autopsies were revealing "thromboembolic events" that had gone unnoticed (or were ignored) before death. Autopsies were mandatory in Germany, despite the WHO mandate that no one was allowed to autopsy people dying of Covid-19. Through post-mortems, Russia discovered that Covid-19 exists not as a virus but as an "extended electromagnetic radiation" that acts as a poison, and that synthesized gain-of-function bacteria can cause a "rotating vascular clotting" (thrombosis).[43]

The chemical nanoparticle is now the favored approach to "modulating" the immune system—for example, silica nanoparticles that penetrate the skin barrier, and titanium dioxide nanoparticles that induce gene expression alterations in the brain. Together, the two induce reproductive and/or liver toxicity.[44]

Barium-Strontium-Titanate (BST)

Barium titanate ($BaTiO_3$) is ferroelectric and ferromagnetic, pyroelectric, and piezoelectric, and used in capacitors (devices capable of storing energy

---

*TiO_2 NPs are in electronic devices and employed as nanothermite in DU (depleted uranium) munitions.

in the form of an electric charge), electromechanical transducers (conversions into electrical signals), and optics (transmission of light).*

Barium-strontium-titanate (BST) is "a complex manmade mineral" used to develop "a wide variety of integrated circuits that create, process and receive microwave frequencies on which communication is based."[45] Thin-film barium-strontium-titanate is soldered onto circuit boards for tuning miniature antennas like the tiny phased array antennas in smartphones capable of having their beams combined into one targeting beam. As German scientist Harald Kautz (previously Kautz-Vella) makes plain, microscope resolution (1000–5000x) droplets of rain falling out of our geoengineered atmosphere are full of such intelligent, exotic nano-crystals that convert body heat to visible light. Under dark field microscopy, point your finger in the direction of the raindrop, and at 1 cm, the raindrop glows like a neon bulb just by converting body heat to visible light, thanks to barium-strontium-titanate doped with europium and other rare earth elements.[46]

The dust in our homes and lungs is loaded with metal oxides, and the nanoparticle mix recently invented to be sprayed on any vertical object such as a tree to turn it into a high-powered antenna or to extend the range of an already existing antenna by a factor of one hundred[47]—probably contains the nanocapacitor barium-strontium-titanate produced by a company like American Elements.

Lithium, a Conflict Mineral

Conflict minerals are those minerals that have been gouged out of the Earth and are blood-soaked as a result of wars fomented by high-tech nations (but fought by the low-tech nations where the minerals are, human cost be damned)—lithium, gallium, selenium, gold, mercury, chromium, niobium, tungsten, molybdenum. The rechargeable batteries of smartphones, computers, and electric vehicles may be emblems of the modern world, but they are powered by men and women, and especially young children, laboring under slavelike conditions in the resource-rich Democratic Republic of the Congo (DRC) under the pall of endless warfare since the Tutsi-vs.-Hutu "Operation Crimson Mist" psyop in the

*Interestingly, the preparation of $BaTiO_3$ entails the very same polyethylene glycol that is in hydrogel

1990s.[48] For example, coltan mining,* which provides the niobium and tantalum for cell phones, computers, PlayStations, and Xboxes. Thus, global corporations in quest of conflict minerals are surely behind the "years of rebel conflict and recurrent natural disasters" that lie behind three decades of millions of deaths and displacement of millions in the Congo as the growth of sophisticated technology requires more and more conflict minerals.[49] According to *Tech Times,*[50] electronics-essential conflict minerals are getting harder to get and may be a goad to globalist corporations to bypass the endless clever devices that need conflict minerals and back whatever brain-computer interface technology it takes to assure remote digital enslavement of Humans 1.0 *and* 2.0.

Lithium is an integral component of laptops, cellphones, and electric car batteries, which makes it a hot potato in the United States versus China competition for rare earth minerals. Like cobalt in the Congo (discussed earlier) and coltan in the Congo, Venezuela, and Bolivia, a "plurinational" state of thirty-eight ethnic groups holds 70 percent of the world's lithium; the Salar de Uyuni salt flat alone holds an estimated 21 million tons.[51] Was President Evo Morales forced out of office in November 2019 because he championed a state-run lithium industry, similar to Venezuelan president Hugo Chavez's elimination for nationalizing oil (and possibly coltan)?

Like aluminum, lithium at the nanoscale is light and stays aloft longer than other good conductor metals like silver, gold, and copper. For example, on January 29, 2013, when the sounding rocket *Orion* took off from NASA's Wallops Flight Facility with lithium rods embedded in a thermite cake, the lithium was ignited. Once vaporized, it left a spectacularly colorful trail of lithium oxide over urban centers below, while the thermite dispersed iron oxide and aluminum oxide. Millions breathed it all in.[52]

We have all heard about the tranquilizing, mood stabilization (via short-circuiting the brain) role of lithium for those diagnosed with bipolar disorder. Dopaminergic neurons are silenced,[53] while lithium's strategic paramagnetic character inhibits oxidative damage to cells and glutathione levels in cerebral cortical cells, proteins, and lipids.[54] The relationship

*Coltan is columbite-tantalum whose rare metals tantalum and niobium/columbium are chemically linked. The synthetic semimetal crystal tantalum arsenide is the 3D analog of graphene, perfect for terahertz lasers.

between lithium-ion batteries and medical implants points to the need for a thorough study of manic depression in electromagnetic terms, as has been done for electrosensitivity and electromagnetic hypersensitivity. Interestingly, when introduced into a magnetic medium, lithium makes magnetic devices inoperable.

Mercury

Mercury is not just found in vaccines as thimerosal and in dental fillings as amalgam. It permeates our atmosphere. In a 2016 paper, nuclear chemist J. Marvin Herndon spelled out the extensive use of ultrafine particles of toxic, mercury-laden coal fly ash as "the aerosolized particulate emplaced in the troposphere for geoengineering, weather modification, and/or climate alteration purposes."[55] The mercury in coal fly ash is fired in a jet's fuel combustion chamber, after which toxic fibrous mesh from methylmercury and ozone-damaging chlorinated-fluorinated hydrocarbons are released into the atmosphere.

"I had all the mercury taken out of my mouth back in the early '90s. The metal creates an antenna around the head that really affects you. The mercury removal dentist showed me with his meters just how much voltage was buzzing around my head. You are a totally different person and think so much better after the mercury removal and heavy-metal chelation removal after that. Everyone laughed at me for having it done." —Greg Abbott, July 4, 2019

Upon exposure to water or body moisture, coal fly ash also contains masses of aluminum nanoparticles in a chemically mobile form, thus contributing to neurological diseases (Alzheimer's, autism, Parkinson's, ADHD, etc.), reduced male fertility, and biota debilitation. The relationship between mercury and Alzheimer's has been known for decades, as has the fact that mercury slowly destroys the blood-brain barrier. Mercury dental amalgams and mercury in vaccines inhibit the efficiency of tubulin, the protein that separates nerves and receptor cells from surrounding tissue. Mercury degenerates the tubulin, which leads to autism and ADHD, *and basically drives the soul out of the body.*

The mercury in coal fly ash makes that substance especially injurious to human health. The small particle size of aerosolized coal fly ash ($PM_{2.5}$) enables particulate intake through inhalation, ingestion, and induction through the eyes or the skin. When inhaled, coal fly ash particles can penetrate and become trapped in terminal airways and alveoli, where they are retained for long periods of time.[56]

Coal fly ash is more radioactive than nuclear waste,[57] yet it is part of the chaff matrix in chemical trails and greatly responsible for weaponizing our weather, primarily by controlling rainfall. Hygroscopic* coal fly ash heats the atmosphere by absorbing solar energy (so much for blaming carbon), melts glaciers, and inhibits rainfall by trapping small water droplets and keeping them from coalescing and growing large enough to form raindrops. The aerosolized coal fly ash particulate also retards heat loss in order to produce an artificial increase in local atmospheric pressure, thus blocking weather fronts and limiting rainfall.

Mercury in the ion propulsion engines of the sounding rockets that deliver thousands of 5G satellites into space is life- and brain-threatening, given that mercury is a strong neurotoxin. Follow the launch arcs of NASA's Wallops Flight Facility in Virginia and Edwards Air Force Base in Southern California and you will see that many launches pass over high-density populations.

With signal-enhancing mercury, along with barium-strontium-titanate, piezoelectric nanocrystals and constant wireless ELF transmissions, conditions like electrosensitivity, autoimmune disorders, and multiple chemical sensitivities are guaranteed.

Aluminum

Aluminum is naturally present in Earth's crust, but it is widely used in industry and has been added to food, water, Big Pharma vaccines, and cosmetics. *Aluminum's multiplying, synergistic effects are vastly compounded by chemical aerosol spraying* to become disruptive to "biological self-ordering, energy transduction, and [cell] signaling systems, thus increasing biosemiotics†

*Hygroscopy is the phenomenon of attracting and holding water molecules via either absorption or adsorption from the surrounding environment.[58]

†Biosemiotics is basically the study of the language that cells use to communicate with one another. In shifting Human 1.0 to Human 2.0, biosemiotics, along with the molecular biology to run AI transmission biology, is crucial—a whole other language for *synthetic* mistaken for *natural*.

entropy. . . [injuring] cells, circuits, and subsystems, . . . catastrophic failures ending in death. Al [aluminum] forms toxic complexes with other elements, such as fluorine, and interacts negatively with mercury, lead, and glyphosate. Al negatively impacts the central nervous system in all species that have been studied, including humans."[59]

The biophysics of water play a crucial role in biodegeneration (lungs being 83 percent water, brain and heart 73 percent, muscles and kidneys 79 percent, bones 31 percent). One study used the toxicity of the water flea to determine how aluminum oxide nanoparticles impact freshwater environments; the researchers concluding that "the toxicity data . . . [was] substantiated by the evidences of internalization of the particles from the transmission electron microscopic images that provided a clear visualization of the disruptions along the digestive tract, indicating a disturbed metabolism inducing cellular death."[60]

That nano-aluminum inundating us from above is now and then mentioned by a few scientists willing to brave the scientism-controlled peer review/publishing gauntlet. One such scientist is the previously mentioned nuclear chemist J. Marvin Herndon. In his 2016 article "Human and environmental dangers posed by ongoing global tropospheric aerosolized particulates for weather modification" (cited above regarding mercury), Dr. Herndon stresses that the ability of coal fly ash—the aerosol particulate favored by geoengineers and the military as chaff for "doping"—to "release aluminum in a chemically mobile form upon exposure to water or body moisture has potentially grave human and environmental consequences."

Toxic, soluble nano-aluminum falls earthward and contaminates the soil (as does coal fly ash), and of course is breathed in. Herndon's study maintains that these heavy-metal nanoparticles are "able to liberate a host of toxins through exposure to body moisture." The toxins, he notes, include aluminum, arsenic, barium, boron, cadmium, chromium, lead, lithium, selenium, strontium, thallium, thorium, and uranium, with their radioactive synergistic spawn and other toxins.[61]

Aluminum nanoparticles are being released on humanity via all three delivery systems: chemical trails and Big Pharma inoculations render chemical synergies in the form of trillions of nanoparticles, while aluminum-tolerant GMO seeds are synergistic with Monsanto's deadly herbicide Roundup (i.e., glyphosate), the idea being that the soil will be so contaminated with aluminum and other heavy metals that the only foods that we will be able to grow are genetically modified organisms. Monsanto's 2009 aluminum-resistant gene patent US-7582809-B2, "Sorghum aluminum tolerance gene, SbMATE," was created at Cornell University and funded by the Bill and Melinda Gates Foundation and the National Science Foundation. The idea behind the nefarious plan to alter our human body chemistry as well as the soil is to kill traditional agriculture on the planet. Monsanto's aluminum-tolerant gene, SbMATE, is "designed to allow genetically modified organisms to thrive in aluminum-poisoned soils [and bodies?] . . . to make it impossible for traditional and organic farmers to grow natural world and heirloom native seeds."[62]

Aluminum nanoparticles are not only conductive and reactive, which is why they're used as catalytic adjuvants in vaccines; they are outright contaminants of all biological neurological systems. Jet chemical deliveries guarantee a wide dissemination in the air, in reservoirs and waterways, in natural forests, and in six inches of topsoil. Aluminum fluoride compounds are found everywhere, and they increase the level of toxicity in the human body to break down the immune and endocrine systems. Hence genetic bioengineers are seemingly in a race to implant in vivo CRISPR technology in all biological organisms in order to advance the transhumanist agenda.

As uptake inhibitors, the tons of aluminum nanoparticles falling to the Earth prevent trees from being able to take up nutrients through their roots. Hence, trees everywhere are dying, starving from the inside out. Case in point: the California fires of 2017–2020 showed dying trees being incinerated from the inside out, as swarms of dry aluminum nanoparticles from aerosols and in the soil served as fire accelerants.

The aluminum content of brain tissue in Alzheimer's* cases as well as in autism spectrum disorder and multiple sclerosis is now significantly elevated.[64] High contents of aluminum have been found in the occipital, parietal, temporal, and frontal lobes of the brains of MS sufferers. The onslaught of aluminum nanoparticles coming down on us from the sky, along with their presence in nearly all childhood vaccinations, is increasing

*Alzheimer's is now being diagnosed in people in their twenties, thirties, and forties.[63]

autism in children logarithmically.[†] In his 2019 book *The Autism Epidemic: Transhumanism's Dirty Little Secret,* author Wayne McRoy builds a convincing case for "autism without intellectual impairment" being purposefully crafted for human brain-computer interface (BCI) with artificial intelligence to foster savant skills while filtering out emotions, empathy, and other undesirable personality traits for entire generations destined for cyborg Human 2.0.

Meanwhile, doctors, scientists, and antivaccine activists churn out excellent studies of aluminum toxicity from vaccine adjuvants, but seemingly do not grasp the fact that literally *trillions* of aluminum oxide nanoparticles are being breathed in daily. The extreme levels of aluminum found in the brains of teenage donors[65] have alarming implications for the entire generation of highly aluminum-vaccinated children. Researchers point out that the burgeoning use of aluminum adjuvant-containing childhood vaccines has been directly correlated with the increasing prevalence of autism spectrum disorder.[66]

Is autism being intentionally developed under a transhumanist mandate? After completing his study of the blood of a hundred autistic children who had been vaccinated, Dr. Jeff Bradstreet, labeled by mainstream media as an "anti-vaccine physician"[67] gave a talk at the 2015 AutismOne international conference and discussed a cure for autism via the highly controversial vitamin D–binding protein GcMAF (Globulin compound-derived protein Macrophage Activating Factor). After his talk, his office was raided by DEA agents in search of GcMAF.† Before his study could be published, he was found floating in a North Carolina river a hundred miles from his home, shot in the chest. The cause, according to corporate media, was "apparent suicide."

---

*From the documentary *Vaxxed* (2016), we learn that 1 in 45 American children is autistic; by 2032, 80 percent of boys may be autistic. A 2004 paper proved a causal relationship between the MMR (measles/mumps/rubella) vaccine and autism, one of whose authors, William Thompson, MD, revealed that the CDC under director Julie Gerberding deliberately omitted crucial data from the study. Gerberding left the CDC in 2009 and became president of Merck Vaccines, which manufactures the MMR vaccine.

†Former CEO of Guernsey-based Immuno Biotech David Noakes, and biochemist Lyn Thyer were arrested in the UK on May 20, 2020, for marketing the "unauthorized" cancer medicine GcMAF. The slander campaign against both in the UK has been relentless, given that cancer is big business.[68]

Numerous physicians around the world who speak out against vaccines and Covid-19 bioterrorism subsequently lose their license to practice medicine as a result. In France, retired French university professor Jean-Bernard Fourtillan was committed to solitary confinement in the psychiatric hospital Centre Hospitalier Le Mas Careiron for his criticism of vaccine adjuvants like aluminum and "the apparition of the SARS-COV-2 virus."*

> Fourtillan has accused the French Institut Pasteur, a private non-profit foundation that specializes in biology, micro-organisms, contagious disease, and vaccination, of having "fabricated" the SARS-COV-2 virus over several decades and been a party to its "escape" from the Wuhan P4 lab—unbeknownst to the lab's Chinese authorities—which was built following an agreement between France and China signed in 2004 . . . Fourtillan himself has said he hopes legal proceedings will allow him to produce evidence he has built up; he is in fact anxious to debate the issues at stake. Now that he is in a psychiatric hospital, the possibility of this happening—in the interest of discovering the truth—is becoming more remote.[69]

Over a hundred holistic doctors have been murdered since 2015,[70] in particular, doctors collaborating on findings regarding autism, such as the discovery that the nagalase enzyme protein was being added to vaccines seemingly to disable the immune system and to prevent vitamin D from being produced in the body, vitamin D being the main line of defense against cancer. Nagalase is found in cancer cells and in high concentrations in autistic children.[71] Dr. Bradstreet's discovery of three DNA markers in autistic children, not two, pointed to a third strand of DNA originating from an aborted fetus cell line in various vaccines that were also to the present *gender dysphoria* among children and youths. Was tucking aborted fetal tissue into vaccines, then, yet another

*Jean-Bernard Fourtillan, Ph.D., chemical engineer, Pharmacist Professor of Therapeutic Chemistry and Pharmacokinetics at the University of Poitiers; expert pharmacologist toxicologist specializing in pharmacokinetics. He has personally filed 400 medical patents. His website is Verite-covid19. Fourtillan is also being sued for the testing success of his hormone patch of valentonin (a sleep hormone) and 6-methoxyharmalan (waking hormones) to help heal sleep disorders and neurodegenerative conditions caused by pollution, adjuvants, and electromagnetics.

Tavistock Institute experiment?* Certainly a more accurate term for autism is *vaccine damage.*

Spraying submicron (nano) aluminum increases morbidity, the slow death that enriches Big Pharma while providing endless "open-air" brain experimentation opportunities—for example, putting fluoride in the public water supply and then measuring how aluminum oxide from aerosols binds with fluoride to form aluminum fluoride compounds that increase fluoride toxicity.

The pineal gland housed deep in the brain—"the seat of the soul," according to seventeenth-century philosopher René Descartes—may be the ultimate target of the concocted chemical/electromagnetic assaults from the sky that humanity has been enduring for the last thirty years. Dietrich Klinghardt, MD, came to this conclusion after studying the combined impact of aluminum, fluoride, and glyphosate (as in Monsanto's Roundup), whose frequencies appear to be electromagnetically pulsed to act together on the human body and brain.† Glyphosate depletes micronutrients of calcium, zinc, magnesium, and other trace minerals, and kills the good bacteria in our gut (our "second brain") by mimicking the amino acid glycine and misfolding the proteins that then become prions.[73] Dr. Klinghardt points out that every American is full of nanosized aluminum, mercury, lead, and organophosphates§ like Roundup, which chelates (binds) trace minerals in food to aluminum, thus preventing nutrients from being absorbed. Roundup in the body binds with aluminum and transports it deep into the brain.

The nanometals found in vaccines are syncretized with the nanometals delivered in chemical trails and are now in our soil and water and the air we breathe: aluminum, barium, silicon, magnesium, titanium, iron, chromium, calcium, copper, lead, stainless steel, tungsten, gold, silver, zirconium, hafnium, strontium, nickel, antimony, zinc, platinum, bismuth, cerium. The Nazi experiments on Jews, gypsies, POWs, twins, and the disabled, supported by the German chemical / pharmaceutical "mother"

---

*Tavistock Institute in the UK has been in the mind-control business since its collusion with the CIA during the early MK-Ultra days.

†Dr. Dietrich Klinghardt expresses the astonishing possibility that organs like the pineal gland are primarily concerned with the spirit—and that Big Pharma is fully aware of this fact.[72]

§Organophosphates were used as nerve gas agents in World War I.

corporation IG Farben, now seem almost tame when compared to Big Pharma's transformation of entire populations into concentration camps for secret, alchemical, in vivo experiments that use conductive nanometals for remote 5G "precision medicine."

Carbon

Ever since the war on carbons was declared at the 2015 UN Climate Change Conference in Paris, carbon experiments determined "to retrace the chemical steps leading to the formation of complex carbon-containing molecules in deep space"[74] have been underway. The carbon-containing ringed molecules (1- and 2-dimensional) called PAHs (polycyclic aromatic hydrocarbons) in emissions and soot from fuel combustion appear to be precursors to the interstellar nanoparticles that account for 20 percent of all carbon in our galaxy (in "the vicinity of carbon-rich stars"). Is this why carbons are being "sequestered"? For learning how to form life's interstellar nanoparticle chemistry in space by connecting and converting 5-sided molecular rings and 6-sided molecular rings?

Without nanoparticles on Earth to work with, such carbon space experiments would not be possible. The IBM discovery of the scanning tunneling microscope in 1981 opened the door to manipulating solid surfaces *the size of atoms*. By 1993, single-walled carbon nanotubes like graphene were found to behave like metals or semiconductors able to conduct electricity better than copper, transmit heat better than diamond, and rank among the strongest materials known. By 1998, an individual carbon nanotube was made into a transistor in Delft, Netherlands. IBM demonstrated that nanotubes could potentially scale up as building blocks for the future of electronics—and Transhumanism.[75]

We have been conditioned to believe that human beings are carbon-based because of the plant photosynthesis that all of organic life on Earth has been dependent upon for their protein base for millions of years. Now, however, if the transhumanists are allowed to have their way, aerosol deliveries of carbon-based nanotube biotechnology will completely replace what is Nature-based ("the singularity"), *carbon nanotubes (CNTs)* being flexible electronics that can be attached to human tissue so as to turn human evolution away from Nature and toward brain-computer interface with artificial intelligence.

"Carbon nanotubes—hollow tubes of pure carbon about as wide as a strand of DNA—are one of the most studied materials in nanotechnology. [Since the 1990s] scientists have used ultrasonic vibrations to separate and prepare nanotubes in the lab . . . Carbon nanotubes are one of the original wonder materials of nanotechnology. They are close cousins of the buckyball, the particle whose 1985 discovery at Rice [University] helped kick off the nanotechnology revolution." —"Tiny bubbles [ultrasonic] snap carbon nanotubes like twigs." News release, Rice University, July 9, 2012

Meanwhile, we are fed the lie that carbons from our vehicles and our bad energy habits—"free energy" being endless, as Tesla attempted to prove for all of humanity—are causing "climate change" as astrophysicists, plasma physicists, and particle physicists call for "extreme weather" energy events necessary to their experiments in decarbonizing and terraforming Earth. Decarbonization requires shutting down photosynthesis and replacing natural forests with virtual forests, and installing megamachines to harvest carbon ("life molecules") out of the atmosphere.*

The CNT "nanoradio" is 1/10,000 of the size of a human hair and 10 nm wide by hundreds of nanometers long, and was created at UC Berkeley's Center of Integrated Nanomechanical Systems in 2007. The new criteria of "absolute limits" and "full-spectrum dominance" are now about tinyness, the *micro, nano, pico,* and *femto* of particles whose extraordinary power is disguised as insignificant, but which actually gives the keys of the kingdom over to technocratic remote control of Nature, human bodies and brains.

Polymer chemist Mike Castle described meticulously in an email to radio host Jeff Rense back on July 14, 2003, how the tiny world implanted from the upper atmosphere—particularly Morgellons—was already being made to work.[77] My brief summary of "Chemtrails, Bio-Active Crystalline Cationic Polymers" in *Geoengineered Transhumanism*:[78]

---

*Direct air capture (DAC) machines suck CO_2 molecules out of the atmosphere, convert it to liquefied CO_2, transport it by carbon capture pipelines, and finally bury it in underground sequestration sites.[76]

> Polymer nanowires far thinner than capillaries are conducting electrical impulses to and inside your brain, and changing shape in response to the nuances of electric fields. Composed of carbon nanotubes (CNTs) made from single layers of carbon atoms and coated with a double wall of oil molecules, the tiny microprocessors encapsulate ion pumps pumping charged atoms of calcium, potassium, etc. in and out of the cell. The hydrotrope adenosine tri-phosphate (ATP) originally powered the tiny ion pumps, but now the wireless network inside and outside our bodies powers them.

Canadian herbalist Tony Pantalleresco studies the carbon nanotubes we're constantly breathing in because many who turn to him for their health issues are among the North Americans whose body composition is still primarily carbon. We are obviously undergoing *synbio* modification, and carbon nanotubes are part of how it's being done epigenetically.

J. Marvin Herndon has written extensively about the type of cloud seeding called *carbon-black seeding* in which carbon black nanoparticles (soot) is released into the atmosphere by aerosols to absorb radiant energy, heat the surrounding air, melt Arctic ice ("dark ice"), or intensify hurricanes.[79] Unfortunately, this practice kills macrophages (immune cells in the lungs responsible for cleaning up and attacking infections) and doubles up cases of lung inflammation (pyroptosis) once the soot is nestled into the deepest parts of lungs, leading to heart disease, premature aging, and death.

Carbon nanotube nanobots are like natural silica-based life forms from a different planetary biosphere. Capable of mimicry and consciousness, *they know when they are being observed.* The filaments that sprout from Morgellons lesions (see chapter 9, "*Synbio*," for a discussion of Morgellons) don't grow, but appear instantaneously and are able to evade detection by Nature's immune system. The filaments appear to be densely packed erythrocytes interwoven with multilayered carbon nanotube strands that have undergone chemical decomposition and fusion and are arranged hexagonally with silicon, copper, and electrical and biological molecular robots smaller than a cell.[80]

Significantly, the main method of making carbon nanotubes is called *chemical vapor deposition*, a process used in the semiconductor industry to produce ultrathin films whereby "solid material is deposited from a vapor by a chemical reaction occurring on or in the vicinity of a normally heated substrate surface. The resulting solid material is in the form of a thin film, powder, or single crystal."[81] Thus, we are not just breathing in carbon nanotubes, but their byproducts as well, the most damaging of which are hydrocarbons. Produce tons of carbon nanotubes and you can expect tons of hydrocarbons—blamed on automobiles, of course—particularly polycyclic aromatic hydrocarbons, the worst.[82] Producers of carbon nanotubes have insisted that the byproducts discharged into the environment are safe, while the effects of their synergies with compounds like Freon refrigerants, methyl t-butyl ether, flame retardants, and the surfactant perfluoroctane sulfanate are ignored.

Here is your Space Fence composition: carbon (more than likely C60 [carbon 60] or C70 [carbon 70]). The Fence just got denser and amplifies more frequencies. Carbon is super-conductive, zero resistance; 3X harder than diamond, 100X stronger than steel. If they are using carbomers (polymers functioning as thickening, dispersing, suspending, and emulsifying agents) or diamines (binds monomers or molecules that can be bonded to other identical molecules to form a polymer), then it is even more than C60 / C70. Carbomer or diamine body armor would resist a .50-caliber and could be programmed to assemble into any form or pattern. Fire it in 6G terahertz, 5G, or a multiple-band frequency, and it will distribute with zero resistance. It also increases visibility and amplifies whatever you're looking at. They're now making lenses out of it. —Email, Tony Pantalleresco, September 8, 2019*

Meanwhile, through its Agenda 2030, the UN is pushing green renewable systems like solar panels and wind farms in tandem with carbon capture and nuclear energy as "sustainable development" in its global "affordable decarbonized energy system" agenda.[83] Go figure. Replacing carbons with carbon nanotubes and carbon black nanoparticles is inversion, not progress, of the natural order.

*The Space Fence increase in density means a greater ability to amplify more frequencies by calibrating all of the Space Fence infrastructure to pulse together.

As for who owns these renewable energy megaprojects, begin with Bonner Cohen, PhD, a senior fellow at the nonprofit National Center for Public Policy Research advocating for free-market solutions to environmental problems. Cohen offers the example of Houston-based GH America Energy as "a hostile foreign power." A subsidiary of the Chinese Guanghui Industry Investment Group no doubt subject to oversight by the Chinese Communist Party, Guanghui could "mess with the nation's power grid and damage our critical energy infrastructure."[84] I have similar thoughts to those of Dr. Cohen when I read peer-reviewed papers clustered around nanotechnology and Covid-19 and encounter one Chinese name after another.

3

They Do It with Smoke and Mirrors

The brain is the hardware through which religion is experienced. To say the brain produces religion is like saying a piano produces music.

C. Daniel Batson,
American social psychologist

Of course, the secret space program has pursued plasma "mirrors" for decades, and now that the sky is loaded with plasma—yes, even plasma clouds—we are hearing about UFOs and strange lighting events everywhere.

Back in 1969, the Environmental Technical Applications Center of the U.S. Air Force published a report titled "Quantitative Aspects of Mirages" about how UFOs were being viewed through the veil of the "temperature inversion" cover story along with "air lenses" postulated by Harvard-trained astronomers like Donald H. Menzel (1901–1976), who claimed that temperatures of several thousand degrees Kelvin would be required to cause the "mirages" attributed to them.[1]

Fifty years on, the public has outgrown cover stories like temperature inversion, balloons, or meteors. Earth's atmosphere is now ionized and loaded with static electricity and plasma, including plasma "mirrors" allegedly for deflecting solar radiation-conducting stratospheric aerosol particles to scatter sunlight back into space, but actually to spread advanced smart dust everywhere for planetary Internet of Things (IoT) and "under the skin" Internet of Bodies (IoB) events. Our Blue Beam sky theater can now be loaded with UAPs (unidentified aerial phenomena) optics, UAPs being the new non-conspiratorial name for UFOs, and phantom images with a seemingly "spiritual," mirage-like quality as plasma TV for the masses.

> The glowing sheath around a UFO could be defined as a nebula . . . The colored, luminous vapor sheath surrounding the UFO, or what is viewed as the fuzzy color of a fireball, is actually a sheath or sphere of air molecules from the surrounding atmosphere, which has been ionized and excited. Air molecules in this state are defined as plasma, hence the term *plasma sheath* is utilized in regard to a UFO.[2]

Real perceptions mixed with Blue Beam perception management technologies make it hard to know if what the eyes are seeing is real or not, rather like the TV coverage of the collapse of the Twin Towers on 9/11.[3] Laser-induced plasma filaments can be used to build a phantom image (i.e., a *plasmoid*, a coherent structure of plasma and magnetic fields often used to explain natural phenomena like ball lightning or objects in cometary tails). In this case, by emitting light of any wavelength (visible, infrared, UV, THz), 3D images can be created by raster scanning, a method of displaying or capturing an image line-by-line, as employed by television and computer graphics, and then projected into the ionized atmosphere.

Ground-based and air-based lasers are as essential to forming 3D images as they are to clearing thick fog a thousand times denser than a cumulus cloud with laser pulses so intense that they change the character of the surrounding air and send air molecules spinning. By means of *quantum cryptography*, the conductivity of the air is changed to "carve out the desired path of least resistance" to protect "sensitive infrastructure" like the plasma channels (*waveguides*) that lasers generate so as to be able to move South Pacific weather systems north along the U.S. West Coast to the jet stream parked over Vancouver Island, British Columbia, utilize the waveguides to attach the weather system to the jet stream, then force it east across the entire continental United States so weather events like storms, tornadoes, droughts, floods, etc., can be manipulated. Lasers also control the ice particle density all the way to optical thickness by orders of magnitude of the artificial plasma clouds our sky is now full of.

Fig. 3.1a and b. Satellite targeting. *Laser-induced plasma filaments* (LIPFs) can be used to build phantom images (*plasmoids*) by emitting light of any wavelength (visible, infrared, UV, THz) to create 3D images by *raster scanning* (rectangular pattern of image capture and reconstruction), the way cathode ray TVs are used to display images. Plasmoids in an electromagnetic field form an auroral column first as plasma orbs, then are flattened into toroids (donut shapes) by magnetic pressure. Evenly spaced sores in Morgellons or on targeted individuals indicate a plasma-drilling attack by a "string-of-pearls nano array." I obtained these images from targeted individual Carolyn Williams Palit's now-removed site *NoExoticWarfareZone* after she was killed.

Fig. 3.2. MUOS satellite (Mobile User Objective System). Satellites do not necessarily look as impressive as their tasks indicate they are. Developed by Lockheed Martin for the U.S. Navy, MUOS is an ultrahigh frequency (UHF) SATCOM system of four orbiting satellites and four relay ground stations of the Space Fence infrastructure. Note the hexagons. MUOS looks like the Wicker Man in the film *The Wicker Man* (1973, 2006).

United States Navy

BLUE BEAM

Which brings us to NASA's Project Blue Beam, widely dismissed as a conspiracy theory, much the way that chemtrails were dismissed as being ordinary jet vapor contrails, or the way that HAARP has been passed off as being about submarine communication.

In deciding to retrieve Project Blue Beam from the conspiracy cover story netherworld, I realized that I was entering the no-man's-land of the 1990s led by former CIA director George H. W. Bush as U.S. president (two years), followed by his MK-Ultra sidekick William Jefferson Clinton (eight years), then his son George W. Bush for the 9/11 blood sacrifice that announced not just a new millennium but the scalar planetary weapon of choice.

By the 1990s, Americans had suffered through forty years of U.S. Air Force BLUE BOOK cover stories created to hide the Paperclip Nazi–inspired secret space program run by card-carrying Nazis like SS aeronautics engineer Wernher von Braun. The thousands of UFO sightings in our skies, the Roswell Crash (1947), Operation Majestic 12 (1947) were real enough, given the Paperclip Nazis' successes in aeronautics, but treated in such a way as to make the American doubt his or her sanity. The news items were seeded with doubt and made into movies and dramas populated by New Age players, cult leaders, CIA agents, men in black and abductees, etc. The 1960s TV series *Star Trek* stirred the psychic pot.

Needless to say, real documentation of the "national security" projects in the years since 1947 (like going to the Moon!) remain classified, other than an occasional reference to a meeting or memorandum, or after the heart attack death at fifty-one of the French Canadian investigative journalist Serge Monast (1945–1996) who wrote proscribed Blue Beam books. So it goes when intelligence agencies—the government version of "secret societies"—and "national security" is made into the litmus for running a nation.

Like the documentation regarding the 1963 assassination of sitting President John F. Kennedy, those who attempt to peer beneath the National Security carpet regarding UFOs / UAPs had to wait until 1994 for the "records" of the 1947 Roswell crash.[4]

Project Blue Beam was mostly about what satellites could be made to do from space, specifically holograms and mind control. For the hologram, interfere three beams (interferometry) from laser-generating electro-optical satellites over the target area, one to illuminate and two to reflect the image in a plasma "sky mirror," so that when the beams intersect, a 3D hologram is formed and observers swear they are looking at the thing itself in the vacuum zone over the target area. For Blue Beam mind control, pulse the temporal lobes and hippocampus of the brain from space with ELF, VLF, and LF radio waves and the magnetite delivered in chemical trails to billions of brains since at least the 1990s* can be made to make the unreal seem real. Tweak the optical fibers and coaxial cables in the ionized sodium layer sixty miles above Earth, and you can play a 3D movie of the Second Coming or an alien invasion.

> We will see [Maitreya's] face on television, but each of us will hear His words telepathically in our own language as Maitreya simultaneously impresses the minds of all humanity. Even those not watching Him on television will have this experience."[6]

Blue Beam's trial run occurred in 1967, during the Israeli Six-Day War and the occupation of the Sinai Peninsula, which coincided with the launch of *infitah*, "economic openness," the neoliberal economics that has brought about a genuine social catastrophe.[7] The popular anti-imperialist Egyptian president Nasser, who'd instituted land reforms, nationalized the Suez Canal Company, and called for pan-Arab unity, had been eliminated and replaced by the pro-West Anwar Sadat. The Egyptian public needed a distraction from the power grab, so the Virgin Mary, in the form of Our Lady of Zeitoun, appeared on top of the dome of the Church of St. Mary in Cairo, Egypt, over a period of three years, beginning April 2, 1968. Virgin appearances proved popular: on May 29, 1971, the Queen of the Holy Rosary Mediatrix of Peace appeared with full coronal discharge east of Necedah, Wisconsin, on Highway 21. Because the air had been overionized in order to create the image, onlookers' ears and hair glowed.

In 1977, *Life* magazine ran a cover story about the sudden flood of spiritual phenomena all over the world—crosses of light appearing in windows from Los Angeles, Bakersfield, and Knoxville, all the way to the Philippines, Japan, and Germany. Maitreya, in the form of a hologram, made guest appearances before large crowds and energized water sources in developing nations. Since 1981, the Virgin Mary has been appearing in Medjugorje, Bosnia-Hercegovina. The original six children who saw her are now adults

---

*Magnetite is a permanently magnetic form of iron oxide.[5]

who pass every truth test, including electro-oculographs, algometers, estesiometers, and ampliphones. Given that these visions can be made to occur in the brain via stimulation of the amygdaloid and hippocampal complexes, similar to what was said above about pulsing the temporal lobes and hippocampus with radiowaves, this shouldn't be surprising. And yet an entire (intelligence-made) cult has grown up around the continuing "revelations."

In June 1988, in Nairobi, Kenya, a man in white appeared out of the blue at a prayer meeting of six thousand. Jesus? Maitreya? Muslims saw Allah in melons, eggplants, tomatoes, potatoes, eggs, and beans. Saints on rose petals! Christ on a rock! Hindu statues drinking milk! Were these signs of technology miniaturized, or simply brain manipulation?

On March 6, 1992 in Manila, capital of the Philippines (93 percent Catholic), the Sun danced amid red, yellow, and blue lights. On Holy Thursday 1998, Dr. Ernesto A. Moshe Montgomery of the Beth Israel Temple in Los Angeles consecrated the "Shrine of the Weeping Shirley MacLaine," as the sky over the synagogue was packed with angels and aliens, Mary and Jesus.[8] The Sardinian *telefono antiplagio contro le truffe dei maghi e delle sette* ("antibrainwashing hotline against the tricks of wizards and cults") was outraged and invoked a 1930s law against *abuso della credulitá popolar*e (abusing the credulity of the people), but the Blue Beam show must go on.

On the military side, the psyops program run by the JFK Special Warfare Center and School has been able to *regionally* order up persuasive messages and 3D pictures of clouds, smoke, raindrops, buildings, flying saucers, whatever, since long before HAARP's control over the ionosphere and the creation of the Space Fence mini-ionospheric ring around the equator, and since their completion 3D hologram "movies" have continued to proliferate over Asian and South American nations—"gigantic holograms that could be used for large-scale deception missions by special operations forces,"* thanks

*"Project Blue Beam: The Technology for Mass Holographic Deception and Psychological Manipulation Has Existed for Decades."[9] The referenced October 19, 2015 YouTube video by Truthstream Media is "Project Blue Beam? 'Floating City' Appears in the Sky over China." I include the 1999 *Washington Post* article that Melissa Dykes references, "When Seeing and Hearing Isn't Believing" by William M. Arkin, February 1, 1999: "The Gulf War hologram story might be dismissed were it not the case that washingtonpost.com has learned that a super secret program was established in 1994 to pursue the very technology for PSYOPS application. The 'Holographic Projector' is described in a classified Air Force document as a system to 'project information power from space . . . for special operations deception missions.'"

to the usual medley of alphabet agencies plus the National Reconnaissance Office (NRO) and the National Geospatial-Intelligence Agency (NGA) previously known as the National Imagery and Mapping Agency.*

Since 2003, the NGA has been in charge of directing the global information grid (GIG)—what we call the smart grid—with CIA contractors like Maxar Technologies aiding in targeting specific populations, while the British corporation Serco, Inc.—"a trusted partner of government"), the world's largest nongovernmental air traffic controller—oversees satellite computer algorithms as well as jet and drone chemical deliveries via air traffic control towers.[10] Blue Beam satellites (laser and electro-optical) serve Space Fence full-spectrum dominance, while geoengineers and ground-based installations assure a highly reflective / refractive sky theater, artificial thought, remote control over individuals, and two-way voice-to-skull (V2K) communication. Remember: neural dust and smart dust (see the chapter 1 section "Dusty Plasmas, Smart Dust, Neural Dust") operate like nano-sized drones. As transceivers, they are delivered in aerosols as nanosensor networks that remain in both the outer and inner environments (the "Internet of Bodies"), GMO foods, and Pharma injections, even as so-called precision medicine delivers neural dust able to monitor the brain from the inside.

Stage magic in the sky for the masses and black magic in the brains of individuals—how will we know to discern sophisticated technology from authentic spiritual experiences? How will we know our thoughts from someone else's? St. John of the Cross (1542–1591) gave this sage advice: *The more exterior and corporeal [supernatural visions] are, the less certain is their Divine origin.*

TRANSMITTING IMAGES AND THOUGHTS TO THE BRAIN

It is difficult for civilian populations to grasp that they too are undergoing 24/7 3D battlefield monitoring that begins with reading their DNA signatures from space.

A recent Cellular Phone Task Force newsletter maintains that 300 corporations and governments are in the process of launching *over*

*Study the websites of these two lesser-known agencies NRO.gov and NGA.mil as they are primary in ordering up mind control technologies from space.

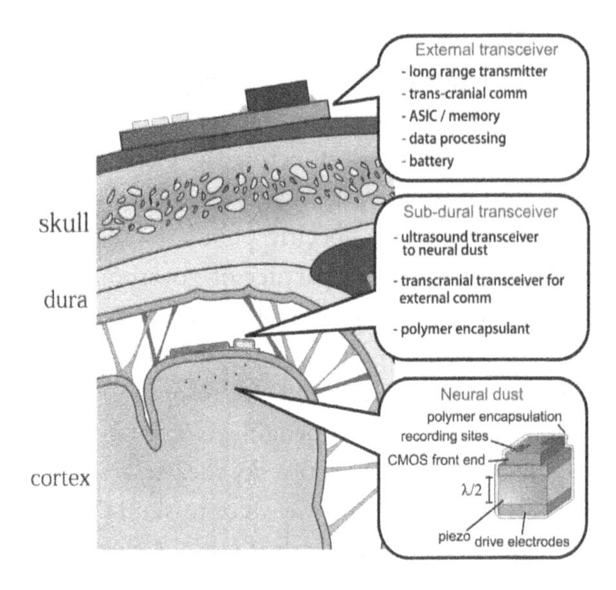

Fig. 3.3. Neural dust—see how it's engineered.
"Intelligent neural dust embedded in the brain could be the ultimate brain-computer interface," *Neurogadget*, July 18, 2013

a million satellites, most of which will be the size of CubeSats. In June 2023, E-Space in France filed a plan for a single megaconstellation of 116,640 satellites after filing the same with the government of Rwanda for 327,320 satellites. According to Cellular Phone Task Force president Arthur Firstenberg:

> The only way to diminish the demand for bandwidth [for more and more cell phones] that is turning the Earth into a giant computer, with all living beings electrocuted inside of it, is to stop using cell phones. Not to use them less frequently, but to throw them away. The *ability* to use them, no matter how infrequently, requires the entire planet to be radiated . . .[11]

Some of the thousands of satellites already up are Blue Beam satellites with electro-optical capabilities for "intelligence, surveillance and reconnaissance (ISR), and electronic warfare (EW)"[12] producing MK-Ultra artificial thoughts and images in brains below. On board each satellite is a *quantitative electroencephalograph (QEEG)* that reads individual brain signatures called *evoked potentials* by tagging a signal onto the target's *nanoantennas* far below. Once the signal attaches, the target's DNA standing wave (scalar) retreats, thus announcing the exact biosignature (evoked potential). This is part of the "under the skin" surveillance scenario promised by Transhumanism advocate Yuval Noah Harari.*

With our individual evoked potential, the agents of agencies and corporations with national security access have entry to however many brains they wish to access. It is no longer a matter of satellite transmitters but of large and tiny transmitters everywhere, from 5G / 6G cell phone microwave towers to cell phones themselves, power lines, smart meters, radar installations, HAARP, the nanotechnology already in our brains, etc. Some people are easier to *entrain* to certain frequencies than others, but we are all in danger of being overtaken by programming à la modular MK-Ultra once our evoked potential is known.

Thus, you discover yet another necessity for maintaining the capability of transmissions from space. The ionized atmosphere with its nanoparticulate doping of metals, the lasers, radar, ionospheric heaters (in the mind control arena, known as *heterodyning* or mixing of signals),† etc., for millions of 24/7 transmissions on Earth (including DUMBs or deep underground military bases) or in space for near-earth orbit (NEO, low-earth orbit (LEO), ISS (International Space Station), and NASA missions in deep space.

Machine-learning supercomputers and quantum computers§ both write their own software and formulate and act on their own responses, but the father of quantum computing, David Deutsch, foresees the day when humans will upload their entire consciousness into quantum computers.[15] Brain-computer interface (BCI) is already so seamless that people who are targeted with directed energy weapons (DEWs) can't tell the difference

---

*Harari is the author of *Homo Deus: A Brief History of Tomorrow* (London: Vintage, 2016) that the *Guardian* calls "spellbinding"—the perfect word.

†For an excellent read on heterodyning and *remote neural monitoring (RNM),* I recommend the 2007 internet-only book *The Matrix Deciphered* by Robert Duncan, who had the experience of working on the very technology for DARPA that was then used on him. He is now a targeted individual.

§"The secret to a quantum computer's power lies in its ability to generate and manipulate quantum bits, or *qubits.* Today's computers use bits—a stream of electrical or optical pulses representing *1*s or *0*s . . . Quantum computers, on the other hand, use qubits, which are typically subatomic particles such as electrons or photons. . . . A quantum computer with several qubits in superposition can crunch through a vast number of potential outcomes simultaneously. . . . The machines' ability to speed up calculations using specially designed quantum algorithms is why there's so much buzz about their potential."[14]

Fig. 3.4. A photon-based quantum computer. Christian Weedbrook, CEO of Xanadu Quantum Technologies, says, "It is only a matter of time before quantum computers will leave classical computers in the dust." Why? Because quantum computers are not just more rapid computation machines than supercomputers but are more like the unique human brain in their ability to shapeshift superpositions and their propensity for quantum entanglement. As a 2015 article in *New Scientist* puts it, "Even if electrical impulses among neurons within the brain—something well described by classical physics—are the immediate basis of thought and memory, a hidden quantum layer might determine, in part, how those neurons correlate and fire."[13]

Hansen Zhong, USTC, public domain

between their own thoughts and incoming words that first were converted into certain frequencies and then were directed back into their brains as words; nor can they tell the difference between an AI program and a human operator when it comes to voice-to-skull (V2K) thoughts. V2K is described by the U.S. military as a "nonlethal weapon which includes (1) a neuro-electromagnetic device which uses microwave transmission of sound into the skull of persons or animals by way of pulse-modulated microwave radiation; and (2) a silent sound device [bypassing the ears] which can transmit sound

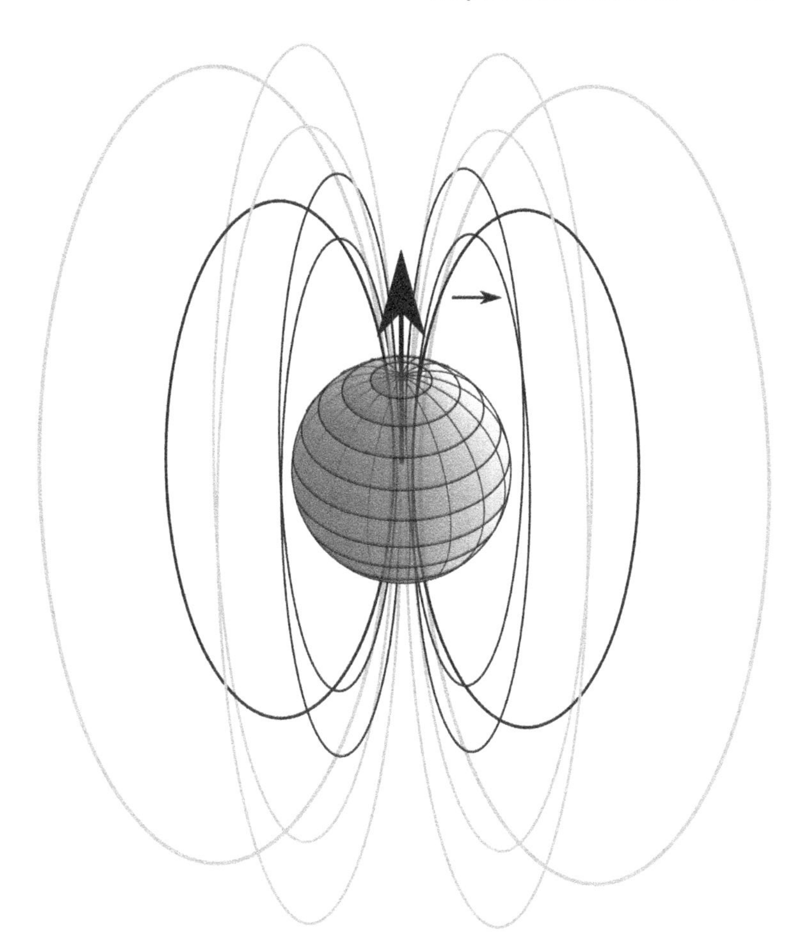

Fig. 3.5. The magnetic field and telepathy. Michael Persinger, PhD: "Suppose you had access to every person's brain, and they had access to yours?" The human brain and geomagnetic field are strongly correlated (and therefore manipulated via directed energy weapons and "dual use" technologies).

into the skull of person or animals. NOTE: The sound modulation may be voice or audio subliminal messages"—the cover story being along the lines of HAARP's submarine communication: that this technology is used as "an electronic scarecrow to frighten birds in the vicinity of airports."*

*See the FAS Project on Government Secrecy web page on voice to Skull devices. James C. Lin's book *Microwave Auditory Effects and Applications* (Thomas, 1978) foresaw all of this and more.

Somewhere between "know thyself" and *caveat emptor* ("buyer take the risk"), we are all now in quest of discernment as AI pushes us toward the Human 2.0 hive mind of Transhumanism. How extensive is Blue Beam's reach? Without a working knowledge of how the technology works, *how are we to know the difference between what is natural, what is supernatural, and what is technological?*

STAGE MAGIC IN THE SKY

With holography, certain comets are real and other comets basically special effects.

In the year 2000, Anthony Hewish, the 1974 Nobel Prize winner for physics, discovered pulses coming from a supernova that had collapsed in Earth's southern sky around 4000 BCE. The story is that the Great Pyramid of Giza was built to commemorate that event and the Egyptian/Sumerian cultural leap that accompanied it. In 1979, science writer and linguist George Michanowsky, in his groundbreaking book *The Once and Future Star: The Mysterious Vela X Supernova and the Origin of Civilizations,* described a piece of Sumerian cuneiform that promised the return of that very same supernova sometime around the year 2000. Then in 1997, Richard W. Noone claimed in his cult classic *5/5/2000: Ice, the Ultimate Disaster* that the Grand Gallery of that pyramid points to exactly where that supernova event occurred in the southern sky.

Comet Kohoutek, in its 75,000-year orbit, made a close approach to planet Earth in 1973. On February 26, 1979, outermost planet, Pluto,* penetrated Neptune's orbit in what comparative mythologist John Lamb Lash succinctly locates as the womb of the Virgin (the constellation Virgo, the second largest constellation in the zodiac), and began its nineteen-year approach to the Sun to eclipse the outpouring urn of the Aquarian Water Bearer and set in motion the oblation of the Moon's shadow passing over Mt. St. Helens in the American Pacific Northwest a year before she gave birth on May 18, 1980, Ascension Sunday.

*In 2006, the International Astronomical Union (IAU) downgraded the status of Pluto to that of a dwarf planet. Now, our solar system is made up of our star the Sun, eight planets, 146 moons, and several dwarf planets, such as Pluto. Comets, asteroids, space rocks, and ice provide sometime phenomena.

It is very much Lash's commitment to give voice to natural phenomena, to return *meaning* to what is being enacted within the precincts of our starry zodiac.* Of course, this is to depart from secret society scientism's attempt to view everything in the world and cosmos as dead forces, and yet even those who control the technologies know that everything has exoteric *and* esoteric meaning, such as the actions of Pluto and Virgo: metaphysical knowledge has not been stricken from their JASON science.†

Early in 1983, while the Reagan-Bush-Cheney troika was busy solidifying the power stolen in Dallas on November 22, 1963, the announcement of the secret space program's "Star Wars" Strategic Defense Initiative (SDI), Jupiter and Saturn conjunct in the solar plexus of the Virgin indicated that the Virgin was in labor during the early "Star Wars" years as the elite followed a script straight out of Revelations—or was it Roman Polanski's *Rosemary's Baby*?

> And there appeared a great wonder in heaven; a woman clothed with the sun, and the moon under her feet, and upon her head a crown of twelve stars; And she being with child cried, travailing in birth, and pained to be delivered (KJV 12:1–2).

Six years after Kohoutek, in the fall of 1985, Halley's Comet, the "Heavenly Dragon," made its twenty-ninth‡ visit in recorded history, executing a backward dive from Taurus the Bull to sever the horns of Capricorn the Goat. The Heavenly Dragon cut an arced scythe through one-third of the zodiac, then once close to the Sun, began its thirty-eight-year

---

*In the spirit of the insight that this book attempts to emulate—"We cannot solve our problems with the same thinking we used when we created them"—see John Lamb Lash's book *The Quest for the Zodiac: The Cosmic Code Beyond Astrology* (Marion Institute, 2007). If you live in the Southern Hemisphere, you will have different "lessons."

†"JASON" is an independent group of elite scientists that advises the United States government on matters of science and technology, mostly of a sensitive nature ["JASON Defense Advisory Panel: Reports on Defense Science and Technology," irp.fas.org]. The group was created in the aftermath of the Sputnik launch as a way to reinvigorate the idea of having the nation's preeminent scientists help the government with defense problems, similar to the way that scientists helped in World War II but with a new and younger generation. It was established in 1960 and has somewhere between 30 and 60 members."[16]

‡Saturn return.

retreat,* pausing for a final attack on Sagittarius the Archer as it skimmed the Scorpion's tail and sped toward the Hydra constellation. As the old German rhyme had it,

> *Eight things there be a comet brings*
> *When it on high doth horrid rage:*
> *Wind, Famine, Plague, and Death of Kings,*
> *War, Earthquake, Flood, and Doleful Change.*

From 1990 to 1994, twelve comets per year were recorded, and yet at the end of 1994, two major comet search programs were mysteriously discontinued at the Palomar Observatory—possibly to avoid scrutiny of the next two Blue Beam–manufactured comets? Comet Hale-Bopp appeared on July 23, 1995, followed by Comet Hyakutake on January 31, 1996, both sighted by amateur astronomer Thomas Bopp in Arizona and professional astronomer Alan Hale in New Mexico.

A mysterious U.S. Naval Observatory survey photo shows pointy-tailed Hale-Bopp mysteriously orbiting Saturn before 1995—mysterious because comets that far out are not usually hot enough to produce enough gases to become visible from Earth. Either something unknown was making it hotter than normal, or it contained unknown gases—or Blue Beam had been invoked.

Hale-Bopp reigned in the sky with spectacular brightness for an unprecedented twenty-two months, from July 23, 1995, to passing perihelion on April 1, 1997. Hale-Bopp was the brightest, best, and longest-observed comet in history. Even the Great Comet of 1811, Napoleon's comet, which comes once every 3,065 years, was visible for only seventeen months. And when the Great Comet's tail split, Napoleon marched into Russia.

Hale-Bopp's orbit was *exactly* perpendicular (89.4 degrees) to Earth's orbital plane, its highest elevation exactly 45 degrees as it crossed the zero degree of the vernal equinox, the beginning of spring, with Hale-Bopp's maximum brightness either on March 22, 1997 (3/22 or 322 is a favored Skull and Bones number) or on April Fool's Day. Media accounts varied. It was as if the comet had been calibrated on a Cray XT4 supercomputer.

---

*Two nineteen-year Metonic Cycles.

"The Field Guide to Comet Hale-Bopp" has been removed from the Internet, as have all other doubts about the veracity of Hale-Bopp, but it is still true that the NASA scientists who use probes to get the best views of target planets could design an orbit for a comet like Hale-Bopp.

After its April Fool's Day perihelion, Hale-Bopp appeared to be heading toward the far reaches of the solar system, between the constellations of Orion and Leo. The three stars in Orion's Belt align with the three great Egyptian pyramids, and Leo the Lion with the Virgin Virgo's head. The Sphinx being Leo-Virgo, I am put in mind of the Freemason obsession with Egypt and what esotericist, philosopher, and Freemason Manly P. Hall stated in *The Secret Teachings of All Ages* (first published in 1928): "The illumined of antiquity entered [the portals of the pyramid of Giza] as men; they came forth as gods."

*As above, so below.** Was this then the Freemason version of the Egyptian journey through heaven's gate? Was Hale-Bopp pointing the way to Freemason death of the old order and rebirth (as in the New World Order)? Was Hale-Bopp celebrating the advent of the "Star Wars" phase of the secret space program?

On January 6, 1997 (Epiphany), a green sphere sped from east to west over Rome—surely a Divine sign! NASA consultant Richard Hoagland and MK-Ultra abductee Whitley Streiber insisted that *something* was trailing Hale-Bopp—a "companion" that would collide with Earth's magnetosphere—i.e., Earth's aura—on March 19, 1997 (the vernal equinox) to enact an aura transformation that would conclude on April 29, 1997 (Walpurgisnacht) to signify the end of an era and the beginning of another.

Then there was the blood sacrifice perfectly timed for the computerized cosmic occasion: the March 26, 1997, suicide of thirty-eight members of the Heaven's Gate cult founded by Marshall Herff Applewhite ("Do") and Bonnie Lu Truesdale Nettles ("Ti") in the San Diego suburb of Rancho Santa Fe. In fact, the Heaven's Gate computer, dubbed "Knowmad,"[17] used the same server (Spacestar Communications) that the highest levels of U.S. security used: the CIA, NSA, DoD, retired

**As above, so below* is the Principle of Correspondence that embodies the truth that there is always a Correspondence between the laws and phenomena of the various planes of Being and Life.

admirals and generals, the military contractor SAIC—even U.S. Navy admiral Bobby Ray Inman, who served on the SAIC board as well as the Council on Foreign Relations. SAIC, headquartered in nearby La Jolla, California, was a neighbor of the Heaven's Gate compound, and three Heaven's Gate members had worked for the defense contractor Advanced Development Group (since renamed ManTech International Corporation), where they developed computer-based instructions for the U.S. Army. Tentacles . . . In April 1996, Heaven's Gate closed shop and sold their Earthship; just shy of a year later, as Hale-Bopp exited, the cult members were all dead.

All discoveries by astrophysicists or amateur star buffs like Thomas Bopp are filtered through nine International Astronomical Union astronomers, one of whom is Alan Hale. Thus Hale-Bopp received the official imprimatur of various organizations: the Central Bureau for Astronomical Telegrams, the national clearing house for information relating to astronomical events; the International Astronomical Union; and U.S. naval Observatory, which influenced public perception of Hale-Bopp based on the 1993 photograph of it orbiting Saturn by posting it to an online interactive program that predicted the comet's position and visibility. The *Sky & Telescope* magazine website, the Minor Planet Center (which operates at the Smithsonian Astrophysical Observatory), the Center for Astrophysics, as well as NASA's Jet Propulsion Laboratory (JPL) all followed the navy's lead, providing canned, monitored, and spun star data.

The devil is always in the details. For example, Hale-Bopp's light could be detected beyond Saturn's orbit and lit the skies of Mongolia during a total solar eclipse, and yet was oddly occluded by the Moon in May 1997, a first in astronomical history. A flaw in the calculations of the "spectacular orbit"? As the comet came closer and photographs on the Web got too clear, NASA's JPL switched to a ten-inch telescope to make the photos appear fuzzier. No Hubble photos after October 1995.

An examination of photographic details indicates that Hale-Bopp and Hyakutake may have been holographic projections of Halley's Comet, developed at Caltech's laser physics lab. The media mentioned mysterious X-rays but icy balls of dirt do not emit X-rays, but light beams—aka interferometry—might.

No astronomer wants to be the first to say that the emperor is stark

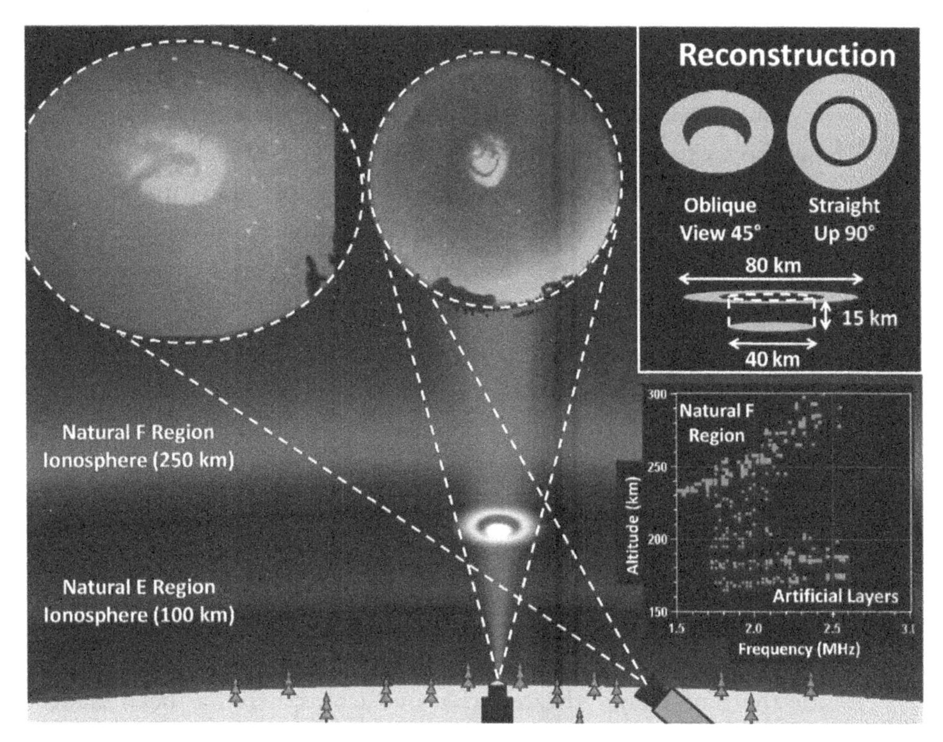

Fig. 3.6. Interferometry in action. The area of the two crossing beams produces scalar rings that heat up due to overpumping radar. Interferometry is also used as a weapon.
Department of Defense, Defense Technical Information Center

naked. The only clue seems to be the 01 in C/1995 01, the official designation of Hale-Bopp, which stands for the first (of many) enhanced comet displays—and perhaps for the brilliant binary computer that created it.

SOLAR SIMULATORS

The glorious golden yellow Sun of yesteryear is now veiled by a white chemical plasma cloud cover ("contrail cirrus") containing masses of nano-metals, as independent scientist Clifford Carnicom pointed out in 2004, along with a warning few have heeded about the façade of "climate change" and "global warming" covering up hundreds of operations in the lower and upper atmosphere:

> . . . the introduction of essentially any metallic or metallic salt aerosol into the lower atmosphere will have the effect of heating up that lower atmosphere. The impact is both significant and measurable. Those who seek and express concern about the so-called global warming program might wish to begin their search with an inquiry into the thermodynamics of artificially introduced metallic aerosols into the lower atmosphere . . . An examination of the specific heat characteristics of an altered atmosphere will provide the path for the realistic conclusions that can be made.
>
> Any claim that the aerosols operations represent a mitigating influence on the global warming problem appears to be a complete façade in direct contradiction to the fundamental principles of physics and thermodynamics. The lack of candor and honesty by government, media and environmental protection agencies in response to public inquiry is further evidence of the fictitious fronts that have been proposed.

Beside the Blue Beam "theater" that our atmosphere has been made into with the help of the conductive nano-metals, there are now artificial suns whose presence is cloaked by plasma cloud cover. At this point, I recommend you view the YouTubes of Jeff P, an extraordinary sky observer.[18]

Look up, and you may see a glaringly white Sun, or even two suns, one golden and one white.[19] Whereas the true Sun is an orb that produces golden light, the solar simulator is angled by its multiple hexagonal lenses and produces white light. The white sun is not really round but has a ragged circumference (the lens "petal" effect). This is the *solar simulator* with blue/red hexagonal honeycomb lenses and xenon arc lamps* reflecting off bent, spinning mirrors as they go through hexagonal apertures. (The hexagonal panels in smaller solar simulators for industry use are lit by LEDs.) Unlike our natural Sun, the 2D white

---

*"A xenon arc lamp is a highly specialized type of gas discharge lamp, an electric light that produces light by passing electricity through ionized xenon gas at high pressure. It produces a bright white light to simulate sunlight, with applications in movie projectors in theaters, in searchlights, and for specialized uses in industry and research. For instance, xenon arc lamps with mercury lamps are the two most common lamps used in wide-field fluorescence microscopes."[20]

artificial sun does not radiate, however much its mirrors spin and its lenses reflect. Heliosynchronous, it is programmed to follow the path of our natural Sun.[21]

Scientists have developed innovative flat optical lenses for secret space program collaboration between NASA's Jet Propulsion Laboratory and the California Institute of Technology.[22] Optical capabilities like the Rochester cloaking effect utilized in simulator operations manipulate light in ways that are difficult or impossible to achieve with conventional optical devices. The lenses are not made of glass; instead, silicon nanopillars are precisely arranged in a honeycomb pattern (again, the hexagon) to create a "metasurface" that can control the paths and properties of passing light waves. Other applications of such devices include but are not limited to advanced electron microscopes, Blue Beam displays, sensors, and cameras.

Another part of the apparatus is the "black cube" spinning *hyperboloid* reflector mirror lens (20 by 20 by 20 feet, according to the patent), which is *not attached to the simulator* but essential to its operation. From the NASA patent US3247367A, "Solar simulator," explaining the role of the 20 by 20 by 20 foot cube "artificial light protector":

> Since no commercially available single light source can approximate the brightness of the sun over a sufficiently large area to illuminate a 20 by 20 by 20 foot cube as required, the invention uses plural light sources and a composite or modular optical system; that is, an array of light sources each having its own optical system. Therefore, the present invention utilizes a large number (say 200 or so) of comparatively small, powerful gaseous or carbon arc lamps, each having a complete optical system associated therewith.
>
> An object of the invention is, therefore, the provision of an artificial light projector for producing substantially the same spectral make-up, parallelism of light rays, and intensity and uniformity of illumination as that of solar radiation in the vicinity of the earth, which projector is capable of evenly illuminating a 20 by 20 by 20 foot cube.[23]

From time to time, the sun simulator is directed to obscure a mysterious large body that is eclipsing the natural Sun and threatening to blot out our natural sunlight. Whether that large body is planet X or Nibiru or a huge, motley plasma "balloon" of plutonium-239—spent nuclear particles

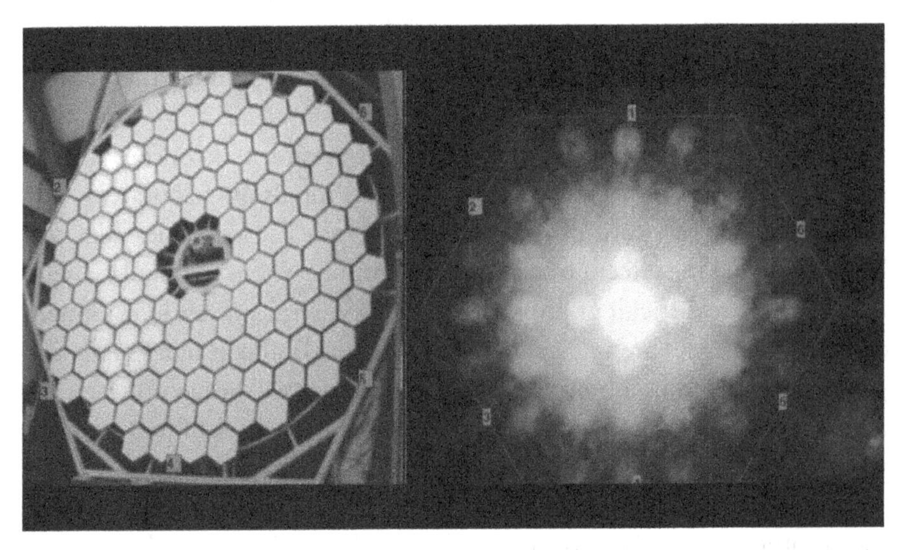

Fig. 3.7. According to Jeff P, the solar simulator is operated by the International Space Station (ISS) 254 miles above us. Solar-collecting satellites with solar panels encircle the natural Sun and provide the power that the solar simulator needs by beaming microwave energy to it as it rotates around the Sun. If the beam is disrupted by a passing object, the solar simulator will *flicker* and malfunction (see "Cell Phones Captures 2" video), possibly requiring repairs at the ISS.

irresponsibly ejected (not "escaped") from nuclear power plants into the atmosphere, a kind of radiation space dump or depot rotating along with Earth—as former SUPO operative (Finnish intelligence à la CIA) and thirty-year nuclear plant veteran Arto Lauri believes—remains to be seen* (see chapter 7, "The Nuclear Behemoth").

One thing is certain, according to Jeff P: When the natural Sun is eclipsed by whatever large mysterious body is passing in front of it, a dark halo surrounded by a rim of natural sunlight forms around the real Sun. The sun halos discussed may actually be evidence of the distance the lens array is from the Earth, given that the halo *is* the lens.

"Some trails went across the whole sky from horizon to horizon, and quite a few ended abruptly near the Sun. A lot of starts and stops, but

*This would not be the first time that the atmosphere has been used as a radiation space dump.

the end result is always the same: MASSIVE SUNDOGS (orbiting halos and rainbows). I believe this is their atmospheric solar energy program: creating structureless solar panels out of particles and artificial clouds. The energy can then be transferred to heliocentric satellites and/or be MANipulated throughout the atmosphere for various purposes."
—ERIC CYPHER, commenting on Facebook, no date

Atmospheric energy harvesting programs are underway, ranging from deriving energy from air[24] (as Tesla envisioned for all of humanity before he encountered the American "energy barons" of the twentieth century), or sequestering carbons from the air.[25]

But there's also the "Death Star" scenario that employs Tesla technology to weaponize lenses and mirrors for plasma laser space-based weapons. *The solar simulator is capable of 30,000-mile transmissions.* At the beginning of "DEW Weaponized Sun Simulator," Jeff P indicates that during the California fires, the simulator was engaged in laser ignition from space.

Tesla's promise that "man could tap the Breast of Mother Sun and release her energy toward Earth as needed, magnetic as well as light" makes beaming solar power down to Earth during the day and using rectenna arrays* to harvest energy from the Sun's infrared radiation at night not so radical an idea as a solar energy plan for orbiting solar farms.[27]

In 2006, the "space weather" corporation Solaren Inc. registered a patent titled "Weather management using space-based power system" (US-20060201547-A1),† a powerful patent for weather *and* warfare:

> . . . Energy from the space-based power system is applied to a weather element, such as a hurricane, and alters the weather element to weaken or dissipate the weather element. The weather element can be altered by changing a temperature of a section of a weather element, such as

---

*"A rectenna array uses multiple antennas spread over a wide area to capture more energy. Researchers are experimenting with the use of rectennas to power sensors in remote areas and distributed networks of sensors, especially for IoT applications. RF rectennas are used for several forms of wireless power transfer."[26]

†Solaren has space solar power patents in the U.S., Canada, European Union, Russia, Japan, India, and China.

the eye of a hurricane, changing airflows, or changing a path of the weather element.

The system uses plasma mirrors ("structureless solar panels") in orbiting solar farms to control weather systems *and* harvest energy in space to maser beam (microwave laser) back to ground stations during the day, while using rectenna arrays to harvest energy from the Sun's infrared radiation at night.* Solar farm mirrors and sun simulators smaller than the massive xenon-powered sun simulator are composed of LEDs, plasma lens arrays, and elaborate mirroring, all of which can be weaponized for igniting fires, melting whole armies, evaporating reservoirs, and turning glaciers into torrents, especially when working with the *LDAL (laser-developed atmospheric lens).*

The LDAL works like a high-powered Fresnel lens to concentrate low-powered light into a tightly focused beam to ignite fires and burn cars like a mean little kid with a magnifying glass trained on a hill of ants. With high-pulsed lasers, portions of Earth's atmosphere are changed into the lens-like plasma mirrors that magnify or change the path of electromagnetic light. Our now constantly ionized air means these space-based lasers mounted on satellites are multipurpose and able to create deflector shields for defense, magnifying lenses for assault, or Blue Beam phantom images.

In 2009, Solaren signed a fifteen-year, 1,700-gigawatt-hours deal with California's number one utility, PG&E, for a "massive floating solar farm" in space.[28] Eight years into the contract, the 2017–2018 geoengineered California "wildfires" occurred, for which PG&E's hand was slapped even as its other hand received $59 billion for a "restructuring plan."[29] On September 19, 2019, the Los Angeles–based Heliogen, a Bill Gates startup, went public with "scientists working in stealth mode" to create "concentrated solar power." Heliogen's AI technology employs a ground-based field of plasma mirrors "to reflect the sun to a single point," to concentrate solar heat up to 1,832 degrees Fahrenheit, plus a way to store solar energy and create clean hydrogen for fuel.[30]

---

*Rectennas are receiving micro-antennas with micro rectifying diodes and micro capacitors on the Earth for capturing infrared radiation and converting it to DC current. These Solaren transmissions may in part explain the California PG&E power surges and outages prior to the 2017–2018 fires.

With this in mind, let's take another look at the enigmatic 20 × 20 × 20-foot black spinning cube referenced in Solaren patents as well as in NASA's patent quoted above.[31] The lens in front of the black spinning cube is not attached to the simulator and appears to be used to focus laser beams for *igniting microwave* fires that firefighters insist is hotter than natural "wildfires." The fires since 2016 or 2017 follow the same pattern (see chapter 4), from the California fires to the 2023 "wildfires" in Lahaina, Maui, and Canada.[32] The black spinning cube "projector" is also able to project Blue Beam broadcasts.[33]

As Jeff P points out in the ""Finally undisputable proof of sun simulator" video, while the cube spins, the *X*-like refractions of two lenses converge and show up on the sides of houses, and sometimes a "sheet lightning" beamed "wall," column, or sphere emanates from the reflecting lenses. As the simulator "eclipses" our natural Sun, multiple artificial eclipses may even show up on the ground![34] As the simulator spins clockwise, its brightness and contrast controls can be adjusted from a nearby, highly classified aircraft like the TR-3B or an orbital spacecraft like the X-37B.[35]

4

Geoengineered "Wildfires," DEWs, and the Climate Crisis Deception

What we can do is focus this very high energy into a very small spot for a very short time, and when that happens, we get the conditions that are very much like inside our Sun.

Physicist Ed Moses, past principal associate director of the National Ignition Facility and Photon Science Directorate, where he directed the largest experimental science facility in the U.S. and the world's most energetic laser

The epigenetic assault on planet Earth's climate is the environmental corollary to undermining Human 1.0 from "under the skin" to create an AI-run Human 2.0. Climate is long term, weather short term, so an examination of patterns of extreme weather as micromanaged chemical and electromagnetic assaults on Earth's living æther reveals that as with the epigenetic assaults on the Human 1.0 immune system, natural weather must first be dismantled so it can be artificially re-created and run via artificial intelligence.

The study of just what has been going on regarding the weather for over a century should begin with World War I (1914–1918) that was both an *air war* and *chemical warfare.* This new type of warfare introduced the petroleum and pharmaceutical industries to a whole new level of profits even as it introduced airplane engine exhaust-induced cloudiness into the atmosphere to show how masses of particulates (including volatile organic compounds like isoprene) might be used to produce low pressures and low temperatures for cloud formation. Forest fires and volcanic eruptions were still natural then, sweeping up into the upper atmosphere to contribute to cloud condensation cloud-seeding. Add electromagnetics and tellingly, there followed a century of weapons development and one war after another in which to test them.

A century after the first air/chemical war, the USDA's Forest Products Laboratory was producing forest-based nanomaterials called *nanocellulose*—organic matter stronger than Kevlar yet lightweight. All of the industries wanted it, including the military. Were the 12.5 million dead trees providers to this industry? Are natural forests being killed off so GMO forests can provide genetically modified pulpwood fiber for the production of nanocellulose substrate, a "crystalline, cellulose-rich waste"[1] similar to the hydrogel "cell growth substrate" for growing human body spare parts in vats and petri dishes?

Nuclear engineer Arto Lauri reports on a story that appeared in an obscure trade magazine about a biopharmaceutical professor in the Finnish town of Rauma who was busy growing human cells on nanocellulose, and that three factories in Finland are now producing nanocellulose substrate.* In mid-January 2016, Lauri collected a phlegmlike cellulose substance from rainwater. As explained in chapter 7, "The Nuclear Behemoth," Rauma is rife with plutonium-239 ion gas leaks, and Finland is deeply invested in the chemical trails and polymerization industries, including nanocellulose/hydrogel production,† as well as plutonium ion experiments.

In 2014, mainstream media had it that trees were dying in the Rocky Mountains for no apparent reason other than the new "climate change" narrative. In 2015, the U.S. Forest Service estimated that 12.5 million trees in Southern and central California forests were dead, with

---

*In 1994, in Oakville, Washington (pop. 665), a gelatinous goo of human white blood cells, two species of bacteria, and live eukaryotic cells fell from the sky six times over three weeks—nanocellulose falling out of the sky.[2]

†From a mechanical perspective, wood hydrogels outperform manmade 3D-printed nanocellulose hydrogels. 3D printing or additive manufacturing constructs three-dimensional objects from CAD models or digital 3D models.

another 120 million trees stressed from years of drought and soon to be dead as well—a total of 20 percent of the state's forests, not including prescribed burns for forest management. Geoengineered drought, geoengineered fires.

"Something very VERY TOXIC was clearly sprayed on the trees on Mt. Ashland this winter.* I drove up there today and was blown away by the death. To have the trees going from looking very healthy this last fall (DESPITE the drought) to now looking very very sick—and with many already dead that were healthy and well last fall is mind-blowing. The tree bark is no longer the healthy brown-red color as last fall; instead, it looks totally grayish white with white or black mold or moss growing in them (an absolute indication of the trees dying, especially when so thick and all up and down the trees). Branches are strangely bending down like they are made of rubber (like NOTHING I HAVE EVER SEEN BEFORE, even with trees under 10 feet of snow), and branches just snapping off. Unbelievable and beyond sad and sick. Absolutely our forests have been sprayed. No way is this quick die-off normal."
—Lucretia Smith, Ashland, Oregon, email, February 29, 2016

In 2015, the globalists moved to make "climate change" their trojan horse that would wheel in global government. Once the climate change narrative was droning on unabated (just preceding the "pandemic" narrative), the 2016 blockbuster film *Geostorm* began preprogramming the world for weather run from space, replete with *microwave plasma fires*, while the controlled corporate media invented new, sexy wildfire terms (*firenado*, *firestorm*, *magnetic storm*, *plume-dominating fires*, *pyrocumulus clouds*, etc.) to cover for endless geoengineered plasma fire experiments. A partial list—all blamed, of course, on "climate change":

*Lucretia's perception made me think of the nanoparticle spray of Chemtech Co., Ltd. (Korea) that acts as a capacitor to turn trees into antennas. Sprayed along the median of roads, the spray produces high bandwidth connectivity with vehicles.[3]

2007 Murphy Complex Fire in Idaho and Nevada, 653,000 acres; same area again in 2011;
2008 "wildfires" in north and central California, 1.15 million acres;
Began February 7, 2009, Australia, 1,100,000 acres;
2011 Wallow Fire in Arizona, 538,000 acres;
2015–2018: Lake County, California (Ranch, River, Rocky, Pawnee, Valley, Jerusalem fires), 280,000 acres
2017 California fires began October 8, 2017,* 9,270 fires / 1,599,640 acres / 10,866 structures;[4]
October 2017: "Wine Country Fire," Napa/Sonoma counties, 245,000 acres
December 2017: Santa Barbara/Ventura counties (Woolsey Fire), 281,893 acres
2018 Camp Fire, Paradise, California; 7,9438 fires / 1,975,086 acres / 24,220 structures;[5]
July 2018: Yosemite, Sierra, Stanislaus national forests (Ferguson Fire), 97,000 acres
July 2018: Shasta and Trinity counties (Carr Fire), 229,651 acres
July 2018: Mendocino, Lake, Colusa, and Glenn counties (Mendocino Complex Fire), 459,123 acres
November 2018: Los Angeles, Ventura counties (Woolsey Fire), 96,949 acres
November 2018: Butte Country (Paradise/Camp fires), 153,000 acres[6]
December 2019–February 2020, Australia, 692,000 acres;
August 2019, Brazil rainforest fires (44,013 outbreaks);
2020, California "wildfires," 4 million acres;
2020, New Zealand bushfires, one a week;
June 2023, Canada from west to east: British Columbia, Alberta, Saskatchewan, eastern Ontario, Quebec province, Nova Scotia—ALL BURNING AT ONCE;[7]
August 8–11, 2023 "wildfires," Lahaina, Maui; 4 fires / 17,000+ acres / 2,207 structures;
2023, Europe (975,431 acres); Turkey, Greece, Croatia, Italy, Spain, Portugal, Cyprus, Algeria, and Tunisia[8]

*At the exact same time on October 8, 2017, the Portugal (Iberian) fires began.

"I left my home shortly before it burned as a 5G tower was erected nearby and then tested. I inferred this was a weapon from my observation of readings of ~1 million times or so over background with my meter pegged, indicating illegal levels of radiation . . . Geoengineering is part of a plasma weapon, a military system now surrounding Earth. The effects of the plasma overhead are the certain thing in the burn; the tower is only a presumed part of the system." —RICHARD LAWRENCE NORMAN, email, April 10, 2020*

While "poor forest management" and "strong Santa Ana winds" were blamed for the thermodynamically created winds of the 2017–2018 California fires, viewers of drone footage on TV news programs watched swarms of "embers" moving fire here and there.[9] But they weren't embers, and no wind was blowing them, as eyewitness James W. "Jamie" Lee clarified in his book *Paradise Lost: The Great California Fire Chronicles* (2020); the "embers" nanobots being directed remotely, programmed to seek out and burn one structure and bypass another.

The fact that trees near thoroughly torched vehicles and buildings didn't burn was a blatant clue that the supposed "wildfires" *were not wildfires at all but microwaved ignitions.* As additionally proved by other trees burning *from the inside out*, with laser strafe marks found everywhere around Paradise.

"The Paradise fire was incredibly selective in what did and did not burn. Tin roofs were spared, but steel guard rails melted. The fire burned incredibly hot, too hot for a mere forest fire. Yep, the nanobots did it, and they were under some kind of intelligent control, whether from within their design or from without, say, via satellite and/or human intelligence. In short, these bots are a new weapon that was being demonstrated in a very public way for various reasons." —DAVID RING, email, June 7, 2022

*Like the scientists of old who studied all of life, Norman writes scientific papers and studies that span philosophy, psychology, neuroscience, and physics as well as verse and fiction. All can be found at the Scientific Research Publishing website, Scirp.org. He is also founder and editor in chief of *Mind Magazine* and the *Journal of Unconscious Psychology and Self-Psychoanalysis.*

"New normal" geoengineered weather depends upon the same mechanisms used to prepare geoengineered fires that consume millions of acres and thousands of structures, whether the frequency dialed in for the "firestorm" is set for forests or homes. It all depends upon the delivery of toxic heavy nanometals, polymers, chemical compounds, nanosensors for remote readouts of temperature, barometric pressure, etc. Next, the relay system of sixteen stationary *Starfire optical lasers* from Kirtland Air Force Base in Albuquerque, New Mexico, northwest to Tonopah, Nevada, and further north to Boise, Idaho, manipulates a north-south laser-induced plasma channel (LIPC "log" or waveguide) off the West Coast to funnel South Pacific moisture north, bypassing drought-challenged California and Oregon to be folded into the jet stream loop waiting over Vancouver Island for moving "weather" destined to be storms, floods, tornadoes, freezes, polar vortices, etc., east over the continental United States—all engineered with chemicals and electromagnetism and run by artificial intelligence (AI) with human assists who tweak algorithms and adjust calibrations of ground installations.

COMBUSTIBLE MICROWAVE PLASMA FIRES

Why are combustible firestorms on a scale of millions of acres and thousands of structures being geoengineered? Sure, for disaster capitalism profits—"asset stripping"—but the larger frame is more like what nanotechnology activist Pete Ramon says:* "While large biomass burning does put vast amounts of CO and particulates in the air, it's ultimately vanquishing Earth's ability to generate her own weather patterns. If you're trying to control global weather patterns, you do indeed have to cut out the middlemen—the Earth and Sun."

As Eugene S. Takle, professor emeritus of agricultural meteorology in the Department of Agronomy at Iowa State University, wrote:

> Biomass burning contributes about 10-20% of the sulfur that leads to carbonyl sulfide (COS), a long-lived form of sulfur in the troposphere. Because it is not easily removed from the troposphere, its distribution is relatively uniform with height and therefore allows COS to

*From his Facebook post of September 12, 2018.

> migrate to the stratosphere where it is converted by photolysis to SO_2 and eventually sulfate particles. Notholt et al (2003) find that biomass burning contributes more COS and hence sulfate aerosols than previously thought. Increase in biomass burning in tropical and subtropical Southeast Asia may have offset some of the decrease in atmospheric sulfur due to emissions controls in North America and Europe.[10]

Experimenting with various ways to "cut out the middlemen" leads to pursuit of Nikolai Kardashev's Space Age and the secret space program doctrine of full-spectrum dominance over the natural rhythms of the Earth and Sun.

One aim of the geoengineered California fires of 2017 and 2018 (in Santa Rosa, Woolsey, Redding, Delta, Malibu, Paradise, Kincaid, and other places) was to clear ancient redwoods for a three-hundred-mile right-of-way for the Great Redwood Trail to converge with the Trans Mountain Pipeline in British Columbia[11] for the Cascadia Innovation Corridor that will connect British Columbia, Washington State, Oregon, and California (maybe even Mexico)*—similar to how the ionospheric heaters are heating the Arctic to melt the glaciers to clear the way for ship lanes: planetary/geophysical engineering for AI-run transhuman "megaregions."

The Woolsey Fire in Santa Barbara and Ventura counties began at the former Santa Susana Field Laboratory in Simi Valley (ceased operation in 2006), home of Aerojet Rocketdyne's research reactor and ten other reactors, amounting to four cores of radioactivity, with iodine-131 radionuclides, trichloroethylene, perchlorate, dioxins, and heavy metals multiplying the contamination. During the Paradise Fire, the local hospital's radionics wing had all the markings of having been intentionally torched to release even more radiation.

While some think of Maui as a piece of paradise, others think of it as the home of the Maui Space Surveillance System (MSSS) at the Maui Space Surveillance Complex run by the U.S. Space Force, with the Haleakala Observatory at the midpoint between Vandenberg Air Force Base in California and the pivotal Space Fence site at the Ronald Reagan

---

*The Cascadia Innovation Corridor and other programs are about merging Washington State and Oregon in the United States with British Columbia in Canada in the name of ecology and "climate change," of course.[12]

Ballistic Missile Defense Test Site in the Kwajalein Atoll. In other words, Maui is essential to "the Air Force's next generation space surveillance system, the Space Fence."[13] Since the Hitachi-sponsored JUMPSmartMaui Project (2011–2017) that promised "40 percent of its electric power generation from renewable energy sources by 2030,"[14] Maui has been touted as "the new Smart Grid in Hawaii." In other words, Maui is electromagnetically wired to the hilt.

Former pharmaceutical industry R&D executive Sasha Latypova's substack asked: "Was weaponized GIS [geographic information systems] used in Lahaina? Click-and-destroy program for sustained development" (August 28, 2023). In other words, government was certainly involved in executing land grab fires.

> Large areas right outside of Lahaina appear to be proposed developments which likely were opposed by the current residents and conveniently positioned to quickly get the go-ahead now that the opposition has been literally wiped off the map . . . Looks like the Hawaiian Homelands Dept. is going to move Hawaiians from the lands that they used to own to housing where the land is owned by the government.

Geoengineered temperatures include the hot summer temperatures during "fire season," so think of the "wildfires" that seemingly arise along the U.S. mainland West Coast as disaster capitalism's "new normal" *prescribed burns* prepared by years of geoengineered deliveries of tons of aluminum nanoparticles then pounded into the ground by scheduled short-lived floods. The Defense Technical Information Center, part of the Department of Defense, informs us that "flammability can be greatly increased by killing all shrub vegetation" and "selecting optimum weather conditions for burning,"[15] but the most powerful preparation is decades of *preloading the environment* with aluminum nanoparticles and then firing masers/lasers to ignite the chemical, statically charged conductive nano-metals.

Retired USDA biologist Francis Mangels[16] measured 4,610 ppm aluminum in water and soil samples in the Mt. Shasta region—25,000 times above World Health Organization (WHO) safety levels. Add to Mangels's findings the statements by retired USAF brigadier general, General Charles

Jones—"These white aircraft spray trails are the result of scientifically verifiable spraying of aluminum particles and other toxic heavy metals, polymers and chemicals"—and former naval officer Denis Mills—"Millions of tons of aluminum and barium are being sprayed almost daily across the U.S. Just sprinkle aluminum or barium dust on a fire and see what happens. It's near explosive. When wildfires break out, the aluminum/barium dust results in levels of fire intensity so great as to cause firefighters to coin a new term: 'firenados.'"[17] The entire United States, in addition to various other NATO countries, is being sprayed.

Wildfires are normally considered to be either a natural phenomenon or the result of arson, but they can also be ignited from low-earth orbit (LEO) and accelerated with nanochemicals like ammonium nitrate (fertilizer) for purposes of warfare. In fact, think of the conductive and combustive chemicals laid by jets and drones over fire target areas a few days in advance as "chempiles," nano-super-thermite metal powder fuels and metal oxides: aluminum, magnesium, titanium, zinc, silicon, and boron. Add phased array "antenna gardens" like the one running north-south on the ocean floor in tandem with the LIPC "log" just off the California coast,"* and you've got a weapon system that just needs ignition.

> Ammonium nitrate (NH4NO3) is an oxidizer [as is aluminum oxide]; it supports the combustion of other materials. When a violent reaction occurs, such as a bomb blast or explosion, the ensuing cloud is made up mostly of nitrogen oxide and water vapor. In essence, ammonium nitrate releases oxygen. Oxygen is the fuel a fire needs to burst and spread.
>
> Now add the coatings of the nanodust metals from years of chemtrails with a new layer of ammonium nitrate and all the Maui firestorm needed as the spark. Yes, the DEW [directed energy weapon] device, whether a laser from a passing plane, satellite-based, or Chinese spy balloon. It wouldn't take much to blow the inferno wide open at night.[18]

---

*Thanks to German chemist Harald Kautz (previously Kautz-Vella) who pointed out this antenna farm in a Google map no longer available on the internet. "This is how the California fires were made to happen," he said, "these antenna gardens are, of course, phased arrays." An internet search of "5G phased array animation" gives various videos that show how phased array antennas work.

IGNITION

It was during the 2017 Santa Rosa fires that eyewitnesses first observed a shaft of blue laser light descending in a column of ionized air and wondered what its role in the "wildfires" was.*

Steve Favis,† an expert in computer science and advanced robotics, has confirmed that China had a satellite ("spy balloon") directly over Maui in earth orbit in the near-infrared (NIR) range for better transmission through the Earth atmosphere on August 8, 2023, at the exact time that all four fires were ignited. (China also has 70-gigawatt (GW) state of the art electro-optical plasma beam-generating lasers in earth orbit.) See the Greg Reese video "CCP Satellites Over Maui at Times of Fires," September 4, 2023, and read his substack:

> The most efficient way to ignite a fire on the surface [of the Earth] from a satellite in earth orbit would be to paint the target in segments by pulsing the laser with an advanced targeting system. To see if this were possible, Favis calculated what it would require to creat a meter-wide mile-long fire.

Fired from a satellite, the Earth's atmosphere will absorb and scatter some of the laser energy. And so the laser would need to be in a wavelength range that minimizes this. The most effective wavelength would be in the near-infrared range . . . invisible to the naked eye, [it] would also have a minimal reaction with objects colored blue on the Earth's surface.

The power of the laser would need to be in the hundreds of kilowatts range. And so Favis based his calculations on a 10-megawatt laser firing from earth orbit. Assuming that the atmospheric loss amounts to 50 percent of the overall power and only 5 megawatts reaches the surface as a one square meter beam, it would ignite a fire almost instantly. If this 50-megawatt beam was pulsed across a one-meter by one-mile-long area in segments, then the time to ignite the entire area would be roughly 2.7 minutes. And it would take approximately 8.8 seconds to melt an aluminum alloy wheel . . .[19]

---

*Laser visibility (or invisibility) is determined by frequency of the laser, strength of the laser, and dust or mist in the air.

†Stephen Favis is CTO (chief technology officer) at Favis Advanced Robotics Corporation.

Citizen eyewitness accounts from drone footage and cell phone cameras* have been invaluable in terms of recording the *sequence* of these "wildfire" events that proves they follow a *pattern* that is not "wild" at all, beginning up to three days in advance with jets and drones strafing the sky over the target zone with nanometal alloys, sensors, and chemicals to determine the indices of refraction† necessary for the operation. Ignition of nanoswarm brigades in the chemtrailed air and already in the drought-compacted soil may precede the laser ignition from the heavens of the actual "wildfires."

As ignition occurs, onlookers may or may not see the telltale shaft of blue light. Thanks to powerlines, underground fiber optic cables, nearby microwave towers and iPhones, and cars acting as conductors and antennas,[20] a microwave energy field is created and plasma fires ignite like they are *combusting*, reminding us of the six explosions—"combustion syntheses"—that occurred between August 3 and August 5, 2020, in Asia, Eurasia, the Middle East, and the United States Were these combustions experimental as well as profit-oriented? If so, surely the geographic—or should I say geomagnetic—locations would have been carefully chosen, rather like the 39° N latitude of both the California and Portugal fires.

- August 3, Hyesan, North Korea
- August 3, Xiantao (outside Wuhan), China
- August 4, Beirut, Lebanon‡
- August 4, St. Paul, Minnesota
- August 5, Ajman, United Arab Emirates
- August 5, Naiaf, Iraq

While the Tubbs Fire in Santa Rosa, California, burned, fire burned for six intense hours in the town of Oliveira do Hospital, Portugal, destroying 93 percent of the surrounding forest, 20,855 houses and buildings, and

*Not everything is picked up by iPhone cameras, due to their filters.

†Necessary to the optics of wireless energy transfers so as to accurately determine the angle by which light is reflected and refracted through different materials.

‡A highly classified black triangle aircraft known as a TR-3B was filmed during this "multiplex" explosion.[21]

Fig. 4.1. Firefighters in Rapa, Portugal, after fires burst back into life over a weekend.
NUNO ANDRE FERREIRA/EPA-EFE/Shutterstock

54 deaths.[22] Drones were spotted when it started, possibly to trigger the laser descending from the plasma cloud cover.

"At the northern border an airplane [or drone] over our forest, everything went dark, five minutes later a fire tsunami, a 20-meter wall of fire arose. Wooden factory in village 3 kilometers away—I saw a beam, heard sounds like glass falling on metal, the sound of electricity, almost like iron in a microwave. We found thousands of empty bottles at the northern border in the deep forest, beside roads, in villages." —Conny Kadia, eyewitness to Oliveira do Hospital, Portugal, "wildfire," email, December 4, 2017

On July 23, 2018, while geoengineered "wildfires" raged throughout California, Greece had its "first event directly related to the climate crisis that affected everyday life on a local scale" when Mati, a holiday resort outside Athens, was incinerated by a "wildfire" that killed 103 people[23]—another little bit of paradise, gone.

COOKING UP THE WIND?

A primary difference between the fires in Santa Rosa and those in Portugal was that the Portuguese fire was accelerated by category-3 hurricane Ophelia, the tenth and final hurricane of a very active 2017 hurricane season. Sweeping up from the southeast Atlantic Ocean, Ophelia arced past Portugal and Spain all the way to Ireland. The hurricane winds acted as a bellows for the fires, with a peak intensity of 115 mph just south of the Azores on October 14. This important fact was not in the National Hurricane Center report.

"This topic involves deeply occulted dark military advanced weaponry science, so open your mind to the incredible possibilities of the best minds limitlessly funded with results bent not for 'free energy' or any good, but primarily for cultish mass murder (Purim sacrifice?) and geopolitical leverage. It was [President Donald]Trump's uncle [physicist John G. Trump (1907–1985)] who bragged to have been involved in the infamous gov-grab of Tesla's papers, particularly regarding his Death Ray—an almost inconceivable energy and particle beam weapon that required high-voltage gradient established over target (by Hurricane Erin), into which particles are injected at high velocity . . . and excited into microwave resonance, such that aluminum, steel, and hard brittle materials are vaporized into nanoparticulate dust." —Michael Brenden, "HAARP Weather Mod Disrupts Earth's Magnetosphere," August 18, 2020, www.toxi.com

---

The presence of hurricanes as the microwave "wildfires" were loosed—including Hurricane Dora (category 4) allegedly 700 miles south of Maui on August 8, 2023—goes back to 9/11 when Hurricane Erin (category 3) was at one point 540 miles east of New York City, for which we have the invaluable 2005 record *Where Did the Towers Go? Evidence of Directed Free-Energy Technology on 9/11* by scientist Judy Wood, PhD. Hurricane

Erin (category 3) was 540 miles from New York City on 9/11; Hurricane Ophelia (category 3) was off the coast of Portugal on October 8, 2017, when the fires in both Portugal (latitude 39° N) and California (latitude 39° N) began; and Hurricane Dora (category 4) was 700 miles south of Maui on August 8, 2023. Why did news agencies outright *lie* or at least avoid referencing the positions of these hurricanes during the advent of the fires? According to Dr. Wood, Hurricane Erin was thoroughly studied, but in chapter 18, "Hurricane Erin," Dr. Wood points out that none of the four major networks showed an icon or mentioned where Hurricane Erin was on September 11, 2001. Dr. Wood includes a satellite photograph of *the eye of Erin*, which "is in the shape of a pentagon and the interior of the eye looks like a maze that winds back and forth. How, and why, are these unique formations there?"[24]

Major airports recorded "distant thunder" as U.S. Navy planes flew into Erin's eye to assess its strength. Erin was geoengineered to generate the high-voltage gradient needed for the molecular dissociation associated with the event: wilting steel, evaporation of metal, cold "heat," dustification, binding of metals and nonmetals.[25]

The path Erin took was most definitely manipulated; Dr. Wood described the magnetometer readings of the magnetic declination of the Earth's magnetic field, especially Erin's eye's dramatic turn northeast away from New York City on 9/11:

> . . . the use of directed energy technology on the huge scale of its use during 9/11 might very well have had weather-altering effects. Such technology might have been able to draw upon the vast energies and field effects of the enormous Tesla Coil known as Hurricane Erin.[26]

Such an insight, coupled with the fact that mainstream media ignored or brushed aside or outright hid the "coincidence" of these hurricanes leads us back to Michael Brenden's turn of phrase above, indicating that Erin was geoengineered to generate the high-voltage gradient needed for the molecular dissociation associated with the event.

Of course, even the jet stream can be used to produce high wind gusts for precision attacks on specific targets. Vic Livingston, independent print/TV journalist and military veteran who monitors weather anomalies, describes what happened to him in Pennsylvania when a weather cell was whipped up to precision-target his home early one morning:

> Using the microwave weapon system to intensify weather patterns, manipulate the jet stream to generate high wind gusts, then operatives can precision-target directed MW [microwave] energy to create damaging winds in the immediate vicinity of a target—thus using weather as a weapon and major weather patterns as a camouflage for local precision-targeted attacks, like outbreaks of "thundersnow" or the tornado-strength winds that pushed against my house this morning shortly before 6 a.m. After I took steps intended to ensure that authorities review surveillance of Lockheed Martin CentCom [unified armed forces command] personnel on duty this early a.m. to determine if this highly plausible theory is true, the tornado-strength winds quickly subsided.[27]

Livingston cites U.S. Patent 7629918B2, "Multifunctional Radio Frequency Directed Energy System" as the possible weapon used. See his account of the November 13, 2012, event.[28]

Tornado-strength winds by land, hurricane-strength winds by sea.

THE VAPORIZING EFFECT

Space launches use aluminum oxide as a propellant for a thermite burn of lithium with aluminum and barium, thermite being a pyrotechnic composition of nanometal powder fuel and nanometal oxide (aluminum powder and iron oxide), meaning it is best used in combination with metals for incendiary purposes, like the nanometals in chemtrail deliveries that end up in the soil and in people's lungs, blood, and brains. Once ignition occurs via laser/maser or other directed energy, firefighters experience it as a *vaporific effect*, a "flash fire" due to the impact of a high-velocity projectile with metallic objects, possibly causing damage so extensive that the target simply vaporizes. For example, nanoparticles of magnesium can burn at 3,600 degrees Fahrenheit, splitting water into hydrogen and oxygen, after which each element undergoes combustion. Magnesium nanoparticles can melt an engine block, just as the exothermic redox of thermite melts iron fences and bolts in creosoted guard rails, as it so obviously did in the case of the Paradise and Maui "wildfires."

The metal nanoparticles inhaled from chemtrails are suitable for a vaporific effect, and I do not need to remind the reader of how much H_2O (water) is in our bodies. The possibility that human beings inside a microwave field can be "cooked" *from the inside out* like a microwave oven cooks food or like the trees burning from the inside out in the California fires, must be considered. In Dr. Wood's chapter 3, "The Jumpers" (*Where Did the Towers Go?*), she goes into a great deal of detail regarding the behavior of "jumpers" on the 105th floor of World Trade Center 1 as they first hung outside the windows, then ripped their clothes off, then let go and fell to their deaths. Did they feel like their bodies were being *boiled alive* in a microwave field?

> At this juncture we must introduce a hypothesis in an attempt to explain this strange behavior. Consider what might be expected if some sort of energy field, such as a microwave field, had been affecting that area just inside the building. Such a field might be part of what comprises the Active Denial Systems (ADS) that are now being used for crowd control. It is equally possible that such a field was part of whatever was destroying the building. In either case, *wet clothing intensifies the pain caused by such microwaves*, as is acknowledged in an article about the ADS: "Wet clothing might sound like a good defense, but tests showed that contact with damp cloth actually intensified the effects of the beam."[29]

In *The Conversation* article "9/11: the controversial story of the remains of the World Trade Center" (September 7, 2021), we learn of how the Staten Island landfill called "Fresh Kills" exudes "toxic vapours proving harmful to the workers on site" and is loaded with tons of pulverized concrete, construction debris, cellulose, asbestos, lead, mercury, and fire dioxins, is also "a graveyard for unidentifiable bodies." Even more shocking is:

> An affidavit filed in 2007 before a Manhattan Federal Court reveals that the [human] remains, mixed with *debris powders known as "fines,"* had been allegedly carried away by city employees to fill ruts and potholes in NYC . . . As the families of victims filed a lawsuit for mismanagement of human remains against the municipality, the authorities objected that the debris had been inspected following a meticulous

> process of classification. In the end, the judge sided with NYC. He said: "the victims perished without leaving a trace, utterly consumed into incorporeality by the intense, raging fires, or pulverised into dust by the massive tons of collapsing concrete and steel."
>
> A marshland in the 19th century, Fresh Kills is now an eco park, including a human-made wetland, secured by a system for the capture and treatment of underground toxic gases that heats 20,000 local homes.
>
> The story of 9/11 provides a stark example of the political economy of waste management, which profoundly shapes the culture of the modern metropolis.[30]

Despite the assurance that debris was "meticulously" inspected, I doubt that the public has been informed as to the studies of the "debris powders known as 'fines.'"

Metal fires burn hotter and longer with less oxygen—to wit, Agent Orange for Vietnam chemical warfare, nanometal particles for California and Maui combustions. The well-known USAF aerospace illustrator Mark McCandlish (1952–2021) told the Shasta County Board of Supervisors Hearing on Chemtrails, July 15, 2014:

> Imagine then how this affects the conflagration that is a forest fire with these materials present in the environment. I have personally spoken to a number of career CDF [Chief of Defense Forces processes air cargo for USAF missions] personnel who have told me unequivocally that fires over the last ten years have become significantly more difficult and costly to suppress. They burned unusually hot, but officials were at a loss to explain why . . . Now as if that weren't bad enough, with aluminum being a conductor of electricity, spraying countless microscopic-sized particles into the sky does something else you might not have considered: It dramatically increases the electrostatic potential of the air, that is, its ability to conduct electricity. So those storm clouds that always seem to follow heavy chemtrailing are primed to produce many more lightning strikes. In late July of 2010 (if memory serves), one such storm produced over 8,000 lightning strikes in our region, many of which created fires. When it was all over a month later, California had totaled over $23M in suppression costs. And since the

chemtrailing started around 1999–2000, the amount of acreage burned and suppression costs have doubled, according to NOAA figures.*

MORE DAMNING EVIDENCE

Investment analyst Catherine Austin Fitts† calls the California fires "9/11 West," primarily for the mortgage fraud that followed from the intentionally set fires so properties could be cleared for wealthy Asians. During a disaster, laws are more flexible and people more desperate. Fitts says that skimming out the back door is common—for example, writing off mortgage fraud[32] and pilfering from disaster recovery funds. Citizen deaths, meanwhile, minimize Social Security outflow. The Red Cross took in $425 million from people wanting to help fire victims, but only provided water bottles.‡

Where did the rest go? Insurance and reinsurance money; catastrophe (cat) bonds§ for money laundering; lucrative real estate deals while devastated and desperate people are crammed into pack-and-stack smart

---

*Most of Mark's clients were top military contractors like General Dynamics, Lockheed, Northrop, McDonnell Douglas, Boeing, Rockwell International, Honeywell, and Allied Signet Corporation.[31]

†Austin Fitts served as assistant secretary of Housing and federal housing commissioner at the U.S. Department of Housing and Urban Development in the first Bush Administration.

‡The Red Cross is notorious for "losing" money, e.g., the geoengineered Haiti earthquake in 2010. From my first book on geoengineering *Chemtrails, HAARP, and the Full Spectrum Dominance of Planet Earth* (Feral House, 2014): "Haiti is the poorest country in the Western Hemisphere, with a third of its population (3.3 million people) 'food insecure.' A wage of $12.50 per day might suffice, but most Haitians live on less than two dollars per day. In June 2009, the Haitian Parliament unanimously passed a 9-cents-an-hour minimum wage pay increase for the assembly zone workers of Hanes, Fruit of the Loom, and Levi's (Levi Strauss). But factory owners refused to pay 62 cents per hour ($5 per day) and were backed by the U.S. Embassy and U.S. Agency for International Development (USAID). Two months later—just five months before the 7.0 Haitian earthquake on January 12, 2010—then-President René Garcia Préval negotiated a compromise, but the U.S. Embassy was still not pleased. WikiLeaks revealed to the local daily *Haiti Liberté* 1,918 U.S. State Department cables insisting upon low wages and threatening factory shutdowns, the fear being that the Dominican Republic and Nicaragua would want higher wages, too. Could the 7.0 Haiti earthquake have been corporate payback, plus a warning to other developing nations to toe the transnational line?"

§Catastrophe bonds are collateralized debt obligations between sponsors and investors.

Fig. 4.2. "FOREST FIRE AS A MILITARY WEAPON,
Final Report, June 1970, U.S. Department of Agriculture, Forest Service."

cities (exactly the plan for Maui); burned and scorched Douglas firs and Ponderosa pines clear-cut for high-priced lumber and polymerization factories . . . The list of "benefits" goes on and on.

With the loss of Paradise, California, we turn once again to the secret societies that many corporate CEOs, foundation donors, political leaders, military brass, intelligence agencies, and scientists dedicated to advanced technologies have given oaths of fidelity to, including the shadow side that is replacing the natural world and Human 1.0 with the transhuman virtual and synthetic world. In the press, violent events (psyops) are couched in *twilight language* so as to program the public subconscious beneath the seemingly rational public mind conditioned to believe what it is told or reads in print. The subconscious is the means by which *a continuous pulse of fear and anxiety* is manipulated, whatever the facts or deceptions in the news.[33] Fire speaks for itself: arising out of nowhere, uncontrollable, insecure, filled with the terror of blood sacrifice. Fire from the sky was once only evident during war, but now . . .

The "Paradise Lost" Camp Fire that utterly destroyed Paradise, California, appears to have been carefully scheduled to follow midterm elections on November 6, 2018, to be in turn followed on November 7 by mind-controlled U.S. Marine Ian David Long, age twenty-eight, shooting and killing a dozen people at the Borderline Club in Thousand Oaks, then killing himself to achieve the favorite secret society number, 13.

*Fire walk with me?**

The Camp Fire followed the fire sacrifice through November 8–9, 2018. Paradise (pop. 27,000) lost 6,700 structures and at least 88 people died (many immolated in metal vehicles), with 1,276 missing and 52,000 displaced or unaccounted for. PG&E trucks and mine-resistant, ambush-protected vehicles filled with militarized police, tactical and sabotage teams, and private security forces were everywhere. Five years later on Maui, it was the same FEMA (Federal Emergency Management Agency) pattern.†

Overhead during the Paradise Fire, TR-3Bs—highly classified, black triangular antigravity aircraft ("the Black Manta")—were spotted hovering in the overarching plasma cloud cover, sometimes captured on cell phone cameras. One acute observer noted three-hole patterns on the hoods and roofs of incinerated cars, in the soil, in stainless steel sinks inside burned-out houses, and in concrete, glass, and even plastic containers. Do these three-hole patterns have anything to do with the three glowing lights on the TR-3B hulls, one at each angle of the triangle? Just as the energy of the Sun can be focused with a magnifying glass to melt rock in fifteen seconds, dirt in forty-five seconds, wood in two seconds, and steel nails in thirty seconds, the evidence of the use of a directed energy weapon (DEW) such as a microwave laser (i.e., a maser) seems obvious. Lasers to ignite, masers to melt? One burned-out car had two antennas, one on each side of the trunk that seemed to have stopped the fire raging over the car

*From the David Lynch / Mark Frost mystery series *Twin Peaks*, April 8, 1990, ran two seasons, then was canceled. Note the "Paradise" twilight language in *Borderline Club* and *Thousand Oaks* with November 6–9 under the zodiacal sign of the Scorpion. If you think twilight language is just accidental, think again.

†High-value murders are often hidden amid chaos and mass casualties.

roof, the trunk being unblemished. Bingo: the Paradise "wildfire" was conducted via a high-powered microwave pulsing weapon.*

Google Earth once exposed a string of underwater microwave transmitters running north and south along the West Coast, but these transmitters have apparently "gone black."† While they may have played a big role in evaporating moisture so as to maintain the recent nine-year California drought (January 2011–2020), Nanowave Technologies Inc. calibrates high-powered transmitters, including those offshore, to prepare the target environments for combustion. Chemical trails deliver the nanoscale oxides, metals, alloys, and sulfides necessary for the combustion synthesis of reactive agents guaranteed to produce thermodynamic-kinetic, "rapid self-sustained combustion reactions" like fires and explosions.[35]

Many public-private partnerships are now dedicated to manufacturing nanocombustibles.

For example, the Department of Energy's Argonne National Laboratory Combustion Synthesis Research Facility in Illinois, has, since 2018, used Flame Spray Pyrolysis "that allows for commodity-scale production of a very broad range of nanomaterials that includes silica, metallic, oxide and alloy powders or particulate films," much like the way spray pyrolysis going on in aircraft engines produces artificially nucleated "contrails."[36]

Beyond the official narrative of "wildfires" are the devil's details proving a microwave plasma fire. Working with our own perceptions, thoughts, research, and study, we take note of:

- Signs of the environment being preloaded for massive spontaneous combustion‡
- Unburned vegetation

*U.S. Navy-trained radar expert Christopher Fontenot says, "The use of MASER (DEW) technology to heat the atmosphere for weather modification, increase evaporation, or even conduct microwave-induced combustion operations in which temperatures reach 950 degrees Centigrade (1742 degrees Fahrenheit) is relatively recent," email, July 2, 2018.

†*Going black* has nothing to do with race; it is an intelligence term meaning to go off radar and disappear into the black budget, etc. The Google map of underwater transmitters is gone from the internet, so this MIT explanation will have to do to give you an idea of how transmission from water to air to land (minus fiber optic cable) might occur.[34]

‡Preloading includes nanochemical deliveries three days in advance of geoengineered events (fires, earthquakes, storms, etc.) and the impacted nanometals and nanobots already in the soil, walls, bark of trees, etc., from years of geoengineering

- Trees burning *from the inside out*, as per a microwave oven*
- Unburned paper
- Unburned plastic
- Melted tempered glass and vaporized wheel-rim metal alloys of cars
- Cars act as conductors and antennas
- Fiberglass boats on land are melted
- Pulverized CMUs (concrete masonry units)
- No molten metal, just melted and twisted steel alloys
- Moving phalanxes that appear as "embers" but are actually nanotechnology engaged in burning some structures and allowing others to remain
- No smell of smoke, only of ozone (O3)
- Physical symptoms of being in a microwave energy field (like having trouble breathing)
- Fire without heat, nor does glowing indicate heat; luminescence without heat points to cold fusion or the Hutchison Effect[37]
- Extreme heat but no fire (microwave fire 2000–3000°F), as described by onsite 9/11 witnesses via cell phone
- Six weeks after "wildfires," microwave fire is still burning underground
- "Wildfires" *explode* into "firestorms"

Begin building a detailed imaginative mental picture, such as how metals in the environment or in our bodies can be triggered with microwave, millimeter wave, or infrared in current satellite systems, in the X-37B unmanned orbital spacecraft or TR-3B Black Manta. And finally, keep in mind that the combustion synthesis is about proximately associated materials (which absorb energy), not the match that is the laser or maser.

"The Great Plume in Paradise: At 6:33 a.m. on November 8, 2018, fire was ignited and immediately traveled WEST from Concow to Paradise with a tight plume that exploded over us within an hour. I was outside until 10 a.m., and it was bluebird. An hour later, complete [chemical cloud] coverage . . . Paradise got hit last year as well; they keep coming

*Do organic creatures like animals and humans also die in this way?

back to the same areas. I've been the only one until recently who has chronicled the worldwide DEW attacks the past 3 years, but it really goes all the way back to the Oakland '91 fires at the same time we were using lasers in Iraq 1." —James W. Lee, videographer and author of *Paradise Lost: The Great California Fire Chronicles* (2019)

THE EARTH IS NOW A CAPACITOR UNDER THE NEW SPACE ECONOMY

The 2020 crown fire in Walker, California, generated enormous pyrocumulus and pyrocumulonimbus clouds visible from space—NASA's "fire-breathing dragon of clouds"[38] injecting masses of nanoparticles into the lower stratosphere ten miles above Earth. *Plume-dominated* means high pressure vertical updraft resulting in strong, erratic winds and extreme fires—fire tornadoes—that sustain and grow themselves, thanks to all the added nanoparticles.[39] The wind that mainstream media kept talking about was not natural wind, but rather blast waves from pressure expanding supersonically (meaning in a scalar dimension) due to explosive cores of compressed gases, followed by blast waves of negative pressure.* Fires moved at an uncanny eighty football fields a minute. Firemen stood in bewilderment.[40]

"I just started reading your new book *Geoengineered Transhumanism* and thought I would get your thoughts on the 'wildfires' here in Colorado—the Cameron Peak fire that began on August 13, 2020, and the one in Boulder on December 30, 2021? The Boulder fire was precipitated by 100 mile-an-hour winds—sounds like microwave tech to me. As far as the Cameron Peak fire, I have been up there this past year to view the damage. Some parts look like the fire was guided, as there were fingers of burned areas that then just stopped; the insides of trees with streaks of blue after the bark peeled off look like chemical poisoning. A lot of pine trees look healthy with green needles but are just falling over. I thought

*Weather engineering technology is capable of compressing the atmosphere to create wind fields.

if the tree has such a shallow root system, surely the needles would show decay before the tree just falls over. I have been living in northern Colorado since 1994, and every time I go up in the mountains, the tree die-off is worse. I was up there the day the Cameron Peak fire started, and the fire crews seemed more interested in getting people out of there than putting the fire out. The darned thing burned until December!" — Christopher Andrina, Colorado Springs, email, September 17, 2022

The "New Space Economy" (i.e., "the rising commercialization of space exploration,"[41]) which includes a manic obsession with decreasing carbon and viewing everyone as "nodes" of energy or information, is much more than we are told. Beginning with the 3D battlefield monitoring of everyone and everything on Earth, the aim is to *read all* DNA signatures and brain-evoked potentials (the electrical potentials in the brain following stimulation by sight, sound, or touch), while cavity magnetrons* read the metal in our brains and bloodstream as well as in the soil, in buildings, in cars, and interference patterns are used to direct phase conjugation, heat nano-metal components, ignite one thing and not another—all from space or from the geomagnetic grid fed by metal nanotechnology now compacted in the soil.[42] Flying, intercommunicating nanomachines—atomic-level swarms of nanobots used as nano-size drone armies in "wildfires"—can shapeshift for whatever programmed task they are commanded to carry out.[43] These are the twenty-first-century nano-swarmtroopers whose hive intelligence is yoked to supercomputers armed with AI algorithms—"programmable matter" that swarms, self-assembles, self-replicates, and destroys or builds their kin in the outer environment as well as inside our bodies and brains.[44] They are the golden key to the geoengineered transhuman.

The Earth is now no longer viewed as the great Natura but as a capacitor that stores electrical charge, while our atmosphere is viewed as a great antenna. Underground ULF lightning discharges are now prolific[45] and can be amped up considerably by the scalar waves sent through the Earth by the ionospheric heater technology. "Lightning" erupts inside trees. Was the fire that moved at 273 mph through Paradise amped up by the town's

*These are high-power vacuum tubes used in early radar systems and currently in microwave ovens and in linear particle accelerators to generate microwaves using the interaction of a stream of electrons with a magnetic field.

gold and silver deposits, then triggered by a laser / maser "kill charge" from space?

In James (Jamie) W. Lee's "Paradise Lost #51" video[46] drone and dashboard footage of what initially looked like masses of embers being blown by the wind are actually phalanxes of nano-"swarmtroopers" flying *against* the blast surges, then swarming up walls and along the edges of buildings, even down highways, intent on target areas slated for destruction.* This was not a random fire but an intelligence operation run from ion-driven "eyes in the sky," directing a swarm intelligence via monitors or military-grade iPhones. Destroying from space with high specificity is the idea. From Jamie's Amazon page for his book *Paradise Lost*:

> On October 8th, 2017 at about 10:30 pm, the winds outside my home were moving the large Redwood trees back and forth, though no storm or wind had been predicted. I went outside and saw several flashes of blue light pulse in the vortex winds forming above me. At 4:30 am, my neighbor was pounding on my door, telling me that emergency evacuations were being conducted due to several fires on the ridges. Outside, I could saw orange skies to the North and East, but no flames and no trees ablaze. Thus began my journey into discovering and uncovering how and why 157 "abnormal" fires began in nine different counties in Northern California. I toured some of the over 4,000 homes destroyed to ash while trees adjacent were left untouched. I saw 3,000-pound cars flipped on their roofs. I saw a 100,000 square foot Kmart building, where the fire had to jump the six-lane 101 freeway, cross a large parking lot, and completely torch the inside while the outside remained untouched. Thus began my journey into learning about what former Governor Brown deemed "The New Abnormal" over the next 10, 15, or 20 years. Never before have firemen seen such fire. They have no context for understanding the advanced weaponry used on not only these fires, but fires that have been set ablaze in over 15 separate areas

*Jamie Lee (aplanetruth3) and the anonymous nanomaterials scientist "Angel" discuss the geoengineered California fires as they watch footage of "ember swarms" of programmable matter, i.e. nanobots programmed to start and spread fires, thanks to the impacted aluminum nanoparticles delivered by chemical trails for years and now in the soil, concrete, walls, etc. Lee is also the author of *Paradise Lost: The California Fire Chronicles* (Self-published, 2019).

of California over the past 1-1/2 years. The deadliest of all California fires was the Camp Fire that occurred in Butte County on November 8, 2018, beginning with an explosion. Again, no weather-related forecast of winds said to travel at "80 football fields per minute," which means these fires were traveling at 273 mile per hour. 52,000 people were said to have safely evacuated Paradise, yet residents who were lucky to escape the fires claim this is nowhere close to being true and many perished while fire departments stood down and military police locked down all of Paradise for a month before allowing the victims who remained to return to see if their homes still stood. There was no warning from public service officials. There were no winds the morning of the fire igntion said to have been caused by (1) falling power poles, (2) bullet holes in transformers, and recently (3) "fire igniting embers" that somehow entered homes and turned them to ash. Each person I interviewed said they received no help from any charitable organizations that have taken in millions of dollars to help victims. Many fled with the shirt on their backs, many becoming homeless or sleeping in their cars with no one to turn to.

This book is a result of some of my 80 YouTube videos I published while investigating the Paradise fires, as well as directly experiencing the 170 California fires in October 2017. This book covers the likely weaponry used to start the fires, the agenda behind the fires, and those who directly benefitted. You can find my work also on YouTube at "Aplanetruth3," "WellHealed2," and websites Tabublog.com, Avvi.info, Aplanetruth.info and Wellhealed.life. This book details like no other the new abnormal and takeover of California's precious resources.

PART II

Tethering the Transhumanist Future

5

Gravity, Æther, and Tesla's Atmospheric Engine

The more we study gravitation, the more there grows upon us the feeling that there is something peculiarly fundamental about this phenomenon to a degree that is unequalled among other natural phenomena. Its independence of the factors that affect other phenomena and its dependence only upon mass and distance suggest that its roots avoid things superficial and go down deep into the unseen, to the very essence of matter and space.

Paul R. Heyl, "Gravitation: Still a Mystery," *Scientific Monthly*, May 1954

Brazilian physicist Fran De Aquino asserts that the global chemtrails-ionospheric heater network that serves the Space Fence promises not just engineered weather and geophysical and staged holographic events, but modification and control of gravity itself.[1]

Gravity is defined as a fundamental physical force that is responsible for interactions that occur because of mass between particles, between aggregations of matter (such as stars and planets), and between particles (such as photons) and aggregations of matter, that is 10^{-39} times the strength of the strong force, and that extends over infinite distances but is dominant over macroscopic distances, especially between aggregations of matter.

Since 1956 and the advent of the secret space program, the U.S. aerospace industry ("the military-industrial complex" that President Eisenhower warned us about) have sought, with the aid of scientists imported from Nazi Germany as part of Operation Paperclip, to conquer gravity so as to get off the planet and through the Van Allen radiation belt, the energetically charged particles that encircle Earth, and out into space. Rockets that require masses of fuel propulsion and nuclear fission have always been about getting satellites up for communication and surveillance, not about space exploration.

THE DESCENT INTO SCIENTISM

Isaac Newton's falling apple didn't really explain gravity, nor did Einstein's "curved space-time" geometry. In the 1920s, under the spell of various secret societies, science went off-course and into *scientism*, the belief that science and the scientific method is the only way to apprehend truth. This when the concept of æther, the universal substance believed for ages to be the medium for the transmission of electromagnetic waves, was removed from the scientific lexicon, along with the idea of the *living universe* put forward by Austrian naturalist and engineer Viktor Schauberger (1885–1958). Schauberger, a contemporary of Rudolph Steiner (and indeed they communicated at length),[2] described the universe as a living, complex interaction of forces that constantly create or reinvigorate matter, and that it is the finer and "higher" energies—i.e., consciousness—that are responsible for creating form and structure, not the other way round, as scientific materialism states. Schauberger was ridiculed by the scientific community for going against the prevailing view that matter is in a perennial state of entropy, but his ideas about the living universe and æther resonate today, in the age of quantum physics.

Electric Universe theory describes a living universe with electric circuits threading through it, an idea promoted by the Thunderbolts Project, an organization devoted to "interdisciplinary research, direct observation, and experimental work confirming the pervasive role of the electric force in nature."[3] Crucial to the theory of the Electric Universe or living universe is the existence of æther, the prime continuum of the fundamentally highest possible frequencies of the universe, producing magnetism, electricity, light, heat, sound, vibration, and mass. Today, science favors the terms *zero-point* *energy* or *zero-point field.** The removal of *æther* and the intentional distortion of nineteenth-century Scottish mathematical physicist James Clerk Maxwell's theory of electromagnetism† has led to a misperception of gravity and fanciful replacement terms like *black hole* and *Higgs field.*

Einstein's special relativity theory was also exploited to hide the existence of æther. We have been led to believe that mass spinning in space drags Space and Time around, creating a curved 4D spacetime continuum, with gravity being the result of an object trying to travel in a straight line through a space curved by the presence of material bodies.

ÆTHER, THE MEDIUM OF FREE ENERGY

A crucial misconception is that the speed of light is an absolute. The autodidact Walter Bowman Russell (1871–1963), a painter, author, mystic, and writer whose ideas reflected the New Thought movement (he was a contemporary of Maxwell), perceived differently—that the stability of planetary orbits indicates that *gravity propagates much faster than light*, at least 20 billion times faster. So how then can the speed of light (in a vacuum, no less) be absolute? And if gravitational interactions are instantaneous, then where does that leave the assumption that nothing can propagate faster than light in a vacuum?

With such different propagation speeds, why wasn't special relativity, abandoned? Einstein's formula $E = mc^2$ only makes sense if *E* stands for *æther / ether.* Was all the obfuscation and lying about gravity just to conceal Tesla's free energy? Except for his work on alternate current (AC), it is true that everything about Tesla was carefully and methodically buried behind "national security," while Einstein's ideas were appropriated (or misappropriated) to hide the light-filled medium of free energy found everywhere: æther.

*Wikipedia states of *Zero-point energy*, "According to quantum field theory, the universe can be thought of not as isolated particles but continuous fluctuating fields: matter fields, whose quanta are fermions (i.e., leptons and quarks), and force fields, whose quanta are bosons (e.g., photons and gluons). All these fields have zero-point energy. These fluctuating zero-point fields lead to a kind of reintroduction of an aether in physics since some systems can detect the existence of this energy. However, this aether cannot be thought of as a physical medium."

†The concept of electromagnetic radiation originated with Maxwell, and his field equations paved the way for Einstein's special theory of relativity, which established the equivalence of mass and energy. Maxwell's ideas also ushered in the major scientific innovation of twentieth-century physics, quantum theory.

Breaking free of Einstein's concept of "curved space-time," which says that gravity arises from the shape of spacetime, is essential to understand that gravity and electromagnetics share fundamental characteristics: (1) they both diminish with the inverse square of distance; (2) they're proportional to the product of interacting masses or charges; and (3) they act along the line between them. In short, apply force to a body, and the force is *electrically transferred* to overcome inertia. Thus, inertial mass and gravitational mass are equivalent, because gravity is a manifestation of electrical force.

Besides Tesla, *antigravity* too has gone black, excepting *electrogravitics* in some quarters, "a synthesis of electrostatic energy used for propulsion—either vertical propulsion or horizontal or both—and gravitis, or dynamic counterbary, in which energy is also used to set up a local gravitational force independent of that of the earth" (Aviation Studies). Other terms that basically mean antigravity are *field-dependent propulsion* and *gravity-shielding*, terms possibly connected to the "Atmospheric fueled engine" (US Patent #6145298A):

> . . . The ion engine propulsion system ionizes a portion of an ambient atmospheric fuel to create a negative ionic plasma for bombarding and accelerating the remaining portion of the ambient atmospheric gas in a focused and directed path to an ion thruster anode . . .

Cal Poly Chris Edwards simplified it on Facebook:

> Where we're going, we don't need roads. Nikola Tesla's wirelessly powered aircraft and "[plasma] cloud maker"—IONOCRAFT? Lockheed Martin Patent US3130945A "Ionocraft" [expired in 1981]—Nikola Tesla's ATMOSPHERIC ENGINE, RCA Patent US6145298A—Atmospheric fueled ion engine utilizing high-altitude ambient [plasma] gas as fuel and producing ozone as a byproduct of propulsion. The ion engine propulsion system ionizes a portion of an ambient atmospheric fuel to create a negative ionic plasma for bombarding and accelerating the remaining portion of the ambient atmospheric gas in a focused and directed path to an ion thruster anode.

Marc J. Seifer, PhD, a recognized expert on Nikola Tesla, wrote in a 2012 letter to *Time* magazine about how a propulsion system depends on æther, and how gravity is transformed for flight:

> The ether, of course, exists. Just look at a picture of a galaxy and you will see it is floating in something. You can call it the Higgs field [i.e., CERN's Higgs bosun "God particle" field] if you want, but it is indeed the medium existing throughout all of space. This ether most likely exists in a tachyonic (faster than the speed of light) realm. It oscillates at such a high frequency that it remains undetectable by modern-day methods. What gravity is, according to this theory, is simply the absorption of ether by elementary particles. During this process, which involves particle spin, the ongoing course of action is converted into electromagnetism. This simple idea explains Einstein's 40-year quest, his dream of Grand Unification, namely, the way to combine gravity (the influx of ether into matter) with electromagnetism.[4]

Chris Fontenot's final word on the atmospheric ion engine is that it has passed experimental testing and *"is the future."* Are these the craft hiding in plasma cloud cover as per Operation #7 of geoengineered operations? *Some even produce their own plasma cloud cover.* (See fig. 5.1.) The following 2017 statement from "The Truth Denied" (in response to a Cloaked Craft article by Jim Kerr) provides excellent observations and analysis of what is being observed in the altered sky above:

> The cloaking cloud formation around these [exotic propulsion] craft is a byproduct of the static energy created by the propulsion system used by this craft. The external electrical static field attracts moisture present in the atmosphere, helping to cloak it. The buildup is the greatest at the corners. The craft has an internal ion generation propulsion system. It is basically like an ion breeze fan with no moving parts, but the ions "blow" in and are contained internally and condense. By changing the polarity of the containment walls, the craft can increase speed and change direction in any aspect. The condensed ions will attempt to repel from the charged wall and "push" the craft in the direction it needs to go. The faster you need to go, the more surface area of the walls is charged. Since you have five walls of a triangle, you can go in all those directions just by charging the proper walls to create repulsion propulsion. The glowing corners are just a necessary fact to release the static energy built up inside the containment vessels created in the process. These craft are actually all electric and only use small, highly efficient frozen methane gas generators to create the

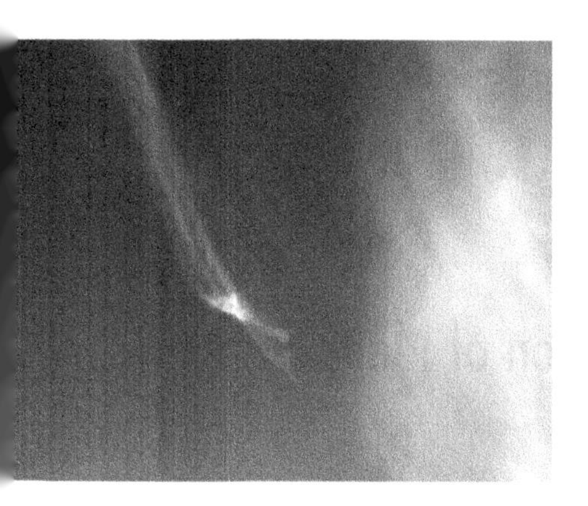

Fig 5.1. Sean Gautreaux, expert plasma cloud photographer and author of the invaluable out-of-print book *What Is In Our Skies, Vol. I, Diagrams: The Study of Cloaked Cloud Craft Above New Orleans* (2014). "The first in a series of books about the cloaked, triangle craft in the clouds above. Or is it all a hologram? Or is it a combination?"

> electricity to power the ion generators. Frozen methane gas is the most abundant power source on the planet. It is easily safely stored and transported. A 2-foot square cube of frozen methane can provide enough fuel to power the craft for a week. Once the ions are sufficiently condensed, very little power is actually needed to operate a craft manned by 3 people who sit inside an enclosed area in the center of the craft. One individual is the pilot, another is the sensor operator, and the third is the mechanical systems engineer. They are all contractors and the craft are capable of landing without a runway and no major ground support. They use closed federal wilderness areas to land when necessary. The craft is virtually constructed entirely out of carbon fiber composite panels and weighs not much more than a family van or small cargo truck. There are a number of diagrams on the web of what witnesses describe as what the undercarriage of these craft look like. Those are the ion generators that are seen. There are multiple triangle compartments as seen by each depressed area.

Hidden in plain sight all these years, like all that scientism has bent and skewed in order to arm Earth's environment for the sake of power. But weaponized or not, Wilhelm Reich was right: orgone (i.e., æther / ether) is God's creative process.

6
The Weaponization of Magnetism

Once World War II was declared over in May 1945 and three atomic bombs were detonated two months later—one in New Mexico, two in Japan—another kind of warfare dubbed the Cold War would run for forty-four years. Operation Paperclip Nazi scientists and engineers were ferreted into the U.S. and Canada, the National Security Act/Agency and Nazi General Reinhard Gehlen's transformation of "Wild Bill" Donovan's wartime intelligence OSS (Office of Strategic Services) into the six-million-strong CIA (Central Intelligence Agency) now multiplied into eighteen intelligence agencies. The fact is that the NSA and CIA have irrevocably transformed politics into what is secret and what isn't—and most everything is now political and secret for the sake of the doctrine of full-spectrum dominance. By the time the Cold War supposedly ended, *the environment itself was being weaponized* and information warfare (IW) was becoming a mainstay of domestic civilian life.

The National Security Act could just as well have been called the Josef Goebbels "Big Lie" Secrets Act, given how it has replaced our constitutional republic with a National Security State. Disinformation, misinformation, lies, and CIA "conspiracy theories" proliferate (not to mention murder and mayhem). In 2010, the Psychological Operations (PSYOPS) of the U.S. Army Special Operations Command (USASOC) was renamed MISO (Military Information Support and Operations). Thanks to Operation Mockingbird mainstream media control and electromagnetic targeting, emotions, motives, reasoning, and behavior of "target audiences" is guaranteed. Three years after MISO, then-President Obama lifted the sixty-four-year ban on perpetrating propaganda on the American people by revising the 1948 Smith-Mundt Act, now known as the Smith-Mundt Modernization Act of 2012. (See appendix 1, "Invisible Mindsets: MISO & domestic propaganda.")

Since the National Security Act and secret entry of Paperclip Nazis into the United States, politics based upon lies have increased. Is it any wonder that Americans are for the most part apathetic and easy to misguide and condition, thanks to endless controlled electronic media? Exposing massive programs like geoengineering requires years of research into events hidden by "national security" all the way back to World War II. Pick at the threads of old "conspiracy theories" like the Bermuda Triangle and UFOs and magnetic cloaking (degaussing) of the *USS Eldridge* in 1943 (the "Philadelphia Experiment") and magnetic MK-Ultra Montauk Project on Long Island in 1983, and one sees what Nazi engineers under Wernher von Braun produced and then hid beneath a context of "New Age" and "paranormal" cover stories. Even after the Cold War ended, "big lies" like the submarine communication cover story for HAARP, the weapon of choice on 9/11, and the coronavirus cover story for acute radiation poisoning transmitted by 5G / 6G, thanks to injections of magnetic graphene oxide have continued pushing the Nazi dream of world domination forward.

Seeing spoons dangling from inoculated upper arms in 2020 made me study magnetism more deeply. Was part of Transhumanism about plugging human beings more deeply into the Earth's geomagnetic grid? I knew that we'd already been inhaling tiny magnetite for years.[1] Was it all connected?

Self-assembling graphene is a carbon nanotube (CNT) whose character is that of a superconductor that is great for transmissions to and from the Cloud (what the Space Fence infrastructure provides) and our Internet of Bodies (IoB), which means the magnetic neural interface among the cells and organs of the physical body *and the bodies around one's body* so as to generate an artificial "shedding." Like a radio module, it is perfect for monitoring and modulating Human 1.0 as well as Human 2.0. It is also capable of amplifying gigahertz into terahertz, which makes it even more toxic. Basically, graphene is a pathogen that generates an inflammatory response and enhances the acute radiation syndrome misnamed covid disease. Dr. Ricardo Delgado, PhD, of La Quinta Columna ("fifth column": a group of people who undermine a larger group or nation from within), references Dr. Roger Lier who explains a highly sophisticated nanotechnology broadcasting / switching system *via scalar wave.* Not only is this system not dependent upon radio waves, but it actually constitutes *the real 5G.**

Joseph P. Farrell, an Oxford scholar and colleague of investment analyst Catherine Austin Fitts, is one of the rare researchers willing to take the road less traveled in order to recover historical truths buried beneath Cold War cover stories. I highly recommend all of his books, but for magnetism and the USS *Eldridge*, read *Secrets of the Unified Field: The Philadelphia Experiment, the Nazi Bell, and the Discarded Theory* (Adventures Unlimited Press, 2008) in which he analyzes what Charles Berlitz and William Moore wrote in *The Philadelphia Experiment: Project Invisibility* (London: Souvenir Press, 1979), and astronomer Morris K. Jessup's 1955 *The Case for the UFO: Unidentified Flying Object, the Varo Edition* (Castelnau Babarens, France: The Quantum Future Group, 2003).†

The USS *Eldridge* experimental objectives concentrated on coupling electromagnetism with gravity to build a strong electromagnetic field that would then be able to deflect projectiles (torpedoes, missiles), refract and absorb radar, and provide optical invisibility ("cloaking") by "bending light"—all of which continued for forty years until the Strategic Defense Initiative (SDI) in the 1980s. The 1928 version of physicist Albert Einstein's Unified Field Theory with its space-time twisting torsion tensor[2] was as central to the Philadelphia Experiment as were the mathematics of Manhattan Project physicist John von Neumann (1903–1957)‡ and the *electron cyclotronic resonance* that HAARP would manifest for increased charged particle density. (See HAARP inventor Bernard Eastlund's 1987 U.S. Patent 4686605 "Method and Apparatus for Altering a Region in the Earth's Atmosphere, Ionosphere, and/or Magnetosphere.")

---

*Thanks to the January 16, 2024, presentation by Ricardo Delgado, PhD, of La Quinta Columna on the video "David Icke was right about 'Covid,' the jab, the Cloud, and manipulation by a non-human force—by the team that identified graphene in the fake vaccine," especially 12:20 to 24:50.

†On April 20, 1959, Jessup (1900–1959) was found slumped over his steering wheel in a Dade County, Florida, park, dead from carbon monoxide poisoning. See Farrell, *Secrets of the Unified Field* bibliography, page 315, for more of Farrell's recommended reading.

‡In the summer of 1955, only a few months after his appointment to the Atomic Energy Commission, Von Neumann became ill with fast-acting cancer. His last public appearance came early in 1956 when, in a wheelchair at the White House he received the Medal of Freedom from President Eisenhower. In April, he was taken to Walter Reed Hospital, where he died on February 8, 1957—or was he moved to Montauk?

Altering the charged density of the Earth's atmosphere impacts biological health, as the work of Robert O. Becker, MD (twice nominated for the Nobel Prize), and biochemist Martin Pall, PhD, stress:

> Cyclotron resonance is a mechanism of action that enables very low-strength electromagnetic fields, *acting in concert with the Earth's geomagnetic field*, to produce major biological effects by *concentrating the energy in the applied field upon specific particles*, such as the biologically important ions of sodium, calcium, potassium, and lithium.[3]

Russian scientist Aleksandr Pressman, who has also made extensive studies of the biological effects of electromagnetic fields, says, "In conjunction with *the broad, seemingly purposeful environmental array of assaults on the endocrine system*, is remote control over ion channels being used to further cut off our communication with our natural world and each other?"[4] (Emphasis added.)

Biological effects acting in concert with the Earth's geomagnetic field. Thus, our transhuman concern regarding the USS *Eldridge* is how biological effects act in concert with the Earth's geomagnetic field. On July 22, 1943, six months after Nikola Tesla's death, the USS *Eldridge* and crew disappeared for twenty seconds. The crew came back disoriented and nauseous. In *The Case for the UFO*, Jessup discounted the electrically charged radar-cloaking paint and degaussing equipment for causing the crew's fried brains because everything would have only *magnetically disappeared* from radar, not physically disappeared. What made the difference was the addition of electromagnetic pulsing, cyclotron resonance, and cloud chambers.

Five Avenger torpedo bombers disappeared off the coast of Fort Lauderdale, Florida, on December 5, 1945, with no debris found. Cover stories of Atlantis and the Bermuda Triangle were trotted out, but what *really* happened magnetically? Did the bombers disintegrate? Become quantum-entangled in spacetime?* Disappear into a magnetic cloud? There were no satellites in 1945, but a pulsed harmonic transmission could have activated a unified field effect, much like what seems to have happened to the USS *Eldridge*.

---

*Harmonic mathematics is the key to understanding quantum entanglement, Einstein's "spooky action at a distance."

The truth is that magnetic, electrical, gravitational, nuclear, thermal, plasmic, acoustic, and other radiative energies are neither decoupled, separated, nor broken down, but in a state of *harmonic unification*. Harmonics means that everything is tuned to everything else. New Zealand pilot Bruce Cathie's books on harmonics are essential reading in order to grasp what it means to live on a planet whose grid system is composed of two electromagnetic fields that interlock and are out of phase.[5] The communications harmonic is 16944, and 695 is the gravity harmonic or reciprocal of light. The 7.82 Hz alpha brainwave and Schumann resonance is tuned to the 695 gravity harmonic.

> The combination of Hilbert space, von Neumann's brilliant calculations, and the advanced Levinson* Recursion for multiple time series revives the question of whether the leading physicists and mathematicians of their day were the secret driving force behind something far more serious than a simple degaussing.[6]

Farrell seems to think that the scientists involved in the USS *Eldridge* experiments were taken by surprise:

> . . . the scientists *expected* . . . ionization of the air, a "boiling" of the water surrounding the ship, and "Zeemanizing"† of the atoms . . . But one unanticipated result of the experiment was an "interdimensional" effect or a mass displacement effect, i.e., an antigravity or even "teleportational" effect . . . the achieved results were wildly beyond what was expected. They were different in degree, not in kind, from the original purposes of the Experiment.

And yet his later comment hints at a change of mind:

> Learn to control and manipulate torsion in the Unified Field Theory context, and one has learned to control and manipulate gravity, time, electromagnetism, and space itself. One can manipulate them for

---

*The role of mathematician Norman Levinson (1912–1975) has been erased from science history, even to the point of obfuscating his name by calling him John Levinson.

†Zeemanizing: induction of a split in the excitation states of atomic particles and atoms.

energy, for propulsion, and of course, for a horrifyingly powerful weapon of mass destruction, a weapon so powerful that it would make the largest thermonuclear bombs look like child's toys.

THE RUSSIAN SIGNAL ("WOODPECKER")

The idea that magnetic fields could transform matter and transport it from one dimension to another dominated much of the classified research from the mid-1950s to the end of the Cold War and beyond.

In 1955, a further inquiry into the weapon potential of the Earth's geomagnetism was launched: the foundation stone of the high-energy particle accelerator known as CERN (*Conseil Européen pour la Recherche Nucléaire*, the European Laboratory for Particle Physics) was laid on the border of France and Switzerland near a town called Saint-Genis-Pouilly, Pouilly referring to the Latin *Appolliacum*, a Roman temple honoring Apollo and therefore considered to be a gateway to the underworld.* Composed of two 1,000-ton superconducting magnets hanging down 300-foot shafts into a cavern through which an underground river once ran until frozen with liquid nitrogen, the Large Hadron Collider (LHC)† was not built until 1998–2008 when HAARP was at last up and running and working on controlling the ionosphere. These powerful magnets were arranged like boxcars around the 5-story, 16-mile (27-kilometer) "ring" circling the cavern. Basically, CERN's massive superconducting magnets enhance and magnify the sustained reach of the Space Fence throughout our ionosphere-fed atmosphere, all the way to the magnetosphere, not to mention the sustained reach of the Space Fence (i.e., "Cloud") into our bodies and brains via the superconducting graphene we've breathed in and been inoculated with.

Speaking of biological experimentation, in 1953 a wireless electromagnetic "Moscow Signal" aimed at personnel in the U.S. Embassy in Moscow began that would continue for the next thirty-eight years, to the very end of the Cold War. U.S. ambassadors fell prey to cancer, one bleeding from the eyes with a rare blood disease; other embassy personnel had

*Leylines point to dowsing and "flowing" magnetism.

†LHC is a synchrotron-type accelerator, a particular type of cyclic particle accelerator in which the accelerating particle beam travels around a fixed closed-loop path.

a 40 percent higher than average white blood cell count. The American public was finally informed of this Soviet assault in 1975 during the Senate Select Committee to Study Governmental Operations with Respect to Intelligence Activities (the Church Committee) that claimed to expose the CIA's Cold War sins. Needless to say, the nation's 1976 Bicentennial celebrations were timed to divert and whitewash what should have been a major housecleaning of what the Nazi-founded CIA (in conjunction with the military-industrial complex) was really up to.

But were the Soviets the only perpetrators of wireless assaults? What about the CIA sending a crack team of American scientists to the Gomel site in Russia to install a 40-ton early SQUID (superconducting quantum interference device) magnetometer capable of generating a magnetic field 250,000X more powerful than the Earth's magnetic field?* At the same time, the Moscow Signal became the "Soviet Woodpecker," an early version of HAARP consisting of over-the-horizon (OTH) broadcasts of 10 Hz on 3–30 MHz bands picked up by power grids and re-radiated into people's homes on their 60 Hz wiring for their unsuspecting brains to be magnetically forced into sympathetic cyclotronic resonance. In *The Body Electric*, Becker tells how the Woodpecker signal was being directed in 1978 to Eugene in Lane County, Oregon:

> . . . The idea was advanced that [the Woodpecker signal] was being directed to Oregon by a Tesla magnifying transmitter. This apparatus, devised by Nikola Tesla during his turn-of-the-century experiments on wireless global power transmission at a laboratory near Pikes Peak . . . reportedly enables a transmitter to beam a radio signal *through* the earth to any desired point on its surface, while maintaining or even increasing the signal's power as it emerges. [Environment journalist]

Becker lists the following symptoms: pressure and pain in the head, anxiety, fatigue, insomnia, lack of coordination, numbness, high-pitched ringing in the ears—all characteristic of strong microwave irradiation.

*The 40-ton American magnet was believed to have been powered by the Chernobyl reactor. Do not forget that the forty-four-year Cold War was a CIA creation for the sake of secrecy and taxpayer support of the covert development of the machinery of social control we are now seeing all around us. Globalists supported the Soviet Union for seventy-two years, just as they have supported China.

Paul Brodeur has suggested that, since the TRW company* once proposed a Navy ELF (extremely low frequency) communications system using an existing 850-mile power line that ended in Oregon, the Eugene phenomenon might have been the interaction between a U.S. Navy broadcast and Soviet jamming.

In short, the Woodpecker was a scalar transmitter, *scalar* being the operative term for sending transmissions *through the Earth*. (HAARP is a scalar weapon system of one of the *two electromagnetic fields* that interlock and are out of phase as per Cathie.) As a quantum potential weapon, the Woodpecker could induce diseases like cancer by mimicking and re-creating their signatures or frequencies in the near-ultraviolet range; now, 5G / 6G is used to force molecular biology to submit to digital biology. Scalar waves are made to penetrate the virtual particle flux that determines the genetic cell blueprint and induce disease symptoms and cell disorder.†

In *Under an Ionized Sky,* I wrote about suicide frequencies and Medford, Oregon, in Jackson County northeast of Lane County:

> In 1973, Medford, Oregon became the suicide capital of the United States overnight, thanks to the ultra-low frequencies being beamed from a nearby military base to people's television antennas. The creation of a standing-wave resonance was connected to depression, *whether the television was on or not.* David Fraser, PhD, of the Department of Toxicology at the University of North Carolina, Chapel Hill, was paid by DARPA to run the Medford experiment.[7]

And what do we hear about Lane County in 2023? "Alarming suicide data released by Lane County": "65 percent greater than the national average . . . Between 2000 and 2020, the rate of suicide increased by 80%."[8] *Not one word about wireless technology or power lines, just about "personal problems."*

Which brings us to the Havana syndrome and Guantanamo Bay Naval Base in Cuba. In late 2016 through early 2017, twenty-four U.S. and Canadian diplomats in Havana, Cuba, complained of hearing loss, loss of balance, and headaches. Then in late November 2017 through April 2018, U.S. consulate officials in Guangzhou, China, complained of similar neurological symptoms. Note that the latitude of Havana and Guangzhou

*TRW, Inc., was an aerospace defense corporation sold off to Northrup Grumman in 2002.
†Thanks to Sine Nomine for this insight.

are almost identical: 23 degrees north and 22 degrees north. These sonic attacks, attributed to *Havana syndrome*, began with a directional loud "chirping" noise and pain in one or both ears or across the broad region of the head, followed by sensations of head pressure, vibrations, dizziness, tinnitus, visual problems, memory lapse, insomnia, vertigo, and cognitive difficulties—in short, the very symptoms that thousands of targeted individuals have complained of over the years.*

In March 2023, the *Washington Post* reported that no foreign power or directed energy weapon (DEW) had produced the "strange and painful acoustic sensations" that constitute "the strange health ailment" studied since 2016 by five intelligence agencies (foxes in the henhouse).[9]

> "Havana syndrome" is a condition some government officials and their family members at US embassies in different countries have reported experiencing since 2016. Symptoms have included headaches, sleeplessness, and other signs similar to those of neurological conditions.[10]

Several years after the initial attack, some of these diplomats have reported brain damage and blood disorders, while two have permanently lost their hearing.[11] The U.S. State Department and the National Academy of Sciences colluded in their cover-up report "An Assessment of Illness in U.S. Government Employees and Their Families at Overseas Embassies"[12] in order to fit the so-called Havana syndrome into the psychiatric industry's *Diagnostic and Statistical Manual of Mental Disorders* (DSM-5). The article suggests that the attacks were done by the United States against its own personnel:

> **Havana Syndrome "Absolutely" the Result of Deliberate Attacks**
>
> For nearly five years, we have been aware of reports of mysterious attacks on United States Government personnel in Havana, Cuba, and around the world . . . This pattern of attacking our fellow citizens serving our government appears to be increasing. The Senate Intelligence Committee intends to get to the bottom of this.[13]

---

*The same symptoms were subsequently reported by U.S. personnel in locations around the world, including Austria, Australia, Colombia, Georgia, Kyrgyzstan, Poland, Russia, Serbia, Taiwan, and Uzbekistan.

Now, detailed medical records are kept on those claiming "Havana syndrome" during the ongoing "top secret investigation into the effects of microwave radiation on humans."[14]

THE MAGNETIC MONTAUK PUZZLE

According to Scottish mathematical physicist James Clerk Maxwell (1831–1879), a magnetic field propagates at 0.4 of the speed of light, an electromagnetic field propagates at the speed of light, and an electric field propagates instantaneously throughout the universe at the rate of c-infinity, making magnetic fields faster than the speed of light.*

Bruce Cathie pointed out that there are eight free energy positions in the geomagnetic grid that can resonate the entire world, three in the sea and one on land in the Southern Hemisphere, and three on land and one in the sea in the Northern Hemisphere.[15]

So what is going on *geomagnetically* along the Atlantic Coast of the United States, from Antarctica and through the South Atlantic Anomaly† to the Bermuda Triangle and north to where the USS *Eldridge* disappeared and reappeared, to Long Island where Nikola Tesla's Wardenclyffe Tower was and Brookhaven National Laboratory (BNL) and Montauk are? The lineup of longitudes at Three Mile Island Nuclear Generating Station (longitude 76°43' W) on the Susquehanna River in Pennsylvania,‡ Guantanamo Bay Naval Base, Cuba (longitude 75° W); Brookhaven National Laboratory, Long Island, New York (longitude 72° W);

*The speed of light at Earth's surface is 143,795.77 minutes of arc per magnetic grid second, or as we learned in high school, it travels at 186,282 miles (299,792 kilometers) per second.

†The South Atlantic Anomaly was discovered in 1958. It is an *expanding* area in South America jutting out into the Atlantic Ocean. It is due to the inner Van Allen belt's proximity (200 km / 120 miles) to the Earth's surface. Besides increasing the flux of ions and exposing near-Earth orbiting (NEO) satellites and the International Space Station (ISS) to higher than usual levels of ionizing radiation, it is the Earth's weakest magnetic field. Wikipedia lays the cause of the South Atlantic Anomaly to "a huge reservoir of very dense rock inside the Earth called the African large low-shear velocity province," which puts me in mind of another *resonance chamber.*

‡The nuclear meltdown at Three Mile Island on March 28, 1979, was a Level 5 "Accident with Wider Consequences." The Chernobyl disaster occurred seven years later in 1986.

and Montauk Air Force Station, Long Island, New York (longitude 71° W), is intriguing.

Brookhaven National Laboratory ("We advance fundamental research in nuclear and particle physics to gain a deeper understanding of matter, energy, space, and time") is located fifty miles from Montauk Point, on the eastern tip of Long Island, which is connected to the town of Montauk by subterranean tunnels and chambers, Montauk being the remnant of an undersea volcanic mountain with bedrock geologically separated from the rest of Long Island. Long considered sacred by the native Montaukett people, the land Montauk sits on appears to be a hyperdimensional energy portal passing into and through Earth's magnetic grid.

Distance is not a determining factor when it comes to geomagnetic fields, nor is time,* so it may not be surprising to learn that Montauk's unique ancient characteristics point to *tetrahedral*† *planetary grid significance*: (1) it is on top of an undersea volcanic mountain or extinct lava tube 10,000 to 12,000 years old; (2) north of Fort Pond is a "bottomless" 300-foot-diameter pond; (3) on the exterior at least the three "Pyramids of Montauk" power spots are three small hills that house two main bunkers and "radar hill."

Montauk Air Force Station was officially decommissioned in 1969, after which it was run by the General Service Administration and renamed Camp Hero State Park. Underground, however, it was (and is) still military, still working with Montauk Point (the subterranean facility extending out under the ocean) and the old World War II navy submarine base on Fort Pond Bay. Not only do triple phase power lines run throughout the area, but a Delta-T antenna continues to transmit *interdimensional* electromagnetic fields. Was this what the child mind control and trafficking underway in the 1980s with thousands of "throwaway" Montauk Boys was about?

Covert radar operations were set up between Montauk and twenty-five other bases around the United States, but Montauk's primary

*Time is a hyperbolic geometric, *hyperbolic* meaning a curve or arc generated by a point moving so that the distance from two fixed points is a constant.

†A tetrahedron is a triangular pyramid composed of four triangular faces, six straight edges, and four vertex corners. Perfect for an energy field.

facility for generating massive electromagnetic force fields is located halfway around the world. With the most powerful transmission power of 1 megawatt in the Southern Hemisphere, the joint American-Australian Naval Communication Station Harold E. Holt station on the northwest coast of Exmouth Peninsula, Western Australia (see fig. 6.6), claims thirteen of the largest very low frequency (VLF) towers. This is known as the Exmouth-Montauk great arc alignment. Why is so much power being generated? According to Valdamar Valerian, a 1990s researcher into technological experiments like HAARP in a newssheet called *The Leading Edge* out of Yelm, Washington, since the Montauk base shutdown, the U.S. Navy's HAARP contingent had taken control of the Montauk underground installation only fifty miles from the Brookhaven National Laboratory's "suite of particle accelerators."[16]

In the early hours of May 1, 1995, a major fireball flew in a north-northeasterly direction toward Perth, Western Australia, situated about 700 miles from Exmouth. It detonated above the eastern side of Perth at approximately 2 a.m. in a huge explosion. Limited Australian press coverage of this event described it as a meteor fireball continuing on toward the Kamchatka Peninsula in far eastern Russia, where a huge Russian electromagnetic weapons complex site lies.[17] The megaton-force of the explosion woke up over half a million people in Australia and demonstrated that Perth could have been obliterated with the flick of a switch. The trajectory of the fireball's origin suggested Enderby Land in Antarctica.

Thirty years later, Antarctica and its huge magnetic field in the deep subglacial Lake Vostok are still buried in secrecy, as are the vast stores of coal and uranium, and the electromagnetic equipment still arriving to control the weather and much more.[18] Now, "whistleblower" Eric Hecker has come forward about the exotic directed energy weapon (DEW) at the IceCube South Pole neutrino observatory, buried in ice to a depth of about 2,500 meters. (Neutrinos are almost massless particles with no electrical charge that in ice produce electrically charged secondary particles.)[19]

The preexisting underground network of tunnels and levels at Montauk has been expanded over the years to as many as seven, according to U.S. Army Corps of Engineers records, which means that Montauk and Brookhaven National Laboratory are both sitting on a huge resonant chamber perfect for advanced esoteric technologies having to do with sound weapons that include quantum wave forms of plasma that exist outside the audio and electromagnetic wave spectrums.*

Don't forget: seawater is an antenna (think of whale communication).

Montauk's unique geomagnetics have greatly contributed to its role in perfecting ultrasonic weaponry—i.e., sound—for mind control. Music has served as the perfect medium in that its emotional component can deeply enlist a victim's psyche and personality to commit to subliminal programming directions. Quantum plasma wave forms can be made to connect music to geomagnetics" by "pocketing" the wave forms inside the music, then directing it at a target to interact with human consciousness and *leave no trail.* The pocketing needs vacuum tubes, which may be why most rock musicians and vocalists now swear by the tonal superiority of vacuum tube amps.

In the 1990s, HAARP began supplementing the psychotronics (brain-computer interface, i.e., electronic harassment) with the super-powerful electrical fields and plasma being generated at Montauk. State-of-the-art mind control psychotronics suggest that the Brookhaven and Montauk particle accelerators are capable of powering particle beams for interdimensional experiments beyond locating and creating geomagnetic quantum access nodes or portals off Mystic, Connecticut (note the name)—exotic particle beam radar systems for quantum information wave packets from human brain pans.

Nazi "Angel of Death" medical doctor Josef Mengele (1911–1979) used the twelve-tone scale for mind control programming whereas at Montauk it seems to have been the magnetic Montauk Chair that was used with various drugs on the Montauk Boys. ITT Inc.,† together with

*Before being imprisoned in federal prison by a Food and Drug Administration (FDA) sting in the early 1950s, psychoanalyst and medical doctor Wilhelm Reich (1897–1957) sent a radiosonde he designed (a telemetry instrument that measures the atmosphere) to Brookhaven National Laboratory for evaluation. The laboratory was impressed and made a compact, lightweight version to be carried by balloon since the radiosonde would not work near metal. The device eradicated radiation (what Reich called "dead orgone" or "dor") and infused the environment with life-giving orgone (æther) by converting electrical energy to etheric energy via two frequencies, 403 MHz and 1680 MHz—suggesting that weaponized etheric / scalar technology can be used to clean up the environment, if doing so is not suppressed.

†Formerly International Telephone and Telegraph Corporation (IT&T), this corporate entity was founded in 1920 and earned a sordid reputation during World War II for its role in keeping careful records of the status of concentration camp inmates, much of which was useful to the CIA's MK-Ultra program.[20]

Mackay Marine in Southampton, Long Island ("We specialize in servicing communication, navigation, safety, and anti-pollution electronic equipment . . . "), built the original chair with three Tesla coils, an ISB receiver, and two ISB detectors—two outputs and one input—the outputs tuned to the hyperdimensional windows at Montauk (41° N latitude*) while a CRAY-1 computer pinpointed the harmonic necessary for transmission into and out of the hyperdimensional window. The perfect microwave length from Southampton to Montauk prevented the transmitter from interfering with the Chair or being subject to incoming fields. A second chair was also built by ITT Inc. but with RCA receivers designed for the Delta T function (discussed above) and standard XYZ Helmholtz coils.†

It is commonly held that Montauk Air Force Station was decommissioned in 1981, but if so it by no means ended the Montauk experiments into magnetic mind control. In the 1990s, new telephone lines and high-capacity powerlines with a gigawatt meter on an equipment maintenance building were installed. After a particle beam radar unit was installed on the Camp Hero bluffs, a Long Island Lighting Company meter indicated that a tremendous amount of electricity was being utilized *below ground*. While it may still be true that the New York State Office of Parks, Recreation, and Historic Preservation took over the base and even the old radar station and renamed it Camp Hero State Park,‡ the federal government retains all rights to the property beneath the surface. Security personnel sporting automatic weapons strolled around the property until the wildlife park finally opened there on September 18, 2002.

I am in touch with Chris Laterreur, one of the Canadian "Montauk

---

*The same latitude as that of Mt. Shasta, California, one of the CIA's favorite MK-Ultra "parking places."

†In Picknett and Prince's *The Stargate Conspiracy: The Truth about Extraterrestrial Life and the Mysteries of Ancient Egypt* are descriptions of strange hyperdimensional occurrences, like the ravens at Livermore Labs in the early 1970s and the nauseous yellow-green "attachment" in the aura, with tentacles grasping at everything.[21]

‡In the 2004 film *Eternal Sunshine of the Spotless Mind*, the main character, Joel, played by Jim Carrey (Mr. *Truman Show* himself), catches a train to Montauk, where he meets Clementine (Kate Winslet), who will eventually have her memory of him and their relationship erased. Efforts to film at Camp Hero State Park were quashed when officials threatened to charge exorbitant filming fees.

Boys" now in his forties who wrote the following in his unpublished manuscript *CIA Terror at Montauk*:

> My next memory is of a group of us walking in the direction of the famous Montauk Chair, then sitting bare ass on its freezing stainless steel. I complained that it was ice cold and Jack P. said, "I know, the other guy forgot to plug the chair warmer in." Ya, right. He then explained the flower of the Epiphyte air plant producing the vapors we were breathing in from the masks we were wearing. The flower grows without dirt in the air. Jack P. said it was discovered around 1980 in the Everglades.
>
> When he finished explaining the air flower, he showed me a white lotus from Egypt and explained that this sacred flower had been used since ancient Egypt to prepare for astral projection. He said it would help get me through my gateway / portal. I just needed to think about it, and it would get me where I needed to be. He said I was the only one among my group of boys who could do it, the only one with access to my own gateway / portal. Everyone has their own gateway, but not necessarily the ability to open it. Once it was open and I went through, he promised he would guide me back so I wouldn't get lost, that I was safe because he'd be beside me.
>
> Before Montauk, I'd had a few LSD micro-dose experiences, but in the Montauk Chair it felt different, almost like DMT (N, N-Dimethyltryptamine) or maybe just different doses. For sure, the buzz effect felt stronger. Less than one percent of the kids sent to Montauk came out alive and still sane. From the first to the second year of the program, I personally witnessed that half the kids weren't there the following year.

THE STAR WARS ERA

The *Star Wars* film trilogy (1977, 1980, 1983), and the 1980 "October Surprise"* election ushered in the infamous *troika* (Ronald Reagan, George H. W. Bush, and Richard "Big Dick" Cheney), and their Strategic Defense

*Referring to the secret deal Reagan's running mate George H. W. Bush made with Iranian leaders to delay the release of American hostages until after Reagan's election victory over then-president Jimmy Carter.

Initiative (SDI) known as Star Wars, cover stories ranging from how missile defense, a shield against nuclear attack, and over-the-horizon (OTH) nuclear submarine communication. In hindsight, it is obvious that the SDI objective was to continue preparation of the atmosphere and geomagnetic grid for wireless control as per the Kardashev model of full-spectrum dominance over planet Earth, Nikola Tesla's discoveries, and experiments like the USS *Eldridge*.

Under the Strategic Defense Initiative, Congress upscaled funding for weather modification, allocating over $20 million for the U.S. Air Force to begin (1978–1982) building 299-foot Ground Wave Emergency Network system (GWEN) towers two hundred miles apart across the entire nation. The 330 feet of underground webs of copper wires radiating from each GWEN tower produced VLF (3–30 kHz) and ELF (1–300 Hz) ground waves and 2,000 watts of power. GWEN arrays alter the Earth's magnetic field within a 200-mile radius, and specific frequencies can be used to control whole populations.[22] Full-spectrum dominance includes magnetic control over the weather as well as over human brains and society.

Both Bernard Eastlund, PhD, in his 1987 HAARP patent, and orthopedic surgeon Robert O. Becker (1923–2008), best known for his extensive research into biocybernetics, emphasize the close relationship between magnetism and cyclotron resonance. Eastlund explains how HAARP excites electron cyclotron resonance heating to increase charged particle density for the billions of earthly wireless transmissions underway at any given moment, military and civilian—a 24/7 operation that all of life is subject to, as well. Dr. Becker explains why this should concern us:

> Cyclotron resonance is a mechanism of action that enables very low-strength electromagnetic fields, acting in concert with the Earth's geomagnetic field, to produce major biological effects by concentrating the energy in the applied field upon specific particles, such as the biologically important ions of sodium, calcium, potassium, and lithium.[23]

Major biological effects? Elsewhere, Dr. Becker states:

> All the evidence points very clearly to the fact that animals (including humans) must intercept a normal magnetic field in order to maintain

> the functional integrity of their central nervous systems. We derive crucial information from the field—information that influences biorhythms, the electrical and chemical properties of the brain, and the growth rate of the organism as a whole . . . It should be obvious by now that abnormal fields cause acute physiological stress, a major predisposing factor toward disease.[24]

Once HAARP had achieved control over the ionosphere in 2013, HAARP temporarily shut down (2014–2015) while the Joint Space Operations Center Mission System replaced and upgraded the hardware and software used for space surveillance, collision avoidance, and launch support, coincided with a number of Space Fence–related dates, possibly because of a well-organized effort to upgrade and synchronize the frequencies of all systems connected with the monitoring and control of geospace (including near-Earth orbit, the ionosphere and magnetosphere) and the harnessing of cosmic processes such as CMEs, solar flares, and solar minimums, for Space Fence optimization.* By 2018, the air force's next-generation space surveillance system, the Space Fence, was operational.[25]

SMASHING AND SPLITTING MATTER

The official story is that the ULF (300 Hz–3 kHz) Space Fence ring around our planet[26] is synchronized with some 30,000 particle accelerators on the ground (International Atomic Energy Agency) to be used as colliders or as synchrotron light sources for studying condensed matter physics. *All can be calibrated to work as one* (see earlier section on the Space Fence and read *Under an Ionized Sky*). Full-spectrum dominance entails weather and climate control as well as human control, thanks to technologies like ionospheric heaters, pulsed wireless technologies, synthetic biology, nanotechnology, satellites, etc.

In 2020, the twenty-three nations comprising the CERN council endorsed the feasibility of building a 100-kilometer "Future Circular Collider" all the way around Geneva, Switzerland, intersecting at two

*As indicated above, the best source for how this "calibration" is done is in *Under an Ionized Sky*, the book that Billy "The HAARP Man" Hayes guided me through.

points with the existing and much smaller Large Hadron Collider, a project to be completed by 2035. The plan calls for a gigantic collider that by 2040 would "smash electrons into their antimatter partners, called positrons, allowing for closer study of the Higgs and possibly dark matter. Initial estimates suggest it would cost approximately €21 billion."[27] Rudolph Steiner pointed out back in 1923 that the action of smashing and splitting matter is a perversion of the natural Divine order: "As soon as division or atomization begins, death enters in. He would not have derived death from the corpse but from atomization, from the division into parts . . . A being that is capable of life, that is in the process of growth, is not atomized; and when the tendency to atomization appears, the being dies."[28]

The rush to smash and split is on. China is building a massive particle accelerator.[29] There are now at least fourteen large particle accelerators around the world, with the one at Brookhaven National Laboratory near Montauk (discussed earlier) being the largest in the United States. More than 160 international groups, including NASA's Ames Research Center and X Development LLC (formerly Google X), a secret research and development organization founded by Google in 2010, are heavily invested in research into high-energy physics, climate science, and genomics, and are using the Energy Sciences Network, "a high-performance, unclassified national network built to support scientific research"[30] run by UC Berkeley's Lawrence Berkeley National Laboratory.

The discovery of the subatomic Higgs boson particle, named after theoretical physicist Peter Higgs, was made public in 2012. The story is that without the Higgs particle, electrons would have no mass and atoms wouldn't stick together. Certainly, *something* holds matter together ("cosmic molasses," Higgs called it), but wouldn't it be more honest to admit that this "something" is the very same æther (or ether) that was banished back in the 1920s, when science became scientism?[31]

What's the real reason behind the current mania for smashing and splitting? Former U.S. Navy nuclear engineer Chris Fontenot clarifies that CERN is not so much engaged in tracking down the legendary Higgs boson "God particle" as in creating a planetary system of electromagnetic power, including ionic propulsion for exotic propulsion crafts like the so-called UFO/UAP. Ken Lebrun, an EM propagation specialist, radar engineer, and software developer, explains that CERN's mission of inducing

Fig. 6.1. Google, now integral to the corporate surveillance state, is far more than a mere web browser.
U.S. Air Force/Joe Davila

voltage into the global atmospheric electrical circuit between Earth's surface and the ionosphere is being used for *global wireless energy harvesting*.*

Think of large colliders like CERN as giant electromagnets working together, like hundreds of nuclear explosions all happening in one second, or hitting the Sun and causing CMEs (corona mass ejections, explosive bursts of solar plasma and magnetic field that fly away from the Sun at thousands of kilometers an hour).†

A bird's-eye view of the Large Hadron Collider (LHC) at CERN. Peter Champoux, a geomancer and the author of the 1999 book *Gaia Matrix* as well as a lifelong student of comparative religion and all things geographic, says, "CERN is a topic ripe for wild speculations as one cannot know its effect until after the fact. Tearing apart the atom and god particles is in the least an abomination and anti-life. CERN's effect on the quantum field will without a doubt/double be Ahrimanic. If anything, CERN is the effect of a weakening magnetic field more than its cause."‡ (The double or Doppelgänger is, at the very least, electromagnetic.)

With the Space Fence up and fully functional, corollary "tests" at CERN (and on the Sun, our star) went live, too—all part of Tesla's planetary surface power system under construction. On December 7, 2015, a "whirling dervish" portal opened above CERN. According to researcher and author Anthony Patch, founder and editor of *Entangled Magazine*,

*DARPA terms energy harvesting *transduction*: Transductional materials convert energy between different forms or domains, such as thermal to electrical energy, or electric field to magnetic field. Devices fabricated from such materials have multiple DoD-relevant applications that include the following: (1) Thermoelectrics (thermal/electric domains) used for energy harvesting, thermal management, and refrigeration; (2) Multiferroics (magnetic/electric domains) used in sensors, antennas, actuators, micromotors, tunable RF and microwave components; (3) Phase Change Materials (various domains) used in transducers, switches, sensors, and control devices.[32]

†"Multiple coronal mass ejections (CMEs) occurred on November 27, 2023. Three of these CMEs appear to have Earth-directed components with the first arrival as a potential glancing blow or near-Earth proximity passage beginning as early as late on November 29 EST (early November 30 UTC)."[33] According to Seifer, *Wizard: The Life and Times of Nikola Tesla* (Citadel, 2016) and *Transcending the Speed of Light: Consciousness, Quantum Physics, and the Fifth Dimension* (Inner Traditions, 2008), Tesla told Joseph Alsop, Teddy Roosevelt's great-nephew, that the Sun absorbs more energy than it radiates. This idea is behind Tesla's dynamic theory of gravity.

‡Email, Peter Champoux, August 21, 2021.

Fig. 6.2. CERN Geomancy.
Image Courtesy of Peter Champoux

CERN was built over an energy node that was the site of an ancient underground Temple of Apollyon, mythic ruler of the abyss in the Book of Revelations.*

When CERN's Large Hadron Collider is in full operational mode, the seventeen-mile-diameter particle accelerator is like an Arctic Circle magnetic field. CERN sits on the French-Swiss border, in geomagnetic alignment with Greenland and the Great Pyramid of Giza. How much

*CERN = Cernunnos, horned god of the underworld. CERN's logo is 666 and a statue of Shiva the Destroyer graces the front of its Geneva headquarters. For more on CERN's occult connections, review chapter 9, "The Temple of CERN," in my book *Under an Ionized Sky.*

do these magnetic lines of force influence us, or as must now be asked, how much do grid manipulators influence us by controlling these cosmic lines of force? How many wars and revolutions have been sparked along strategic geomagnetic alignments? Similar to the effects produced by the Montauk experiments, airplanes may drop from the sky when the Large Hadron Collider is running full tilt: to wit, in 2015, Germanwings flight 9525 was cruising at 38,000 feet, just 127 miles from CERN headquarters before crashing in the French Alps. Then there were two incidents in 2014: Malaysia Airlines flight 17's disappearance and Indonesia AirAsia Flight 8501's crash in the Java Sea. Is CERN implicated in the cover story that the South Atlantic Anomaly, an area where Earth's inner Van Allen radiation belt comes closest to Earth's surface, leading to an increased flux of energetic particles, is responsible for a possible geomagnetic reversal?[34]

Does the South Atlantic Anomaly have anything to do with the north-south alignment of Montauk, Brookhaven National Laboratory, and Guantanamo Bay, and the workings of the largest particle accelerator in the world? Masses of electric power are pouring into Earth, thanks in no small part to technological control over the ionosphere and the ongoing particle physics experimentation. On April 23, 2016, the magnetosphere collapsed—actually disappeared—for two hours; and on December 18, 2022, a plasma "shockwave" blamed on a coronal mass ejection (CME) cracked the magnetosphere.[35] Was the magnetosphere similarly affected by the multiple CMEs on November 27, 2023, mentioned above?

Fired up, the Large Hadron Collider becomes a magnetic Arctic Circle all its own, generating a hundred times the magnetic energy of the entire planet due to its geographical alignment directly over the ancient Temple of Apollyon, along with HAARP's earth-penetrating tomography (i.e., scalar earthquake technology) synchronized with the entire Space Fence infrastructure.

Anthony Patch (mentioned above) sets forth another major concern: namely, that CERN is in league with the entire Space Fence *apparat* in the race to plug transhumanist Human 2.0 into AI by means of brain-computer interface technology:

> The mind operates on multiple frequency levels . . . The luminosity resulting from the collisions of subatomic particles in the collider

> inside the detectors at CERN produced identical frequencies as the mind. They are reproducing through collisions of particles two of the frequencies within which the mind operates, and you'd better believe that they're operating at *all* of the various frequencies, alpha and beta ranges, of the mind. Does that mean that CERN is directing mind control systems? No. It means that they're doing the research and development that is then applied to the supercomputer systems like D-Wave for the purpose of—in the environment of the *sentient world simulation**—controlling people's minds.[37]

While particle accelerators—once *called* cyclotron particle accelerators, as in cyclotron resonance heating defined earlier—are smashing atoms, legions of swarm-conscious nanotechnology built at the atomic threshold are now being breathed in and implanted in humanity. Thus, even more insidiously than during the Manhattan Project (1942–1946), we now stand at the atomic/subatomic quantum threshold of Earth's existence. Are we ready to contemplate the relationship between those dedicated to scientism instead of science and their ancient secret societies' quest for access to "the final frontiers"‡ of parallel universes described by LHC physicist Mir Faizan as "real universes in extra dimensions"—the quantum scale of what D-Wave† chief technology officer Geordie Rose calls "the Old Ones"?

As antimatter mass production alters the very fabric of Space,§ CERN is able to electrically connect via Birkeland currents with Saturn in the Saturn-Venus-Mars-Earth alignment. Developments such as these must be reconsidered in the light of a new, less materialistic way of thinking, as must the secret space program.** While harvesting Earth's living energy,

---

*Sentient world simulation, a "continuously running, continually updated mirror model of the real world that can be used to predict and evaluate future events and courses of action."[36]

†A reference to the *Star Trek* television series preprogramming we underwent in 1966–1969.

‡For more information see D Wave Systems website.

§Professor Jeffrey Hangst of Aarhus University, Denmark: "Antimatter is just the coolest, most mysterious stuff you can imagine," he told me. "As far as we understand, you could build a universe just like ours with you and me made of just antimatter."[38]

**Mark Skidmore, professor of economics at Michigan State University, and investment analyst Catherine Austin Fitts discovered $21 trillion in government transactions, primarily in defense, missing between 1998 and 2015. Fitts maintains that much of it has gone into the secret space program.[39]

including human energy, could NASA's secret society of scientists known as the Saturnalian Brotherhood be bent on turning Earth into an AI generator replete with synchrotron-generated "rings" so as to "restart" Saturn, a supposedly dead planet whose rings—if we are to believe Norman Bergrun's book about the Voyager 1 and 2 flybys of Saturn after their launch in 1977—are still maintained by artificial intelligence machines?[40] Certainly, such possibilities are along the lines of the Kardashev mandate to control Earth's solar system.

Tesla averred that the future is his. We are indeed living in Tesla sci-fi times, but sadly all of his technology has been coopted and weaponized. Become aware of the global environmental infrastructure being geoengineered and tell me that it simply can't be.

MAGNETISM AND OUR CELLS

Up until electricity and the present electromagnetic era, our physiological rhythms and collective behavior synchronized with natural solar and geomagnetic rhythms. Back then, disruptions in the natural electromagnetic field of the planet greatly affected our health, behavior, and even our ability to reason. The same is true now in our wireless geoengineered world, albeit from different causes and with longer lasting effects. I cover this in greater detail in the book on synthetic biology.

Most important here is to recall the entry of graphene oxide into our bodies, whether from aerosols or inoculations or processed foods, given that every cell in our bodies is bathed in an external and internal environment of fluctuating magnetic forces that affect every cell and circuit of our biology. As Robert O. Becker says in his book *Cross Currents: The Perils of Electropollution*, "All biological cycles are directly related to the planet's magnetic field (which averages about 1/2 gauss, with a daily change in strength of less than 0.1 gauss*—compared to a refrigerator door magnet at 200 gauss strength)."[41] In fact, every living organism emits radio waves, which makes every cell in living organisms an electromagnetic resonator that emits and absorbs high-frequency radiation, each cell's nucleus with its own oscillating frequency. New Zealand harmonics mathematician and

*Electrical fields are described in kilovolts per meter (kV/m), and magnetic fields in gauss (G or Gs).

former military pilot Bruce Cathie, author of books on the harmonics of light and the geomagnetic grid, writes:

> The geometric makeup of the cell causes it to act as an electric circuit which has self-inductance and capacity. The natural oscillation of energy in the cell I believe to be due to the constant interaction of the matter and antimatter cycle . . . A pendulum-like pulsing occurs between the physical and non-physical substances [i.e., cyclotron resonance]. When stronger radiations are imposed upon the cell by outside influences, then the natural rhythm of the cell is affected and it begins to break down. If the radiation [frequency] of the cell can be restored to its original rhythm, then it will resume its healthy state.[42]

Russian-French engineer, author, and inventor Georges Lakhovsky (1869–1942) agreed that "every living being radiates and emits EMF," that our cells are connected to the frequencies of the cosmos, and that disease is the oscillatory disequilibrium of the cell due to external causes.[43] We have seen that all living cells possess oscillating circuits composed of neural filaments. Amazingly, these cells are propelled into motion by the motion of Earth, which moves at a velocity of 17 miles per minute at the equator. The terrestrial electromagnetic field is thus swept along in the same rotatory motion as the cells, which means that our cells' movement in variable electromagnetic fields is generated by a source *external* to Earth within a vast field of cosmic radiation emanating not just from our Sun, but from the Milky Way and beyond, from the immensity of celestial space.[44]

Magnetometers are used to measure slight or extreme deviations in Earth's magnetic field, from the signatures of mountains, rivers, streams, lakes, and plains, to disturbances created by such natural geological features as upthrust dikes, fault lines, caverns, and various types of mineral deposits, as well as human-created disturbances caused by dams, fracking wells, tunnels, and deep underground military bases (DUMBs). Whether used on the ground or via satellite, powerful magnetometers like SQUID (superconducting quantum interference device) and MEG (magnetoencephalograph) neuroimaging scanners, as well as the nanosized smart dust microelectromechanical (MEMS) magnetic field sensors in chemtrails, scan and read and measure the lines of force that surround the human head and body.

Our neurons, muscle cells, and touch receptor cells all use ion channel receptors to convert chemical or mechanical messages into electrical signals. This action is subverted by the clusters of heated magnetic nanoparticles delivered aerially, and now by mRNA inoculations. By exposing us to a magnetic field similar to that of an MRI (magnetic resonance imaging), and then adjusting the heat factors to deliver the desired results, our ion channels can be remotely controlled to provoke certain physiological and psychological responses, thus subverting our sensory, nervous, and endocrine systems. For example, 93.2 degrees Fahrenheit initiates an avoidance response.

Millions of brain experiments are being conducted on us via satellite, tower triangulation, and 5G as we daily breathe in trillions of MEMs, NEMs, nanobots, and conductive metals like magnetite, graphene, and other iron oxides needed for *remote* transcranial magnetic stimulation (TMS) and deep brain stimulation (DBS):

> By injecting magnetic nanoparticles into the brain, researchers have found that they can manipulate neurons by applying external magnetic fields . . . [A team at MIT] developed a system that involves an injection of iron oxide particles that are subjected to an external alternating magnetic field . . . Particles capable of deep penetration of brain tissue rapidly heat up under the influence of the magnetic fields, stimulating nerve cells.[45]

Wireless magnetothermal deep brain stimulation depends on the concentration of magnetite and graphene nanoparticles in our brains. Once the superparamagnetic nanoparticles warm up, an ion channel opens up and activates neurons. This "noninvasive" approach is regulated not by applying electrodes or tiny fiber optic cables (electrodes) to the brain, the old-fashioned way, but by activating jet-delivered chemical aerosols loaded with nanobots via remote 5G/6G microwaves from cell phones and Internet of Things technology, or the next tier of towers, radar installations, fiber optics, and satellites.

Israeli researchers have confirmed "the abundant presence in the human brain of magnetite nanoparticles that match precisely the high-temperature [iron oxide–rich] magnetite nanospheres formed by combustion and/or friction-derived heating, which are prolific in urban, airborne particulate matter."[46] Entering first through the nose and then into the olfactory nerve to the brain, nanoscale magnetite particles respond to external magnetic fields, producing reactive oxygen species (ROS) that are linked to a host of neurodegenerative diseases such as Alzheimer's. Professor David Allsop, an Alzheimer's disease expert at Lancaster University, says, "There is no blood-brain barrier with nasal delivery. Once nanoparticles directly enter olfactory areas of the brain through the nose, they can spread to other areas of the brain, including hippocampus and cerebral cortex—regions affected by Alzheimer's disease . . . An impaired sense of smell is an early indicator of Alzheimer's disease.[47]

Notably, loss of the sense of smell has also been associated with Covid-19 symptoms as well as a side effect of Covid-19 "vaccines." Could nose-to-brain transport of nanoparticles into the brain via the olfactory pathway be the objective behind the PCR swab "test"?[48] The chemistry on swabs has been analyzed multiple times since 2021.[49]

Thanks to Lockheed Martin's Space Fence lockdown infrastructure, our atmosphere is now fully ionized and loaded with compressed plasma clouds suspended in a condition of magnetic resonance. We are breathing in a constant aerial delivery of iron oxides coated with biological polymers, which assures ionic binding and guarantees that our bodies and brains interact with external manmade magnetic fields.[50] In short, the constant delivery of nanometals ensures that we and our bodies can be remotely manipulated via brain-computer interface (BCI) to interact with 5G/6G transmissions.

THE PRECISION MEDICINE OF CELL TRANSFECTION, MAGNETOFECTION (MAGNET-ASSISTED TRANSFECTION OR MATRA), ELECTROPORATION

The medical tyranny over nations by the dubious World Health Organization (WHO) has already begun under the covert cover of a wireless convenience world. New terms like telemedicine and precision medicine are making their way into the populace who are being conditioned to think of computerized medicine as "progress." Thus, it may neither shock nor concern people to realize that many of the wireless transmissions invisibly shooting through our ionized atmosphere are dedicated to the new synthetic biology of the transhumanist overhaul now underway. In this section on magnetism, let's take a look at *transfection.* I will go much deeper into these transmission matters in the synthetic biology book.

Under precision medicine, transfection can be remotely accomplished. No need for a lab or hospital setting. Transfection, like much that has to do with digital biology, has to do with frequency and 5G/6G transmission. For example, "caged"* sodium, aluminum, and carbon can now be remotely transfected into the cells of all living organisms as nanoparticles-become-pathogens.

Transfection is the process of "infecting" a cell with isolated nucleic acid so as to transform the nanoparticle into a pathogen via frequency transmission. Transfection is basically how genetically modified organisms (like GMO foods, Morgellons, and other synthetic biology) are produced, whether the DNA is transfected *in vitro* (in a lab) or *in vivo* (directly into an organism). *Magnetofection* is more of the same but particularly powerful in that it employs superparamagnetic iron oxide nanoparticles with unique magnetic characteristics once their surface is modified by polyethylenimine, at which point they are able to smoothly deliver plasmic DNA in the form of a "vaccine" into cells.[51] The inclusion of graphene oxide in the Covid-19 genetic injections that left the inoculated with magnetic reactions indicates that magnetofection is the preferred *in vivo* transfection candidate.†

> Recent efforts combining nanotechnology and magnetic properties resulted in the development and commercialization of magnetic nanoparticles that can be used as carriers for nucleic acids for *in vitro* transfection and for gene therapy approaches including DNA-based vaccination strategies. . . . It is possible to combine superparamagnetic nanoparticles with magnetic forces to increase, direct and optimize intracellular delivery of biomolecules.[53]

---

*With the caging technique, a target molecule can be rendered biologically inert (or caged) by chemical modification via a beam of light. Photomanipulation of cellular chemistry using caged compounds means control over a cell or cells. Is this another use of optogenetics?

†"While magnetofection does not necessarily improve the overall performances of any given standard gene transfer method in vitro, its major potential lies in the extraordinarily rapid and efficient transfection at low vector doses and the possibility of *remotely controlled vector targeting in vivo*" (emphasis added).[52]

In a lab (*in vitro*), a nucleic acid (DNA) solution is "incubated" over a universal magnetic plate (i.e., a magnetic field), but for inside the body (*in vivo*), magnetic nanoparticles like magnetite and carbon nanotubes (CNTs) of graphene oxide (GO) work quite well for transfection. U.S. Patent 20110130444A1, "Methods and compositions for targeted delivery of gene therapeutic vectors," was filed in 2008 and then abandoned, to be replaced by U.S. Patent WO2008137114A1 wherein a "transgene" and nucleic acid "with perfluorocarbon gas-filled microbubbles" create a mixture then introduced into the bloodstream where ultrasound pulses "disrupt" the microbubbles to release their payload, "thereby enabling uptake of the transgenic nucleic acid into the cells."

Is *magnetofection* just a more recent term for MATra (magnet-assisted transfection) benignly described by J. Bertram in the footnote above, or is it the seemingly weaponized "Drug-loaded nano-microcapsules delivery system mediated by ultrasound-targeted microbubble destruction [UTMD]: A promising therapy method"? Given the hundreds (and thousands in large smart cities like New York) of microwave "cell phone" towers providing "external manmade magnetic fields," it is not difficult to foresee mass-scale transhumanist experiments (or crowd control) by activating the payloads in UTMD magnetic nano-capsules delivered by aerosols and already in vivo.[54]

On March 24, 2016, the *Guardian* newspaper ran a story announcing "Magneto," described as "a magnetized protein that activates specific groups of nerve cells from a distance."[55] Whereas optogenetics and chemogenetics* require a multicomponent system, magnetogenetics does not, now that the human brain is loaded with magnetics. As the *Guardian* article states:

> Magneto can remotely control the firing of neurons deep within the brain, and also control complex behaviors. Neuroscientist Steve Ramirez of Harvard University, who uses optogenetics to manipulate memories in the brains of mice, says the [magnetogenetics] study is badass. . . . Previous attempts [to use magnets to control neuronal activity] needed multiple components for the system to

***Optogenetics* switches related neurons on and off with pulses of laser light; *chemogenetics* uses engineered proteins activated by "designer drugs" to target specific cell types.

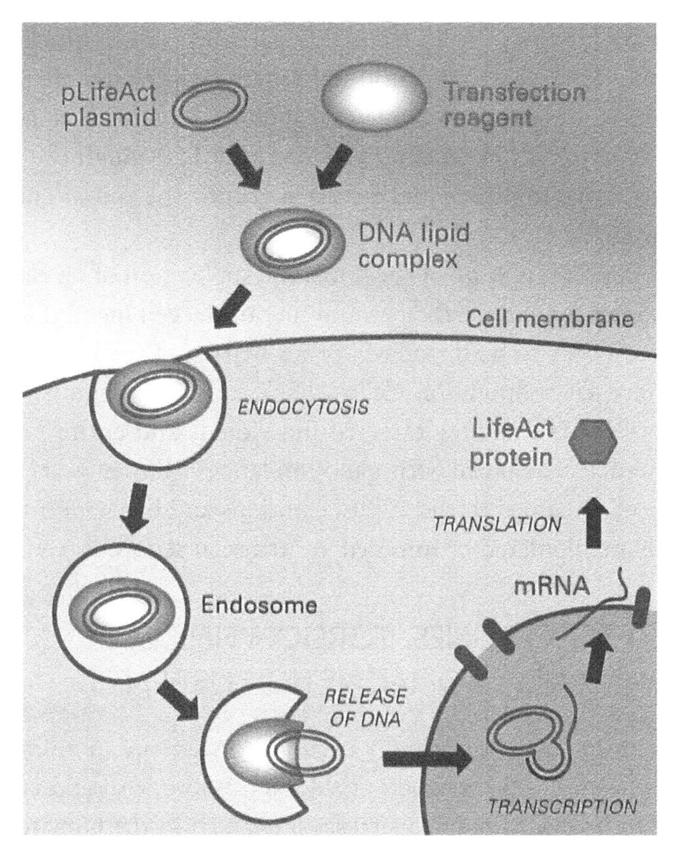

Fig. 6.3. The process of transfection. "Prior to the transhuman shift of biology from molecular to digital biology (5G/6G transmissions of frequencies), *gene therapy* was about injecting a tiny needle loaded with DNA into cells,* usually by means of electroporation. Now, electroporation remotely shoots the nucleic acid / magnetic metal nanoparticles "Transfection reagent" / "DNA lipid complex" into cell nuclei.
Society for Mucosal Immunology

work—injecting magnetic particles, injecting a virus that expresses a heat-sensitive channel, [or] head-fixing the animal so that a coil could induce changes in magnetism . . . The problem with having a multi-component system is that there's so much room for each individual piece to break down.

*Bacteria can be engineered to "infect" cells with DNA.

Better by far than "head-fixing" is when nerve cell proteins sensitive to radio waves and magnetic fields remotely attach themselves to the magnetite nanoparticles being delivered by chemtrails, GMO foods, and mRNA serums. The Magneto current travels through the cell membrane, then nerve impulses make their way into the spinal cord and up into the brain. *Badass . . .*

Electroporation is more of the same: the application of an electric field to cells in order to increase the permeability of the cell membrane to allow chemicals or DNA to be introduced, thus bypassing the blood-brain barrier. In effect, electroporation electroshocks an inoculation site to force DNA into cells. Is this how targeted individuals and entire populations are implanted at a distance with nano-implants?[56] Thirty years after neurocircuitry was mapped in the 1990s, emotions and behavior are now easily remotely manipulated or imposed as "artificial senses."[57]

LEYLINES, TORSION FIELDS, AND GEOMAGNETIC CURRENTS

It is well known that the naturally occurring magnetite in animals allows them to navigate Earth's magnetic field.[58] Therefore it should come as no surprise that the magnetic nanoparticles implanted in the human body and brain, such as the magnetite and graphene loaded into chemical aerosols as well as injections are useful for tracking bodies and brains in different geomagnetic regions, and may even change one's consciousness in geological "power spots" while affecting moods and intentions, behaviors and psychic abilities. Telekinesis, plasma entities, and UAPs/UFOs have been observed on such electromagnetically sensitive land. As neuroscientist Michael Persinger put it while observing the temporal lobes of the brain, "Transient and unusual phenomena should occur in areas where tectonic stress is accumulating." The temporal lobes seem to mediate parapsychological phenomena, whether naturally occurring from the tectonic strain within the Earth's crust producing luminous phenomena attributed to unidentified aerial phenomena (UAP) or from the weaponized use of electronics like buried fiber optics.*

*Persinger (1945–2018) was director of Laurentian University's Consciousness Research Laboratory. He and Ghislaine Lafreniere authored Space-Time Transients and Unusual Events (Nelson-Hall, 1977). He is the inventor of the famous "God helmet."

Our atmosphere once rang like a Schumann resonance bell when struck by solar flares, but now solar flares, sunspots, solar (plasma) winds, solar minimums, and Earth's "brainwaves"* (i.e., the Schumann resonance) are being manipulated by technocrats seeking full spectrum global dominance over not just economies and technology and diseases but over a hybrid humanity increasingly *biomagnetic*. Meanwhile, chemical trails follow torsion fields, ancient leylines, and tunnels that connect deep underground military bases (DUMBs) with 15-minute Smart Cities. The air is now viewed as a dielectric conductor and humans as dielectric biometric transceivers. Human consciousness is no longer just being environmentally influenced by telluric energies but by endless 5G/6G wireless on Earth and from space.

Ancient megalithic stone circles emit high and low levels of radiation and ultrasound, their placement calibrated to correlate with Earth's energies and forces. Primitive superstition? Hardly. We would do better to try to harness these sources of free energy instead of blindly submitting to energy barons bent on controlling us via the conductive nanometals in our brains and in the geomagnetic grid. For thousands if not millions of years, ancient magnetic lines of force have been generating cosmic energy for astronomy centers, pyramid transceivers, roads, cathedrals, forts and military bases, nuclear installations, and particle accelerators, with magnetrons now taking the place of dowsers for probing powerful geomantic alignments such as

Mexico City to Mount Carmel in Israel and Nova Scotia-Belfast-London-Brussels-Kosovo.

What do geomantic alignments have to do with the oldest American Smart Cities like New Orleans at the mouth of the Mississippi River (site of HAARP-driven Hurricane Katrina in 2005); Atlanta, Georgia (with its open lattice pyramid under an obelisk on top of the Bank of America's Plaza skyscraper); Washington, DC (built according to Masonic design esoterica); Baltimore, Maryland; Philadelphia, Pennsylvania; New York City; Boston, Massachusetts; etc.?

Grid manipulators are crucial to shaping society by controlling cosmic lines of force. Wars and revolutions along strategic geomagnetic alignments are like "geographic acupuncture." As geomancer Peter Champoux says,

---

*By reading the magnetic H wave (the psychoactive component), it was determined that Earth's brainwaves at that time were identical to human brainwaves.[59]

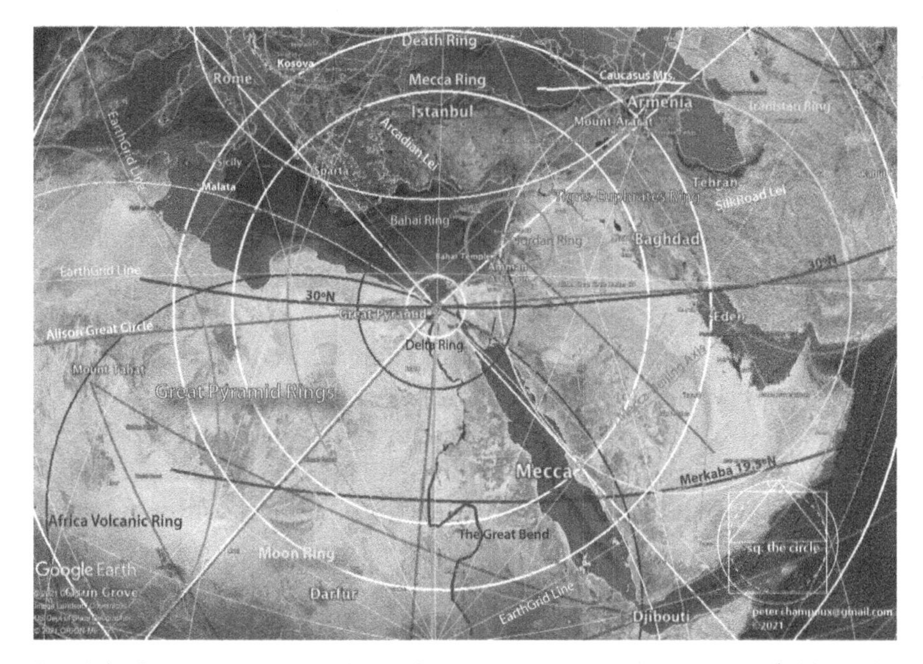

Fig. 6.4. Geomancy expert Peter Champoux: "The consciousness field set by the Great Pyramid is under the control of Ahrimanic forces, dark lords as such. This conclusion was reached by the manifestations of genocide radiating in its EarthRings. With Mecca's Kabba stone on its SE ley and the Shrine of the Bab on its NE ley, the hierarchical model / worldview is embedded in our top-down world control template. To shift this, the pyramid needs to be seen for what it originally modeled: a water molecule."

Image courtesy of Peter Champoux

> So many forces are at play here, both natural and Ahrimanic [Ahriman, a spiritual force of evil that human beings must grapple with]. A pat answer to the questions of our time is not as simple as blaming a group of mystery men behind the screen, although without a doubt they are trying to pull the levers of anti-awakening, as the sovereign human soul is the greatest threat to their earthly power. Ley are fought over for control. The more violence on a ley, the more violence it attracts, and vice versa. Elemental collaboration with human consciousness or a concordant harmony sung at these sites would, in effect, change the frequency and that of the manifestation of its mass.*

*Peter Champoux, email, January 25, 2023.

These lines of force and their "power points" of energy are employed for rituals, as well—good *and* evil—even rituals having to do with military objectives, most of which are kept from the public, as determined by secret societies like Mithraism that have been central to governments and institutions for thousands of years.* In Chapter 23, "El Camino Real and the Waters of Babylon" in my fictional series about America since President John F. Kennedy's assassination *Sub Rosa America: A Deep State History*,[60] I go into geomantic detail about the beginning of the American rocket program and the roles of "Satanic rocket fuel genius" Marvel Whiteside "Jack" Parsons at the Jet Propulsion Lab by the Devil's Gate vortex in Pasadena. For example:

> The California Institute of Technology (Caltech) had been around since the late 19th century, but the Jet Propulsion Laboratory (JPL), NASA's rocket design center, was born on Halloween 1936 at the entryway of the Arroyo Seco. Jack and Ed and the rest of the Caltech Suicide Club posed for the Nativity Scene that night, a ritual that continues to this day with mannequins standing in for the JPL's John Dee and Edward Kelley . . .
>
> The Linda Vista-Annandale area extends from the west bank of the Arroyo Seco to Linda Vista Hills and from Devil's Gate Dam to Colorado Boulevard. In late 1977, the Hillside Stranglers would ritually arrange their victims' bodies near Devil's Gate. Oak Grove Park (now Hahamongua Watershed) lies between the Jet Propulsion Lab (JPL) and Devil's Gate Dam. Oak groves are sacred to Druids, as are rivers and their tributaries. Annandale is an old Scottish family name, as is Stewart (Stuart). King of Scots Robert the Bruce was the 7th Lord of Annandale, and Robert Bruce Stewart was president of Annandale Corporation from 1977 to 1991, after which he founded Acacia Research Corporation (ARC), then was CEO of Arrowhead Research Corporation (ARC).

*London was once the early Roman town of Londinium seven meters below modern London, where Roman soldiers worshipped Mithras. Mithraism preceded the arrival of Christianity as the State religion in the fourth century CE. The ancient Temple to Mithras (*mithraum*) has now been refurbished by Michael Bloomberg (see London Mithraeum website) and Western Christianity's Christmas cleverly blends the feast of Saturnalia with the birthday of Mithra on December 25.

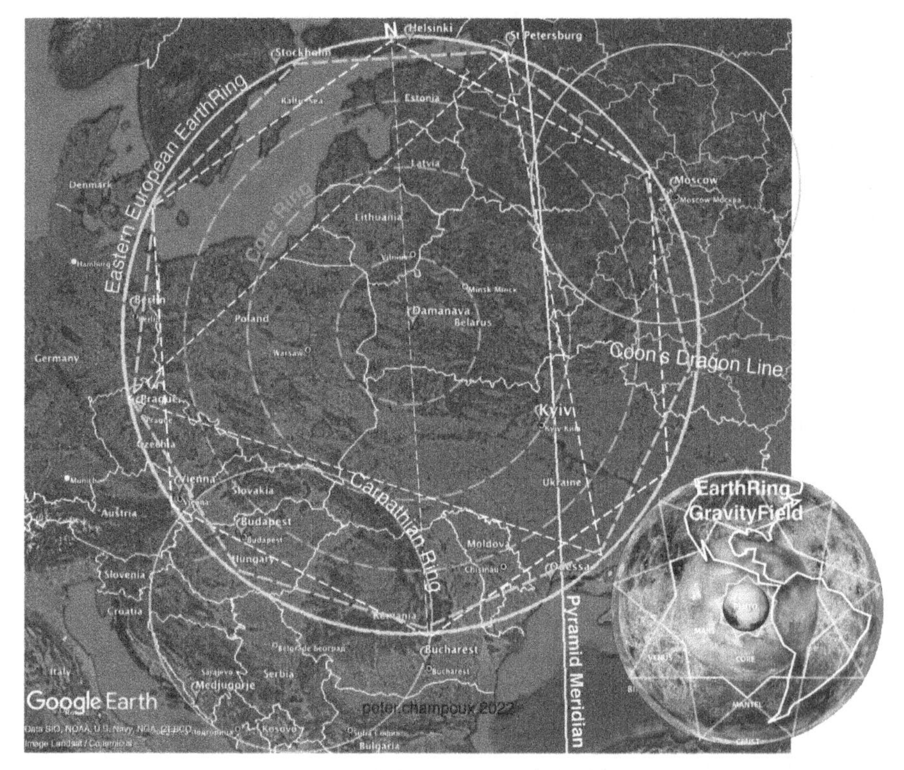

Fig. 6.5. Geomancy expert Peter Champoux: "To the exact North of the Great Pyramid is a proverbial line in the sand between Russia and NATO. Russia's western border follows this ley with the exception of Ukraine. Kiev (Russian cyrillic Киев) being the spirit heart of the Russian Orthodox Church furthers the multilevel tension along the ley's course. Consciousness coalesces into dissonance through repetition of the top-down pyramid model myth resulting in the modern era's Ahrimanic manifestations, [the war in] Ukraine being the most recent."

Image courtesy of Peter Champoux

Speaking of leylines, where is the line between paganism and Satanism? Surely, this is a question that the American military must finally ask itself.

Grid Manipulators

In our high-tech era getting progressively higher and higher, dowsing wands no longer locate and measure leylines and the geomagnetic grid. Now, the elite scientific cabals have created a network of grid manipulators armed with particle accelerators and ionospheric heaters, massive radar installations, phased array antennas, power lines, underground cables, microwave towers, etc. The following list offers examples as per installation, location of the installation, and locations affected by manipulation of the particular grid / leyline:

- **Fermilab** (Chicago particle accelerator): Greenland and Teotihuacan
- **Triumf** (Vancouver, BC, particle accelerator): Greenland and Kiribati*
- **DESY** (Hamburg, Germany, particle accelerator): magnetic north and Giza
- **CERN** (French-Swiss border particle accelerator): Greenland and Giza
- **EISCAT 3D** (Norway ionospheric heater): magnetic north and Giza
- **HAARP** (Gakona, Alaska, ionospheric heater): magnetic north and Kiribati
- **HIPAS** (Fairbanks, Alaska, observatory): magnetic north and Olgas, Australia
- **Poker Flats** (Alaska rocket range): magnetic north and Olgas, Australia†
- **Millstone Hill MISA** (Westford, Massachusetts, steerable antenna): Giza and Kiribati
- **NCAR** (Boulder, Colorado, atmospheric research and development): synced with HAARP/Triumf/Teotihuacan
- **SRI International** (Menlo Park, California, R&D): magnetic north and Greenland
- **SuperDARN** (Super Dual Auroral Radar Network), consisting of thirty low-power, high-frequency radars, in a collaboration between Virginia Tech (the lead institution),‡ Dartmouth College, the University of Alaska Fairbanks, and the Johns Hopkins University Applied Physics Laboratory: Kashina (Mongolia and Kiribati); Kansai/Kobe (magnetic north and Olga); Yamagawa (Tibet and Teotihuacan); Okinawa (Giza and Kiribati); Goose Bay (Greenland and Cuzco); Iceland (magnetic north and Giza); Kapaskasing

*Thirty-two Micronesian atolls and reef islands; Midway Island falls on this line, as well.

†HAARP, HIPAS, and Poker Flats work together as a unit.

‡Note how extensive Virginia Tech's SuperDARN grid is—much like the successor infrastructure, the Space Fence, created by Lockheed Martin.

(Greenland and Teotihuacan); Saskatoon (magnetic north and Teotihuacan); Antarctic (Giza and Kiribati).*

An aerial view of the Australian–United States joint Naval Communication Station Harold E. Holt in Exmouth, Western Australia, the most powerful transmission station in the Southern Hemisphere. Harold E. Holt was seventeenth prime minister of Australia when he disappeared while swimming December 17, 1968, (solstice), three months after the station was commissioned on September 20, 1968 (equinox). The base is currently operated by Raytheon Australia.

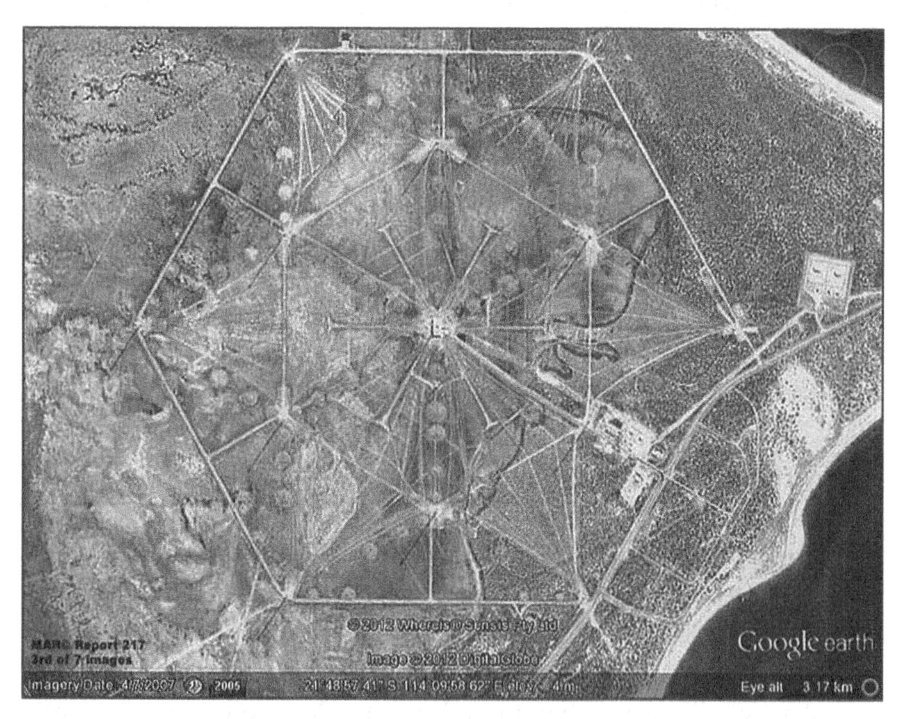

Fig. 6.6. "Hex," or hexagon?
Map data © Google

*The Arecibo, Puerto Rico, ionospheric heater collapsed along with the telescope in 2020. Nor is the Sura ionospheric heating facility in Russia (56.13°N / 46.10°E) included in this list of grid manipulators. Sura is active with an ERP (effective radiated power) of 190 MW, compared with HAARP's 5,000 MW. Other known ionospheric heaters in Russia are at Tula, Khabarovsk, and Novosibirsk, where the mind-control cult Ashram Shambala was founded in the 1980s.

The thirteen towers are tiered in TWO INTERLOCKED HEXAGONS around Tower Zero (387.4 meters / 1,271 feet), the outer hexagon made up of six odd-numbered towers T1–T11 (each 358 meters / 1,175 feet) around Tower Zero, the inner smaller hexagon of six even-numbered towers T2–T12 (each 303.6 metres / 996 feet) around Tower Zero. Buried in the ground beneath the VLF spider webs of wire antenna array is 386 kilometres of bare copper ground mat.

"For the previous two weeks, I stayed in Exmouth, a small coastal tourist village originally built to house the U.S. Navy and Raytheon personnel who constructed and currently operate the Harold E. Holt Naval Communications facility there. I could look out my window and see the antennas in the distance. I had a pronounced ringing in my ears while in 'Exie,' as locals like to call it, and wondered if the uniform mentality exhibited by much of the mostly white population was due not only to the constant surf and sun, but also an 'apathy' or 'zombie' wave. HEH is probably the most powerful microwave/phased-array radar transmitter in the Southern Hemisphere outside of Antarctica, and an integral part of the global HAARP network."
—Jeff Wefferson, March 5, 2022

The helix, vortex, spiral, and torsion are everywhere in Nature and the cosmos, from the spiral at the top of our skull evident in our hair, to our helical heart[61] and the unique whorls in our fingertips. Billions of proteins surge through a newly fertilized egg's surface in a spiral pattern, like vortices in quantum fields. As well, wave forms in our pre-geoengineered atmosphere and the oceans once oscillated outward "as tiny, hurricane-like spirals."[62] These same electrical signals propagate in the human heart and brain, and can be experienced as telepathy and psychokinesis. According to Gerald H. Pollack, PhD, water has a fourth phase beyond solid, liquid, and vapor (gas), namely a *hexagonal geometry* able to draw electric current from sunlight.[63] The molecular structure of the melatonin produced by the pineal gland, the so-called third eye organ whose highly active dream-like and imaginative picture consciousness we now call clairvoyance, is a hexagon; the molecular structure of DMT (dimethyltryptamine)—which some think is produced in the pineal gland—is also a hexagon. Even the north pole of Saturn is covered by an extraordinary hexagonal polar jet stream cloud pattern.[64]

Does the geometry of the hexagon announce the presence of doors of perception between dimensions? Are the occultists in the military aware of the power of the geometry of the hexagon? I am sure they are, and beyond the actual *fact* that "The mystery behind the [hexagonal honeycomb] pattern is that natural forces work to keep an electric charge moving in an interrupted circuit."[65]

The invisible vortices created by crossing magnetic lines of force are called *torsion fields.** These pure fields of coherent force have been studied by many inventors (other than Tesla) seeking to finally free people from the greed of controlling energy barons—inventors like Karl Schappeller (1875–1947)[67] and his system of uniform deflection of electrons by magnets, and India's chief nuclear scientist, Paramahamsa Tewari (1937–2017)[68] and his electrostatic field of atoms and electrons interacting to produce invisible energy. Vacuum in rotation, the quantum spin of empty space producing the motive power of magnetic motors, the overunity of free energy! Whereas Tesla's method was a wave system, Schappeller's and Tewari's systems entailed the flow of æther and, as Tewari put it, "Each electron is tied up with the whole ether stream."† In fact, with just a narrow stream of water and a crystal plate, a torsion field and torus ("donut")‡ of glowing plasma called by Tewari a plasma torus and Schappeller "glowing magnetism" can be created without powerful electromagnetic fields or a vacuum.[69]

Those investigating the relationship between rotating magnetic fields and strain-related "earth lights"—or as Michael Persinger coined them, magnetic "anomalous luminous phenomena" (ALP)—use Geiger counters, methane detectors, infrared sensors, and radiofrequency detection equipment loaded with amplifiers and oscilloscopes.[70] They pay attention to

*Also called axion fields, spin fields, spinor fields, or microlepton fields, these are debunked by the mainstream media as pseudoscience.[66]

†See the excellent 2016 Austrian film *Aus Dem Nichts (Out of the Void)*. The first half is about Schappeller, the second half about Tewari.

‡The yin-yang symbol is a toroid. A torus is contraction-expansion in balance due to vacuum energy fluctuations.

lunar tidal cycles, railroad tracks laid in alluvial soil, power lines stretching for miles, the colors of released lights, the radon gas creating ionized pockets of air that seemingly produce plasma orbs, and the effects on the humans living nearby, including UAP sightings and subsequent psychological and physiological effects.

Clues as to how torsion fields affect us emotionally may lie buried in the many leyline or crossroads anecdotes about spirits and deals with the devil, like the nursery tale "The Three Billy Goats Gruff" about the troll under the bridge over a moving river of energy, and how American blues man Robert Johnson sold his soul to the devil in exchange for musical talent. Rudolf Steiner lays out an *intraterrestrial* dimension that humanity is subject to, beyond all the different levels of *karma* with which individual lives are bound up—a dimension seemingly having to do with *unredeemed moods of soul* like quantum entanglements are bound up with the history of human evolution and geographic torsion fields. It is from this dimension that black magicians (now working primarily technologically) derive power over souls *via the electric currents in our bodies*.[71] This is why electromagnetic technologies should be studied from both a physical *and* metaphysical vantage point, such as this book about geoengineering and the following book about synthetic biology are attempting to exemplify. Rudolf Steiner called this approach "geographic medicine."

To the following list of qualities of the intraterrestrial dimension to which humanity's development is bound—the nine levels of *subnature* counterbalanced by nine *supranature* hierarchies (Angel, Archangel, Archai, Exusiai, Dynamis, Kyriotetes, Thrones, Cherubim, Seraphim), I have added mathematician Robert Powell's astronomical time computations of Christ's Second Coming.[72] From the Nazi takeover of Germany in 1932 and Hitler's appointment as chancellor on January 30, 1933, to the three atomic bombs in 1945 and onward to our arrival in the realm of the Earth Severer, the Etheric Christ has moved inexorably through the subnature spheres in a twelve-year Jupiter heliocentric pulse, meeting resistance at every level that is then magnified in our human sphere.

Levels	Date	Qualities
Mineral Earth	April 1945	Fixed
Fluid Earth	April 1, 1945–February 10, 1957	Mechanical
Air Earth	February 10, 1957–December 21, 1968	Cold feelings, a death sphere, destroys sensation and feeling
Form Earth	December 21, 1968–November 1, 1980	Stereotypes, caricatures, warped fantasies
Fruit Earth	November 1, 1980–September 11, 1992	Magnetic forces, unlimited growth like fungus; possesses soul
Fire Earth	September 11, 1992–July 23, 2004	Unbridled passion, darkened will, devastating effects in earthquakes and volcanoes, material kingdom of Ahriman-Mephistopheles
Earth Mirror	July 23, 2004–June 3, 2016	Love of evil that perverts virtue and sacrifice; transforms Nature into its opposite; happy over others' downfall
Earth Severer	June 3, 2016–April 14, 2028	Dissonance and chaos; fragments all that is moral; hatred; brought the substance of evil into the world
Earth core	April 14, 2028–February 24, 2040	Source of spiritual evil; anti-human

THE SHADOW BIOSPHERE

Then there is the hidden world that astrobiologists call the *shadow biosphere*. A 2006 paper in the *International Journal of Astrobiology* by Carol Cleland and Shelley Copley explains how we coinhabit Earth along with "microbial life forms that have a completely different biochemistry from the one shared by life as we currently know it."[73] Astrobiologists call it "weird life."[74] "All the micro-organisms we have detected on Earth to date have had a biology like our own: proteins made up of a maximum of

20 amino acids and a DNA genetic code made out of only four chemical bases: adenine, cytosine, guanine and thymine," says Cleland. "Yet there are up to 100 amino acids in nature and at least a dozen bases. These could easily have combined in the remote past to create life forms with a very different biochemistry to our own. More to the point, some may still exist in corners of the planet."[75]

The shadow biosphere calls to mind the infrared photographs of "macroorganisms" in the American Southwest desert taken by Wilhelm Reich and later by former U.S. Merchant Marine electronics officer and early UFO researcher Trevor James Constable (1925–2016).* Their photographs reveal life forms in various shapes and sizes, with invisible plasma bodies that only infrared could detect. The photos featured in Constable's 1976 book *The Cosmic Pulse of Life: The Revolutionary Biological Power behind UFOs* look decidedly like giant single-cell microbes with close-set eyes navigating our atmospheric sea, peering down at us much like the pulsating, diaphanous orbs that flocked around NASA space shuttles, not so much threatening as exhibiting a petlike curiosity.

Is it time to cast our net of consciousness toward *intraterrestrial* thresholds like Skinwalker Ranch† in the Uintah Basin at the edge of the Uintah and Ouray Indian Reservation in northeastern Utah, stretches from Vernal, Utah, to past Fruitland, Utah. The ranch is crowded with diaphanous, shape-shifting plasma skinwalkers, hologram drones, UAPs dropping plasma from their hulls, flashing lights (red, white, and blue), and strobing laser weapons. The question is, how much of this is military holographic interface with tectonic stress and leyline highways, and how much is actual shadow biosphere? David Gessner drops a clue in his *Mother Jones* article (originally in *OnEarth* magazine):

> I started seeing UFOs, something I had never really ever seen before. The most peculiar thing is many of the odd situations came about around electronics, especially manifested through cell phones. I rested quite a bit on the couch and noticed my phone would heat up when

---

*The ancient Vedas of India (far older than the 3,500 years often quoted) referred to the atmosphere and deep space as a "cosmic ocean" populated by large creatures and vimanas, flying craft that are thought to have been UFOs.

†In Navajo culture, a *yee naaldlooshii*, or a skinwalker, is a witch able to turn itself into, possess, or disguise itself as an animal.

> near my foot. Not a big deal, until I noticed my phone actually showed it was charging.[76]

A particularly powerful natural geomagnetic field? Or was a SQUID (superconducting quantum interference device) magnetometer buried at Skinwalker Ranch in order to experiment with anomalies and brains?[77] Geomagnetism can do much more than influence the weather; it can also be used to influence multiple human minds to see what may not normally be visible,* particularly now that our brains are suffused with magnetite. (Was the underground Montauk Chair that influenced the brains of the Montauk Boys designed to function like a SQUID magnetometer?)

Originally, Robert Bigelow of the National Institute for Discovery Science and founder of Bigelow Aerospace† was in charge of this "hotspot" ranch, along with biochemist Colm Kelleher, who was in charge of biological testing, and retired U.S. Army Colonel John B. "Nonlethals" Alexander, a leading advocate for military applications of the paranormal.‡ Alexander hired ex-military Buffalo Soldier Freemasons skilled in Haitian Vodou (voodoo) to deal with the psychic battles between whatever was manifesting at the Skinwalker Ranch and the shamans of the Uintah and Ouray tribes.

The Uintah Basin is a geomagnetic anomaly loaded with the mineral gilsonite, a very pure, resinous rock discovered in the 1860s and formed from a complex combination of different kinds of hydrocarbons originating from the solidification of petroleum. In 1888, Samuel H. Gilson, one of the prospectors associated with the mineral's discovery, succeeded in pressuring Congress to take back seven thousand acres from the Uintah and Ouray Indian Reservation so that the mineral could be legally mined.

---

*For example, magnetic storms have been known to increase psychiatric hospital admissions because of the ELFs present in the micropulsations of Earth's geomagnetic field connected to the brain.

†Bigelow Aerospace in 2011 was second in scope only to NASA.

‡Alexander served with the U.S. Army Special Forces and the Army Intelligence Command, and at Los Alamos National Labs as manager of nonlethal weapons. His PhD was in thanatology. In 1985, he founded the Advanced Theoretical Physics Project, an informal group that included government officials as well as people from the army, navy, and air force, plus representatives from the defense aerospace industries and some members from the intelligence community.

Gilsonite resembles shiny black obsidian, but its shadow biosphere qualities may be more like the famous "black goo" (i.e., graphene, discussed in chapter 9) that German scientist Harald Kautz claims arrived in the Texas oil fields as a massive meteorite 20,500 years ago, and that "black goo" is not only *conscious* in an intraterrestrial way, but is the source of *unlimited* petroleum.[78] (Gilsonite is only found in the Uintah Basin in Utah, in Colombia, and in Iran, and no doubt accounts for Big Oil's presence at Skinwalker Ranch.[79]

Is gilsonite actually graphene? Gilsonite is not magnetic in itself, but its status as a hydrocarbon means it is involved with the observable magnetic anomalies induced by iron-bearing minerals (ferromagnetic), particularly where petroleum seepage is occurring.[80] Its unusual geologic origin should be noted: In the Uinta Basin, gilsonite veins are vertical and northwest-tending in a 60-by-30-mile area of Tertiary (57 to 36 million years ago) sedimentary formations.[81]

7

The Nuclear Behemoth

A column of smoke and dust rose into the air like a column of smoke issuing from the bowels of the Earth. It rained sulphur and fire on Sodom and Gomorrah, and destroyed the town and the whole plain and all the inhabitants and every growing plant. And Lot's wife looked back and was turned into a pillar of salt. And Lot lived at Isoar, but afterwards went to the mountains because he was afraid to remain at Isoar. The people were warned that they must go away from the place of the future explosion and not stay in exposed places; nor should they look at the explosion but hide beneath the ground . . . Those fugitives who looked back were blinded and died.

DEAD SEA SCROLLS, THIRD CENTURY BCE

I believe that there have been civilizations in the past that were familiar with atomic energy and that by misusing it, they were totally destroyed.

FREDERICK SODDY (1877–1956),
THE INTERPRETATION OF RADIUM,
ORIGINALLY PUBLISHED IN 1909;
SODDY COINED THE WORD *ISOTOPE*

At a 1958 congressional hearing, Major General Walter Dornberger—former commander of Nazi Germany's V-2 rocket research at the Peenemünde Army Research Center (Bell Aircraft Corporation sponsored his Paperclip entry into the United States)—insisted that America's top priority must be *space*, in order to meet astrophysicist Nikolai Kardashev's protocol for a true space age, "to conquer, occupy, keep, and utilize the space between the Earth and Moon" (the Earth-Moon gravity well) with bases on the Moon and orbiting battle stations (i.e., satellites) controlling the pathway on and off the planet. Domination of the pull of gravity between Earth and the Moon requires domination of Earth's natural geomagnetism and frequencies. At a national missile industry defense conference, Dornberger informed participants: "Gentlemen, I didn't come to this country to lose the Third World War."*

In 2014, NASA engineer Kelly Smith, in the sales pitch video "Orion: Trial By Fire,"[2] about the Orion Deep Space mission goal of getting to Mars in 2030, admits that the perilous Van Allen radiation belts are deadly to astronauts, which is why Orion is loaded only with sensors and not Human 1.0: "We must solve these challenges before we send people through this region of space," Smith declared.

In the future, the trail of the development of the secret space program will be viewed as *the* major influential secret buried in the twenty-first-century historical record. It began with the Manhattan Project and atomic bomb, then leaped into high gear with the Paperclip Nazis in service not to the U.S. government but to SS astrophysicist Wernher von Braun. Since then, our atmosphere has been constantly radiated by both ionized and nonionized radiation.

In 1958, Explorer 1—the first satellite launched by the United States—discovered the Van Allen radiation belts operating 5,000 to 32,000 miles above the Earth like a net, capturing charged particles from solar and galactic winds and spiraling them down along the magnetic lines of force that comprise the magnetosphere, until they converge at the Poles as the extraordinary Arctic and Antarctic auroras. Immediately, the U.S. Navy detonated three nuclear bombs in the Belts under Operation Argus.

In 1961 under Project West Ford, the U.S. Air Force launched 480 million tiny copper needles half the wavelength of 8,000 megahertz microwaves that began the creation of a permanent metallic ring of dipole antennas around the planet to serve as an artificial ionosphere. At

*"Dornberger declared that in Europe, the military devised ideas and asked industry to fulfill them, whereas in this country it was industry which made proposals for acceptance by the military."[1]

2,299 miles above Earth, the copper needles created a donut-shaped toroid cloud approximately nine miles wide and nineteen miles thick. When the dipoles dispersed and three satellites and the innermost Van Allen belt were destroyed, Hawaii blacked out and the sky over the Arctic Circle caught fire. The following year, Starfish Prime detonated three more nuclear bombs while attempting to storm the Belts, but managed only to further irradiate our planet—and form a new Belt.

> The inner Van Allen Belt will be practically destroyed for a period of time . . . [and] the ionosphere . . . will be disrupted by mechanical forces caused by the pressure wave following the explosion. At the same time, large quantities of ionizing radiation will be released . . . On 19 July . . . NASA announced that as a consequence of the high-altitude nuclear test of July 9, a new radiation belt had been formed, stretching from a height of about 400 km to 1,600 km [300–1,200 miles].[3]

After 200+ U.S., 200+ Soviet Union, 20 UK, 50 France, 20 China atmospheric "tests"[4] (and a multitude of underground "tests"), the natural electron fluxes in the lower Van Allen belt were forever changed, and the cosmic radiation ("killer electrons") in the inner belt, 435 to 6,214 miles above the equator, remains damaging to humans, satellites, and sensitive spacecraft instruments to this day. Passenger jets fly at an altitude of 31,000 to 38,000 feet (5.86 to 7.19 miles) through radiation clouds whose dose rates are double what they should be, which means that the radiation is not coming just from cosmic rays.[5]

> Sometimes dose rates skyrocket for no apparent reason . . . All of the surges observed occurred at relatively high latitudes, well above 50 degrees in both hemispheres . . . The ARMAS [automated radiation measurements for aerospace safety] flight module recorded a 2X increase in ionizing radiation for about 30 minutes while the plane flew 11 km (36,000 feet) over the Antarctic Peninsula. No solar storm was in progress. The plane did not abruptly change direction or altitude. Nevertheless, the ambient radiation environment changed sharply. Similar episodes have occurred off the coast of Washington state.[6]

◆

> According to the patent documents, the walls of the engine's combustion chamber would be lined with a fissile material like Uranium 238, which would react with the high-energy neutrons produced by the nuclear reaction to generate huge amounts of heat. On the other side of the chamber wall, a coolant would pick up this heat and be sent through a turbine/generator to produce electricity . . . used to power the engine's lasers.[7]

Dose rates skyrocketing "for no apparent reason" should mention military-industrial-intelligence experiments that mainstream media are not informing the public about, like Boeing's plasma pulse engine,[8] which uses high-powered lasers to, in the patent language, "trigger a small nuclear reaction" for jet engine thrust. A stream of hydrogen isotopes (deuterium or tritium) triggers a small thermonuclear explosion, from which heat energy is harvested.

Perhaps human beings are not intended to physically explore their solar system and beyond. What is certain is that the national security objective of many of the "tests" was about commandeering "the space between the Earth and Moon" (Dornberger), beginning with opening a pathway through the Van Allen radiation belt so that human beings in non-lead-lined spacesuits could come and go from the planet without massive cell destruction by radiation,* and so the United States would be the first to control entry and exit to planet Earth.

The problem of the radiation belts certainly influenced how the secret space program proceeded—from experimentation on human beings of ionized versus nonionized radiation to the distant hope of genetically engineering a less vulnerable, more *cybernetic human body* impervious to radiation—and of course continuing to work on the Nazi *Die Glocke* (The Bell) exotic propulsion craft still emitting a pale blue glow that was unfriendly to human tissues and blood:

> During the tests, the scientists placed various types of plants, animals and animal tissues in the Bell's sphere of influence. In the initial test

*Even in 2024, the "wild boar paradox" of high levels of radiation in wild boars across central Europe was mounting, not because of the Chernobyl catastrophe in northern Ukraine in 1986—mutant wolves now wander there—but because of the two-decade assault on the Van Allen belt.[9]

> period from November to December 1944, almost all the samples were destroyed. A crystalline substance formed within the tissues, destroying them from the inside; liquids, including blood, gelled and separated into clearly distilled fractions.[10]

The Strategic Defense Initiative (SDI) Space Fence was never about missile defense. It was always about installing the conductive metallic ring around the equator and satellites patrolling above Earth, which was why the whole "Star Wars" SDI had to be tabled until HAARP was successful in wresting control over the ionosphere so the atmosphere could be thoroughly ionized and the Space Fence at last be up and running. Meanwhile on the ground, construction of a massive militarized electromagnetic infrastructure continued apace, the ultimate objective being synchronized control over "under the skin" self-replicating nanotechnology interfacing with AI algorithms via millimeter (5G) and terahertz (6G) transmissions.*

OUR CONSTANTLY RADIATED ATMOSPHERE

With so many cover stories for so many covert programs, how to ferret out the truth? Military strategists claim that orbiting space trash is impeding their laser weaponry. On August 27, 1998, it was reported that a powerful blast of stellar radiation—gamma and X-ray radiation from the magnetic flare of a compressed neutron star (magnetar) 20,000 light-years away—severely affected radio transmissions and shut down seven spacecraft. Was the blast *really* from a magnetar, or was it some kind of experimental radiation "event"?

Even storms are not what they used to be. In late December 2022, the Tokyo Electric Power (TEPCO) began dumping one million tons of Fukushima wastewater packed with cesium and tritium off the Japanese coast along the North Pacific Gyre.† This one internationally illegal act

*The World Economic Forum's Yuval Harari uses the term *under the skin* to describe the transdermal contamination of all of humanity with nanotechnology via aerosol, GMO pharming, and inoculations.

†After the 2011 Fukushima disaster, hundreds of tons of spent nuclear fuel rods were dumped along the Japan Trench, the Philippines Trench, and in the Southern Sea south of New Zealand and Australia.

unleashed radioactivity-powered "atomic storms"* as far away as the Sierra Nevada and Central Valley "bread basket" of California. Autumnal lightning storms in 2023 caused power outages across the Americas. In his March 8, 2023, article, "Fukushima Year 12—Radioactivity Powered Storms Blast the USA," Yoichi Shimatsu clarified:

> These events are not acts of God nor the outcome of so-called "climate change" due to carbon dioxide emissions but instead are generated by free-floating nuclear isotopes, which are the final outcome of financial investment and misplaced trust in nuclear power.[11]

As the Pacific Ocean heated up, the heated vapor rose as pink, gold-tinged super-heated radioactive clouds.

In 1972, physicist Wilmot N. Hess, then-director of the National Oceanic and Atmospheric Administration (NOAA) Earth System Research Laboratories in Boulder, Colorado, announced the modification of the near-earth space environment and expressed the prevailing Kardashev military mindset bent on conquering space at an international meeting of the Society of Engineering Science in Tel Aviv, Israel:

> In the last few years, experimenters have artificially modified the space environment. We can now produce artificial aurorae. We can change the population of the Van Allen radiation belt. We can artificially modify the ionosphere from the ground and our other ideas about artificial experiments for the future stretch as far as trying to copy the sweeping action being carried on naturally by Jupiter's moons.†

Ionospheric heaters now beam microwaves into the ionosphere, DSX (Demonstration and Science Experiments) satellites monitor the magnetosphere‡ for the very low frequency (VLF) transmissions of the

---

*These storms are often blamed on "atmospheric rivers" (tropical storms originating in the Philippines) without clarifying that they have become radioactive since Fukushima, generated as they are by free-floating nuclear isotopes.

†Several of Hess's NASA papers on the Van Allen belt are available on the internet, at NASA's History Collection portal.

‡The Sun's radiation creates and maintains both the ionosphere and plasma in the magnetosphere.

growing metallic ring around the equator,[12] satellites beam energetic particles to produce electromagnetic waves in space, while the "power beaming" satellite PRAM (Photovoltaic Radio-frequency Antenna Module) converts solar energy into microwaves for greater power density,* injects thermal plasma at high altitudes, and transmits electromagnetic waves beamed from Earth to affect particles trapped in the Van Allen belt[13]—for example, to control geomagnetic storms in the outer belt (8,000 to 40,000 miles above Earth).

As you can see, ionized and nonionized experiments and operations continue unabated. Certainly, maintenance of communications and intelligence for the U.S. Army's Combat Capabilities Command Center C5ISR operations† has been achieved.

But at what cost? We are now basically living in a condition of *radiation brinkmanship.* The risk is no longer just borne by suborbital spaceflight crew members subjected to increased occupational radiation exposures not so much from cosmic or solar radiation or thunderstorms as from radiation clouds loaded with plutonium-238 and -239; all of us on planet Earth are now being subjected to radiation from our constantly radiated environment. This is not just the cost of living in an electromagnetic Space Age, but is due to how all of us have, for a few generations, been eating genetically modified foods grown from seed irradiated in outer space to induce mutations[14] while being lab rats in secret environmental genetic alternation experiments intent on breaking down the natural, organic, Human 1.0 nervous system and body to make way for the transhuman model that will easily navigate the Van Allen belt.

In the stratosphere, LEO (low-earth orbit) satellites beam radiation down on all of life in the name of "communications." When the public is informed about new versions of SpaceX satellites with laser communication capabilities being launched to provide "space-based internet services"—as with Starlink's Mini V2 satellites followed three weeks later by NASA's ILLUMA-T (Integrated Laser Communications Relay Demonstration Low Earth Orbit User Modem and Amplifier Terminal)[15] payload to the International Space Station 250 miles above us, orbiting the Earth every

*Power density is the amount of power per unit volume (W/m3).

†C5ISR stands for "command, control, communications, computers, combat, intelligence, surveillance, reconnaissance."

90 minutes—I cannot help but wonder what kind of "space-based internet services" are at hand. For example, two-way, end-to-end laser relay communications—could this have anything to do with coupling human minds with computers via laser (light) instead of radio frequency (maser)? Laser optical communications promises

- Increased bandwidth in unregulated portions of the spectrum
- Small beam size (PAT: pointing, acquisition, tracking)
- Hard to intercept, detect, or jam; low probability of interference
- Vast amounts of data (terabytes) at uncanny speeds*
- Ultra-secure networks for netcentric warfare (cognitive warfare) and covert operations

As NASA claimed in "Communicating via Long-Distance Lasers":

> An infrared laser can carry information in a similar way to radio waves. By modifying the invisible beam a certain way, the varying modulation can transmit a digital signal. Space is a perfect use for the technology, because there's no atmosphere or buildings to impede the beam's path, and compared to other communications standards, lasers offer a wide range of benefits. Light waves can support a high data rate and take less power to run.[16]

In the 1996 *New World Vistas*, a 14-volume U.S. Air Force study of future weapons development, we read in the fifteenth ancillary volume under "Biological Process Control":

> One can envision the development of electromagnetic energy sources, the output of which can be pulsed, shaped, and focused, that can couple with the human body in a fashion that will allow one to prevent voluntary muscular movements, control emotions (and thus actions), produce sleep, transmit suggestions, interfere with both short-term and long-term memories, produce an experience set and delete an experience set.[17]

*Invisible infrared packs data into tighter waves, which means more data can be transmitted in one downlink.

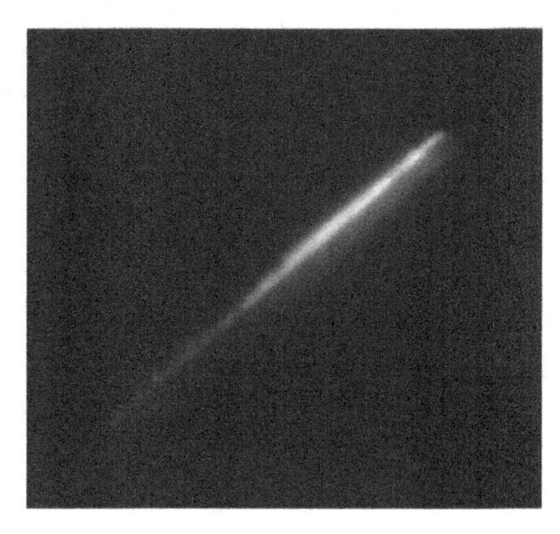

Fig. 7.1. A Star Wars laser "bullet."

Thirty years later, "space-based internet services" mark the arrival of the global transhumanist promise that we will all be living *on or in the internet* one way or another—in sentient world simulations (SWS),[18] as digital twins or *doppelgängers* in their own metaverse,* and as *wireless body area networks (WBANs) or human biofields.* Miniature optical beam steering from space means "free-space laser optical communications and light detection and ranging (LiDAR) for use in laser communications"[19] *with brains* via photoacoustic computerized tomography (PACT) instead of fMRIs (functional magnetic resonance imaging machines) which use radio waves and magnetic fields, whereas PACT shines a pulse of laser light into the head that is then picked up by ultrasonic sound waves that build an image of the brain.[20]

STAR WARS WEAPONRY

The claim regarding the Mach 5† offspring of the 1985 *Aurora* is that the top-secret hypersonic SR-75 *Penetrator* reconnaissance aircraft does

*A simple example is Google Maps. The U.S. Space Force has a digital twin of space. SpaceX has a digital twin of its Dragon capsule spacecraft.

†1 mach = 768 mph; the ratio of the speed of a body to the speed of sound in the surrounding medium. The existence of the *Aurora* has, of course, been denied, as has most projects labeled "top secret." By 2011, suborbital space shuttles were phased out, as were stratospheric passenger flights for the rich and famous ($20 million per flight)—or did it merely go black?[21]

not really exist.[22] But Boeing's Phantom Works orbital test vehicle, the X-37B—another stealth Star Wars weapon—is now declassified and circling the planet every ninety minutes (at 5 miles per second). Launched by either a carrier plane or orbital rocket like SpaceX's Falcon Heavy, the X-37B is essentially a phased array orbital HAARP kinetic weapon* that can precisely and rapidly scan or focus energy on the ionosphere, steer the jet streams (see the chapter 1 section "The Controlled Jet Streams"), create and steer hurricanes and tornadoes, and trigger earthquakes (land) or tsunamis (sea). Onboard is earthquake/tsunami-inducing tactical high-energy laser weapon technology, while far below on the ground, phased array antenna downlinks are everywhere. When a camera caught the destruction of a SpaceX rocket loaded with a Chinese satellite destined for orbit but still on the pad, the perpetrator was immediately recognized as being an X-37B, but since then has been covered up as a UFO.[24]

Before the X-37B, earthquakes and tsunamis were induced by beaming the earthquake frequency (2.5 Hz) into the ionosphere from the ground-based HAARP installation in Alaska or from the U.S. Navy sea-based version of HAARP, the SBX-1. The ionosphere then bounced the beam to a preselected longitude-latitude target on Earth's surface or on the ocean floor. With the X-37B in orbit, surprise earthquake-inducing radio frequencies can be beamed from directly over the target[25] (think Haiti, Japan, Iran, waters off the coast of Puerto Rico or California, Yellowstone National Park, or the New Madrid fault line).†

Then there are the high-powered microwave weapons that are also suborbital and armed to the teeth. In 1999, in Mumbai, India, the Bhabha Atomic Research Centre (BARC) assembled an electron-accelerating, rapid-fire beam weapon system appropriately named KALI-5000 (Kilo Ampere Linear Injector), Kali being the triple Hindu goddess who creates, preserves, and annihilates ("I am the dance of death that is behind all life, the ultimate horror, the ultimate ecstasy. I am existence, the dance of destruction that will end this world, the timeless void, the formless devouring mouth"). Blood sacrifices, skulls and cemeteries appease the dark goddess.

---

*"Rods from God" are another kinetic Star Wars weapon consisting of orbiting "battle stations" armed with tungsten steel rods.[23]

†An aircraft alert radar (AAR) can automatically shut down a HAARP X-Band transmission when aircraft are within or approaching facilities. Will this work with the X-37B?

KALI-5000 shoots thousands of microwave pulses, each 4-gigawatt burst lasting 60 billionths of a second, between 3 and 10 gigahertz. The cluster of high-powered microwave pulses travel in a straight line that does not dissipate. Unlike laser weapons that eat through metal, KALI's accelerated electrons make electronic systems go haywire—a "soft-kill" approach, you might say. KALI generates flash X-rays for ultra-photography to trace projectiles at lightning speed, and throws up electrostatic shields hardened to withstand thousands of volts per centimeter for light combat aircraft.

At present, every nation with the means is researching how to survive electromagnetic Star Wars and accustoming itself to puzzling out the truth from lies in the information warfare / cyberwarfare* we are now subject to, soldier or civilian. Though the saber-rattling of nuclear war is now and then resorted to in order to scare the public into conformity, nuclear war itself as in Hiroshima or Nagasaki has been politically abandoned. Harmonics mathematician Bruce Cathie (1930–2013) made a thorough study of how atomic bomb pilots had to drop their payloads at precalculated times and harmonically tuned places because the nuclear bomb is geometric and the electrons have to fly off at a tangent that depends on harmonic latitude, longitude, and angle of the Sun. Is this why nuclear submarines cruise from geometric point to geometric point? Due to that angle of the Sun, a hydrogen or plutonium bomb can only be ignited on two dates of any year. Some world leaders are aware of this these dates, some not, but public fear continues for political advantage. In his 1990 book *The Energy Grid* (pages 59–60), Cathie explains:

> . . . The bomb is, in short, a geometric device which can only be detonated in accordance with the unbreakable laws of geometry. The device is detonated by the manipulation of the relative motions of the atomic particles enclosed within its casing; and this can only be effected by placing the bomb on, under or over a specific geometric point related to

*According to Wikipedia, "*Information warfare* is the battlespace use and management of information and communication technology (ICT) in pursuit of a competitive advantage over an opponent. It is different from cyberwarfare that attacks computers, software, and command control systems. Information warfare is the manipulation of information trusted by a target without the target's awareness so that the target will make decisions against their interest but in the interest of the one conducting information warfare. As a result, it is not clear when information warfare begins, ends, and how strong or destructive it is."

> the earth's surface at a specific time. The relative motions of the earth and sun, at this instant of time, cause the disruption of the unstable particles of uranium, plutonium, cobalt or whatever unstable matter is used to trigger the explosion. Every test of nuclear devices since World War II has been designed to discover all the geometric combinations possible for the detonation of the atom . . . The computers of each nation concerned are fed with this information so that calculations can be made ahead as to where and when a nuclear device might be triggered. In this way the spiral of insecurity and distrust has increased its momentum until, at this present time, the world has become a madhouse with ourselves its inmates. Unless something is done, a system set up for pooling of all knowledge, we shall end up as have so many other civilizations, intent on annihilation of the race . . . The geometric nature of the bomb makes a nuclear war completely illogical—and perhaps this explains the present concentration of research in the United States, England, Switzerland and elsewhere into bacterial and chemical warfare techniques . . . An even more dangerous and complex game has now been initiated by the same group. Knowledge obtained from the bomb has opened a door to secrets of anti-gravity; if one group or another obtains supremacy in this field, they could, if they desired, rule the world.[26]

Cathie was right about "bacterial and chemical warfare." As we have learned since 2019, digitized synthetic biology is being developed and tested on global populations. As Todd Kuiken wrote at *Slate* in 2017, "It's possible, then, that DARPA's work is bending the entire field of synthetic biology toward military applications,"[27] which goes far to explain why radioactivity is being broadly used, given the scientific interest in "the effects of exposure to long-term, low-dose ionizing radiation" and "continuous environmental radiation exposure on a large-bodied mammalian species," as has been evidenced in the feral dogs that have been wandering and breeding since the disaster at Chernobyl, Ukraine, in 1986.[28]

- 1989, Galileo space probe, 49.25 pounds Pu-238 (half-life 87.75 years)
- 1990, Ulysses space probe, 25.6 pounds Pu-238 (half-life 87.75 years)
- 1997, Cassini space probe, 72.3 pounds Pu-238 (half-life 87.75 years)

- 2006, New Horizons space probe, first spacecraft to explore Pluto, 24 pounds Pu-238 (half-life 87.75 years)

Multiple space probes have employed plutonium-238 (Pu-238) for space travel rather than solar panels. NASA claims this is because plutonium-238 generates 745 watts of electricity to run instruments. But what about the greater risks of blowing up during launch or a space probe slingshotting around planet Earth in order to use its gravity for velocity, and the probe burns up in the atmosphere and disperses deadly plutonium? Or blows up and spreads it in space? NASA's *Perseverance* rover was apparently successful in landing on Mars on February 18, 2021,[29] but will that always be the case? More than twenty-five years ago, Karl Grossman wrote *Wrong Stuff: The Space Program's Nuclear Threat to Our Planet* (Common Courage Press, 1997), and it's still going on more than ever.[30]

An interplanetary shock wave that occurred on April 19, 2020, was said to be a "supersonic disturbance in the gaseous material of the solar wind usually delivered by coronal mass ejections (CMEs)," causing blue (nitrogen) auroras in the northern latitudes and red, yellow, and green (oxygen) in southern auroras over Tasmania.[31] Was the supersonic disturbance due to CMEs, or were ionospheric heaters cooking up an aurora borealis? And why nitrogen in the Northern Hemisphere and oxygen in the Southern, when both hemispheres have both?

WASTE, LEAKS, LOSS, AND "ACCIDENTS"

Peter A. Kirby's *Chemtrails Exposed: A New Manhattan Project* (second edition, 2020) maintains that high-tech geoengineering constitutes a second national security Manhattan Project mired in state secrets, untested before being loosed on the environment and populations, ignored future problems of waste and health, and sold to a public traumatized by World War II and 9/11. In both cases—and, one could argue, in the *third* "Manhattan Project" of 2020–2023 stealth biowarfare known as Covid-19—"public safety" was invoked as a rationale for secrecy and duplicity.

Finnish nuclear engineer Arto Lauri—former SUPO (Finnish CIA) operative and thirty-year nuclear plant worker—lives on the border of the Arctic Circle and therefore has a front-row seat when it comes to observing the atmosphere/ionosphere over HAARP (Alaska), KAIRA (Finland),

LOFAR (the Netherlands), EISCAT (Sweden, Norway), and SURA (Russia) ionospheric heaters. Lauri can read the colored chemical signatures of high-latitude clouds bloated by dry radiation pressure, not moisture, to determine the presence of nitrogen ionization (light blue), nitrogen tetraoxide (yellow), oxygen (pink), blasts of plutonium (red), and so forth.* Strong cloud boundaries indicate strong ionization. Also, he states that every reactor in the nuclear industry *purposely* leaks 3,000 kilograms of plutonium (Pu-239 ion gases, 24,131 half-life) per year. Why purposely? Lauri agrees with me: to weaken Human 1.0 and our organic planet.

In this same video at 6:00 is an FAA weathercam capture of what seems to be a massive bubbling planetary body or balloon much closer than the Moon. Lauri posits that the discolored, pockmarked sphere is not a planet, but a huge ionized plasma balloon of plutonium-239, perhaps the 3,000 kilograms per year jettisoned waste from nuclear power plants, secret deployments, nuclear "accidents."[33] "Parked" so as to rotate

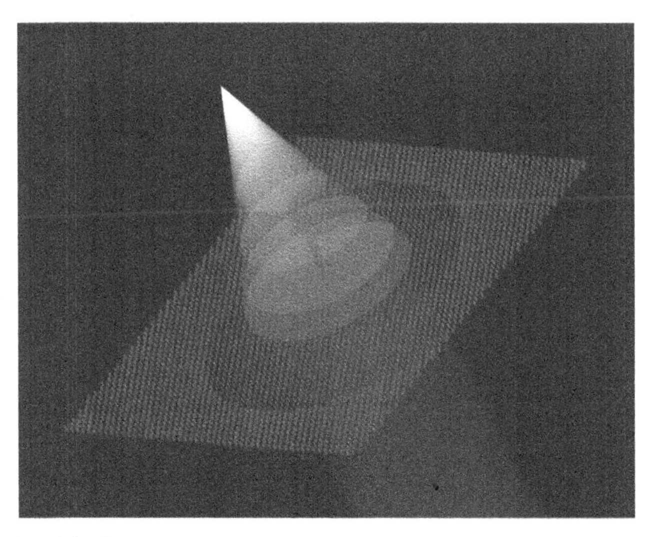

Fig. 7.2. Silicon nanopillars arranged in a hexagonal pattern create a "metasurface" that generates and focuses radially polarized light. Think solar simulator.
Amir Arbabi/Faraon Lab/Caltech

*In "Arto Lauri 251: Latest Chemtrail," Lauri describes a plasma cloud mass as iodine reddish brown, with a metallic sheen. In the 1960s, 1970s, and 1980s, chemicals like trimethyl aluminum, barium, and triethylborane respectively created red, green, and purple clouds over Wallops Flight Facility on the East Coast of the United States.[32]

with the Earth, the "balloon" blocks our natural Sun for as much as four hours daily in certain northern latitudes, which is why the solar simulator is used to hide this bizarre lethal weapon—capable of setting our nitrogen-oxygen atmosphere on fire!—helped along by internet gossip about Planet X, Nibiru, etc.

Lauri points out that the carefully crafted chemical compounds in chemtrails play a crucial role in reducing the risk of these radioactive balloons of plutonium exploding and setting our atmosphere on fire, the proof being that the EU-US Open Skies Agreement (March 30, 2007) makes NATO nations designate 15 percent of their budgets to help defray the cost of endless chemical trails.* (Chemtrails also obscure the overlapping lenses of the solar simulator.)

"The main danger from geoengineering is the radiation the spray blocks. DNA is unable to replicate at certain levels of UVC radiation. Such radiation, typically at around 245nm, is used to sterilize biology and destroy DNA's ability to replicate. That dangerous radiation is produced as sunlight shines through energetic plasma." —Richard Lawrence Norman,† email, July 27, 2020.

Under the rubric of "industrial tourism," the energy-conscious public is being wooed to think of "green"‡ and nuclear energy as "Disneyland for adults."[35] Back in 1999, I read a short *Wall Street Journal* article (no longer online) about a million French citizens relaxing in the warm waters used to

*If radiation coverup—more accurately, the ongoing experimentation with synergies between radiation and various chemicals—is the primary purpose of the chemical trails we've had for twenty-five years, the nuclear industry would have to drop from full capacity, if not close down entirely, if chemical trails were discontinued. The two deadly industries work in tandem.

†Besides being the founder and editor in chief of *Mind Magazine* and *The Black Watch: The Journal of Unconscious Psychology and Self-Psychoanalysis*, Richard Lawrence Norman is co-founder of Future Life Net, a group of advanced quantum physicists and artists who share a new human vision based on quantum physics.

‡By making carbons the enemy, the nuclear industry is viewed as "an affordable decarbonized energy system [zero-carbon generators]."[34] Wind and solar can never provide enough power in a timely manner, even with weather control.

cool reactors. As if they were vacationing at a spa, they flocked to nuclear, hydroelectric, tidal, and thermal power plants of state-owned Électricité de France, but especially to the uranium enrichment factory. In Prague, doctors recommended the radioactive waters of the Jáchymov former prison camp built into the side of an old uranium plant* for a "Hormesis effect," a therapeutic low dose of an otherwise toxic substance. The International Atomic Energy Agency "Radiation Protection and Safety of Radiation Sources: International Basic Safety Standards, No. GSR Part 3,"[37] proscribes *any* quantity of radioactivity, with nuclear scientists agreeing that radiation weakens cells for cancer takeover, with "accidents" at nuclear plants being more of the same.

And then there is the jettisoned waste of nuclear plant emissions released into the atmosphere from nuclear power plants all over the world. The half-life of a neutron† leak is seventeen minutes, after which the ionized gas will explode with the power of one million electron volts (eV). If the ventilation is cut off or does not occur, everyone in the nuclear plant dies. Nuclear emissions since the 1930s have increased fifteen-fold, but because the waste is blasted hundreds of kilometers into the atmosphere at the speed of light, radiation readings at nuclear plants remain low.

In the United States, nuclear waste, whether under Yucca Mountain or in a North St. Louis County landfill, is chemically bound in solid glass (silica) at 2,100 degrees Fahrenheit (1,149 degrees Celsius) in a process called *vitrification*. It is an expensive and dangerous problem given that the nuclear fission of just one uranium atom can destroy 20 million cells in a human body. The North St. Louis County nuclear waste problem goes all the way back to 1942, when Mallinckrodt Chemical Company was purifying tons of uranium for the Manhattan Project. Today, high levels of benzene and hydrogen sulfide are in the air over metro St. Louis, where an underground landfill fire continues to burn 8,700 tons of nuclear weapons

---

*Post-World War II, Jáchymov, Czech Republic, hosted prison camps whose inmates (life expectancy forty-two years) were forced to mine uranium until 1964. "The radioactive thermal springs arising in the former uranium mine are used under the supervision of doctors for the treatment of patients with nervous and rheumatic disorders. They make use of the constantly produced radioactive gas radon (^{222}Rn) dissolved in the water."[36]

†It takes one neutron of uranium-238 to produce plutonium-239. A *neutron* is a subatomic particle of about the same mass as a proton but without an electric charge, present in all atomic nuclei except those of ordinary hydrogen.

waste. A *dirty bomb*—a nondetonated release of radioactive particles—could occur anytime.[38]

Is our nitrogen-oxygen atmosphere in danger of being ignited by plutonium? Arto Lauri maintains that Finland's version of HAARP, known as KAIRA, drives ionized clouds loaded with radiation to chosen locations to sterilize designated populations. Once, he even saw "something" slip out of the Loviisa power plant near where he lives, then explode.

Besides intentionally expelled waste, *nuclear reactors leak at least three tons of radiation a year.*

Finland plays a powerful role in the international nuclear industry. It has the only active uranium mine in Europe;* Talvivaara mine is the only active uranium refinery in Europe;[39] and Posiva, a Finnish company, is the only deep-geological nuclear waste "final disposal" facility in Europe[40] since a 2005 law made European producers of uranium—namely, Finland—responsible for *all* nuclear waste. The big question is how much waste is being buried in "deep-geological" facilities, and how much is being blasted into the atmosphere?

Depleted uranium (DU) means uranium "depleted" to 70 percent of the original reactive uranium. During uranium enrichment, some of the dust is released into the atmosphere, but most is recycled for weaponry. When the Germans ran out of tungsten in 1943, they used depleted uranium for their armor-piercing munitions. *The half-life of depleted uranium is 4.5 billion years.* It is highly pyrophoric, meaning it ignites and burns hot like magnesium (3,092 degrees to 5,432 degrees Fahrenheit). Once vaporized, it condenses into nano-sized hollow spheres that float on the wind and the water and lodge in the lungs. Water will not put it out.† Roughly 8 percent of the Earth's total land mass (57.3 million square miles) was already severely contaminated with depleted uranium years ago.

> Since 1991, the United States has staged four wars using depleted uranium weaponry, which is illegal under all international treaties, conventions and agreements, as well as under the U.S. military law. The continued use of this illegal radioactive weaponry, which has already

---

*All of Lapland is reserved for uranium mining.

†The U.S. Navy's anticruise missile Phalanx pumps out 60 to 120 shot burst rounds of uranium-238 at 2,000 per minute.

> contaminated vast regions with low-level radiation and will contaminate other parts of the world over time, is a serious international issue.[41]

Some 300 tons of depleted uranium were used in the first Gulf War in 1991 after which 200,000 American GIs returned home, ill down to their genes with Gulf War Syndrome, their lungs riddled with heavy-metal nanoparticles from depleted uranium debris, their families and even their pets ill as well due to radioactive "shedding." Despite the cost to humanity and in defiance of all standards of ethics, let alone legality, depleted uranium is now common in conventional warfare. Some 2,000 tons of it were used during Operation Desert Fox, the four-day bombing campaign in Iraq in 1998. Doctors in that country have since reported sky-high rates of cancer, leukemia, and birth defects. As General Leslie Groves, who oversaw the construction of the Pentagon and directed the Manhattan Project, put it in 1943, "One millionth of a gram accumulating in a person's body would be fatal."[42]

In the twenty-year American war in Afghanistan (2001–2021), nondepleted uranium, a processed form of pure uranium more toxic than depleted uranium, was used. As its metallurgy burns through concrete and steel, depleted uranium and nondepleted uranium bombs such as the GBU-28 "bunker busters" are converted to micron-size particles that sicken and kill and cause genetic mutations to successive generations exposed to this weaponry. The newest weapon in the nuclear arsenal, used by the Israeli army against Palestinians in Gaza beginning in 2006, is:

> Dense Inert Metal Explosives, or DIME, a so-called LCD ("low collateral damage") weapon developed by the United States Air Force. DIME bombs blast a superheated "micro-shrapnel" of powdered heavy metal tungsten alloy (HMTA). Studies indicate that HMTA embedded in the body disrupts biochemistry and rapidly causes cancer. Like depleted uranium (DU), HMTA is genotoxic—it is capable of inflicting genetic mutations. . . . DIME is part of the U.S. Air Force's Focused Lethality Munitions (FLM) program, which is expected to "allow" the targeting of "terrorists" wherever they are, even in places "previously off limits to the warfighter." The ideal of FLM is to reliably kill every human within the blast zone—one way or another. It is "total war" on a 50-foot circle, within which deaths are not admitted as collateral, but

> purchased as insurance. Israel's new weapon "slices" off its victims' legs, leaving "signs of heat and burns near the point of the amputation." It's "as if a saw was used to cut through the bone," according to Dr. Habas al-Wahid, head of the ER at Gaza's Shuhada al-Aqsa hospital.[43]

Not only do governments continue to have "accidents," but they also continue to "lose" nuclear elements—enough "material unaccounted for" (MUF) to bomb Nagasaki eight hundred times over. For example, the March 2017 theft of two disks of plutonium and cesium were supposedly stolen from a Ford Expedition owned by employees of the Department of Energy, who were staying in a Marriott Hotel in San Antonio, Texas. An oversight? In all, six tons of radioactive materials—gone.[44]

Three Mile Island (March 28, 1979, latitude 40.15° N/longitude 76.72° E), Chernobyl (April 25, 1986, latitude 51.27° N/longitude 30.22° E), and Fukushima (March 11, 2011, latitude 37.38° N/longitude 140.47° E)—all "accidents," all in the open air, all in the spring, and all at relatively close latitudes. Was radiation being tested per the Sun's UV radiation and galactic cosmic rays?* The Three Mile Island "accident" ten miles south of the Pennsylvania state capital of Harrisburg yielded a thirteen-year study of 32,000 people, but the data was skewed to reassure the public that there was absolutely no connection whatsoever to the cancer deaths of residents living five miles from the reactor. Just before Three Mile Island, the results of a fourteen-year study of the 1951–1962 aboveground nuclear tests in the Nevada desert were released, *but only after a fifteen-year delay.* Farm Belt exposure to iodine-131 showed up in 11,300 to 212,000 cases of thyroid cancer. Many had drunk fallout milk in childhood.[45]

The nuclear destruction at Chernobyl in Ukraine occurred during Western Christianity's Easter, April 27–30, 1986. I mention this synchronicity only because the radioactive death cloud was the equivalent of a hundred Hiroshimas and Nagasakis, and *Chernobyl* means "wormwood," which references both the helpful herb *Artemisia absinthium,* known to kill cancer cells, and the fallen star in Revelations 8:10–11: "And the third angel sounded, and there fell a great star from heaven, burning as it were a lamp, and it fell upon the third part of the rivers, and upon the fountains of waters; And the name of the star is called Wormwood: and the

---

*Tesla was convinced that radium could transform cosmic ray energy.

third part of the waters became wormwood; and many men died of the waters, because they were made bitter." Meanwhile, thirty-five years after the disaster, neutron counts are rising at Chernobyl like "embers in a barbeque pit," signaling possible fission.[46]

The Chernobyl tragedy occurred just as Mikhail Gorbachev's campaign of *glasnost* and *perestroika* was winding down the Cold War and what had been done in Russia, East Germany, and Eastern Europe under atheistic Bolshevism.* The nuclear fallout contaminated all of Eastern Europe. Vegetation in hot, radioactive soil is now returning, but has been transformed by plutonium, cesium-137, strontium, and radioactive iodine. The river flowing past the reactor and feeding into the Dnieper River, which in turn feeds into the Black Sea, is the watershed for nine million Ukrainians. It was not until the end of 2000 that Reactor No. 3 was finally shut down.

The 2011 Fukushima nuclear reactor disaster that released iodine-131, strontium-90, cesium-134, and cesium-137 took place twenty-five years after Chernobyl and one year after the Deepwater Horizon oil spill in the Gulf of Mexico. Fukushima was about a long-term study of the impact of radiation on the North Pacific Gyre,† whose debris plume was projected to reach the U.S. West Coast by 2014. Radiation opens Human 1.0 cells to greater vulnerability and triggers the immune system to go into overdrive, which ultimately ends in exhaustion. Releasing reactor radiation into an atmosphere already altered by geoengineering weakens the immune system in preparation for genetic alteration, sterilization, mind control, and consciousness attacks

*Rather than blame the Russians for Sovietization, it is more accurate to blame what we now call globalists (in those days, "internationalists"). An extraordinary book written by a Swede in 1998 sets the record straight: *Under the Sign of the Scorpion: The Rise and Fall of the Soviet Empire* (Stockholm: Referent Publishing, 1998) by Juri Lima: "What happened in February (March) 1917 was not a revolution, but a coup d'etat organized from without. The Bolsheviks themselves, however, did not carry out a coup d'etat in October (November) 1917, as we have learned in the West, but simply took over power. It was an internationally controlled conspiracy. . . It was also the Germans who put down a revolt among the cadets at an army training school in Petrograd, captured the Kremlin for the Bolsheviks in Moscow, fought back Krasnov's Cossacks and performed other similar actions vital to the survival of the Reds. General Kirbach promised that Moscow and Petrograd could be occupied by German troops if the Bolshevik government was threatened. The weak Soviet regime was protected by up to 280,000 disciplined German soldiers."

†A gyre is a giant circular oceanic surface current.

in the form of direct assaults on the etheric bodies of human beings.* Why else would the total gamma radiation in the United States, as confirmed by the EPA's gamma radiation monitors, register 41.832 billion counts per minute (CPM)† between January 1, 2010 and February 29, 2020?

What happens when ionized radiation is released on such a global scale? First of all, charged particles are trapped by Earth's magnetic field, rotating around the magnetic lines of force thousands of times per second, producing a mirroring spiral that makes sure the ions and electrons remain in the atmosphere to undergo a slow drift around the Earth's magnetic axis.[48] The truth is *it has all been intentional*—the "experiments," the waste, the leaks, the loss, the "accidents"—all of it, and it's been about atmospheric electron-cyclotron resonance heating that increases charged particle density. It's right there in the language of Bernard Eastlund's 1987 HAARP patent.

In the 1990s I was contacted by a targeted individual named Clare Louisa Wehrle. She wrote that she had been harassed and gang stalked / gaslighted since her PhD thesis on radiation at Stanford University in the 1970s. Her then-boyfriend, who worked for the CIA, was murdered, after which Clare was used for a host of radiation experiments. She writes,

> The scope of the effects which can be caused by these substances is much wider than mere "harassment." A major question in the early MK-Ultra research was the need for a method of propulsion of substances such as chemicals, microbes and odors through the air to a target. In the past few decades, they have found such a propellant in nuclear energy supplied by radioactive substances and radiation-emitting chips.‡

---

*See the article "The Three Occult Purposes of the Fukushima Operation," on the Montalk.net website. Since 1945, there have been 2,053 atomic explosions. Radiation prompts the manifestation of the Asuras, the demons of Hinduism and Buddhism, which are particularly useful to shadow technologies. The formation of noble gases allows invisible antihuman entities to materialize. Since 2011, consciousness has been attacked by increasing waves of radioactivity.

†Gamma radiation is penetrating electromagnetic radiation arising from the radioactive decay of atomic nuclei. Counts per minute (CPM) is for Geiger counters, the ionization rate in a given quantity of decaying radioactive material.[47]

‡Clare Louisa Wehrle wrote "The Use of Synthetic Body Odor for Mind Control," a typewritten account of "harassment substances" she was being subjected to while on the run. She was finally killed by a car driven by Raymond Peters of St. Petersburg, Florida, on January 6, 2006. All references to these events have been wiped from the internet.[49]

Over the years since Clare's certain murder, I have learned through other targeted individuals that nuclear aerosol agents make good invisible delivery systems for nanosized lethal drugs or implants. Clare grasped what was up decades ago as she wrote in her typewritten "Cybernetic degradation" (November 2003):

> High levels of radiation can break bone, metal, move objects, or destroy brain or body tissue . . . Furthermore, the CIA styles the reduction of IQ as an almost trivial effect, even claiming it will "feminize" women. But adaptation to the loss of IQ would be almost impossible. Were the wealth of the mind to be lost, one would be in an alien land, a mental desert. The destruction of the mind's wealth is actually a goal of some CIA experiments. Some experiments even seek to damage basic skills which exist in animals, such as spatial orientation, transmogrifying the subject to a sub-animal existence, as if by sorcery. The extinguishing of the light of the mind can actually be cruelly sensed by voyeuristic mind-reading technology—the technology used to enhance sadistic pleasure in the crude, low-tech murder of the physical vehicle, as well.

As if by sorcery . . .

The channels between the human brain and external electronic devices are being purposely opened by high concentrations of radiation. Unbelievably, even human waste containing radioactive isotopes is being "mined" from sewage for nuclear energy production by means of the nanosized sensors now ubiquitous in the environment and in our bodies and brains in quest of endless data to be reported back to supercomputers and artificial intelligence, while nanochips and MEMS measure pressure, temperature, acceleration, magnetic fields, radio frequency signals, GMO pharming, "precision medicine" drugs in vivo, computer monitors, video games and TV, geoengineered weather, even UFOs.

During the March 28–29, 2011, Carnegie International Nuclear Policy Conference—seventeen days after the Fukushima disaster on March 11—Gregory B. Jaczko, the former chairman of the U.S. Nuclear Regulatory Commission, recommended that all 104 U.S. nuclear reactors be taken out of commission and replaced, and that additional twenty-year extensions be stopped. Did it happen? Of course not. The United States is the world's biggest producer of nuclear power. Carbon sequestering was never the real goal of the 2015 United Nations Climate Change Conference in Paris. Carbons were "shaped" into a problem so that the predecided "solution" ("affordable decarbonized energy system") would send the public not toward actual renewable energy, but straight into the waiting arms of the nuclear industry.

8
Bios

The matter and energy in our vicinity will become infused with the intelligence, knowledge, creativity, beauty, and emotional intelligence (the ability to love, for example) of our human-machine civilization. Our civilization will expand outward, turning all the dumb matter and energy we encounter into sublimely intelligent—transcendent—matter and energy. So in a sense, we can say that the Singularity will ultimately infuse the world with spirit.

Ray Kurzweil, *The Singularity Is Near: When Humans Transcend Biology*

We began *The Geoengineered Transhuman* with a chapter on nanoparticles (NPs) and nanotechnology and how cosmic nanoparticles like dusty plasma have been copied, retro-engineered, and weaponized into nanotechnologies like smart dust, MEMS (*microelectromechanical system*), GEMS (global environmental MEMS sensors), NEMS (nanoelectromechanical systems), etc. It is now time to take a look at βίος, Greek for *bios* or living microbes that have no doubt played a cosmic role from long before being thought of as part of "life on Earth." Synthetic biology is now being forced to redefine life on Earth and therefore a recategorization of life that will include an artificial *metaverse** subject to algorithms created by human

*According to Wikipedia, metaverse: virtual worlds in which users represented by avatars interact, usually in 3D. The term originated in the 1992 science fiction novel *Snow Crash* as a portmanteau of "meta" and "universe."

scientists as well as supercomputers and quantum computers* instead of Nature.†

Our highly technological era has seamlessly followed fast upon the industrial and postindustrial eras, and Human 1.0 health has become more and more fragile and fleeting as agrobusiness genetically alters the mass food supply and diseases like cancer and "long covid" (aka radiation flu) are churned out of biowarfare labs while Big Pharma alchemists whip up the latest injection for "national security" purposes not shared with the public, and American Medical Association (AMA)-obedient doctors are paid off to push whatever drug they're told to push (or their license to practice medicine may be revoked). Generations of parents since microbiologist Louis Pasteur (1822–1895) have been conditioned to think that their newborn must be vaccinated and doctor "wellness" visits faithfully attended for yet another childhood "vaccine."[2] As Human 1.0 ages, specialists and an army of prescription drugs are sought out and obeyed so as to save oneself from cancer and death. People have become docile before the phalanx of MD and medical PhD "experts," obediently lining up as guinea pigs for experimental inoculations, "medicines," jabs, pills, treatments, and surgeries—oblivious to the jet aerosols that have been chemically and genetically prepping them for medical tyranny for decades.

Under *Bios*, we'll take a look at the Earth's ancient survival kit of *water*, the cosmic substance of life itself; *bacteria*, accused since Pasteur of causing disease while being exploited for building gain-of-function bioweaponized microbes; *Mycoplasma*, the ancient anaerobic bacterium preferred

*While a conventional computer works on binary, quantum computers rely on a unit of information called qubits (subatomic particles such as electrons or photons) with far greater processing power. The qubits only work in a controlled quantum state–under sub-zero temperature or in ultra-high-vacuum chambers. Quantum computing is predicated on two phenomena: superposition and quantum entanglement. Superposition is the ability of qubits to be in different states simultaneously, allowing them to work on a million computations at the same time. However, qubits are sensitive to their environment, so they can't maintain their state for long periods. As a result, quantum computers can't be used to store information long term. Quantum entanglement is the ability of two or more quantum systems to become entangled irrespective of how far apart they are. Thanks to the correlation between the entangled qubits, gauging the state of one qubit gives information about the other qubit, which accelerates the processing speed of quantum computers.[1]

†In 1980, Diamond v. Chakrabarty established the patentability of human-made life-forms, except those encompassing the human organism.

by the military; *biofilm*, bacteria's pseudo-skin survival tactic; and *soil*, the living Earth's foundation of agriculture—seed and crops, livestock, and forestry—now being transitioned from its bedrock of five types of natural worker microbes (bacteria, actinomycetes, fungi, protozoa, and nematodes) to a planetary agrobusiness of conjured genetics, chemistry, and electromagnetics so as to transform what the Transhumanism guru Ray Kurzweil calls, in the quote above, "dumb matter and energy" into the metaverse.

WATER

> *And God made the firmament, and divided the waters which were under the firmament from the waters which were above the firmament: and it was so.*
>
> Genesis 1:7–9

The human body consists of 70 percent water. Water as an energy source connotes tremendous power, whether it's a geoengineered storm or a tidal wave or a flood or in how rivers are harnessed to dams. And water is also a critically important source of energy in the human body, as it regulates metabolic function, body temperature, hydration, and the absorption of nutrients.

Harnessing the power of water goes back to ancient times, when early forms of hydraulic energy were used: aqueducts and water wheels, then dams and steam power. Fast-forward, and we are now hearing that Turkey is harnessing the power of ocean waves to build the world's largest tidal power station,[3] while on the other side of the world, the Fukushima Daiichi Nuclear Power Plant will be releasing more than a million tons of radiated water (mostly tritium) into the Pacific Ocean.[4]

But harnessing water for electricity such as with dams and "wave power" is not why water itself is so powerful. It is because water is *the* magical transducer of all of life, primarily due to its dielectric constant or ability to insulate. While water may be a poor conductor, it can store an electrical charge and act as a solvent that can dissolve ionic compounds. The water in our bodies, oceans, rivers, rainfall, and geoengineered storms of rain and ice acts like a sponge when it comes to electromagnetic radiation. Once irradiated, water in the environment or in our bodies turns acidic, due to an increase in the concentration of hydrogen ions; as in the oceans, the pH scale of our blood is inverse to hydrogen ion concentration: the more hydrogen ions, the higher acidity and lower pH.

We have heard for decades how pollutants acidify rain, aquifers, the oceans, and the atmosphere, but we've heard precious little about how the acidity of pollutants, coupled with electromagnetism, affects the water in our bodies for profound health consequences. Health effects of lower pH and higher acidity for greater conductivity is of little concern to the military-industrial-intelligence phalanx, which is obsessively focused on increasing Space Fence (smart grid) electromagnetism around the Earth and in human bodies for the greater and greater levels of conductivity needed for full-spectrum dominance and control over cyborg Human 2.0.

Water pollution with a *terraforming* (transforming the Earth) agenda on a grand scale occurred with the "accident" known as the 2010 Deepwater Horizon oil rig catastrophe in the Gulf of Mexico. Not only was it no "accident," but it was a Big Pharma / geoengineering operation to test the environmental and biological effects of three million liters (800,000 gallons) of the oil dispersant Corexit via British Petroleum and Synthetic Genomics Inc. The synthetic bacterium *Mycoplasma laboratorium* (aka Synthia) was added to the Gulf waters to purportedly gobble up the oil spill, but it actually self-replicated so much that it spread what locals call "BP Syndrome." Gulf residents and disaster workers reported blood in the urine, heart palpitations, kidney and liver damage, migraines, multiple chemical sensitivities, neurological damage resulting in memory loss, rapid weight loss, respiratory and nervous system damage, seizures, skin irritation, burning and lesions, and temporary paralysis—all symptoms of neurotoxicity and an immune system under attack. Lung damage, depression, and anxiety plagued the 50,000 people directly involved in the spill's cleanup[5] as the genetically modified strain *Vibrio vulnificus*—a species of pathogenic bacteria of the genus *Vibrio* present in estuaries, brackish ponds, or coastal areas—entered wounds and cuts, lodging between muscle and skin, where it released a toxin that destroyed tissue, and in some instances led to amputated limbs. This environmental disaster continues to this day; the Gulf will never be what it once was.[6]

Wells both private and public, watersheds, and aquifers are now acidified. In the past decade, 85,000 fracking wells* from Big Oil's "shale

*By 2040, the projected figure for fracking wells is 675,000.

revolution" of horizontal drilling and hydraulic fracturing have gone through 72 trillion gallons of water and 360 billion gallons of chemicals that are suspiciously similar to the chemicals being delivered in atmospheric aerosols. Human and animal pharmaceuticals end up in public water supplies and watersheds, and so do the residues of medicines for pain, infection, high cholesterol, asthma, epilepsy, mental illness, heart problems, antianxiety, and hormones. Then there is PVC (polyvinyl chloride), dry-cleaning solvents, pesticides like Syngenta's paraquat (whose use doubled between 2006 and 2016, according to the National Water Quality Assessment Project), bleach in the manufacture of paper, chlorine gas, chlorine-based DDT, dioxin, and PCBs—all of which find their way into our water. And lest you think drinking bottled water is safe, corporations that bottle water are not required by the government to test for pharmaceuticals,[7] and the same goes for home filtration and sewage treatment systems—reverse osmosis can help—and then there's industrial agriculture and all of its synthetics.

> Cattle, for example, are given ear implants that provide a slow release of trenbolone, an anabolic steroid used by some bodybuilders, which causes cattle to bulk up. But not all the trenbolone circulating in a steer is metabolized. Water sampled downstream of a Nebraska feedlot had steroid levels four times as high as the water taken upstream. Male fathead minnows living in that downstream area had low testosterone levels and small heads. . . . Our bodies may shrug off a relatively big one-time dose, yet suffer from a smaller amount delivered continuously over a half century.[8]

In 2014, the city of Flint, Michigan, switched its drinking water supply from Detroit's system to the Flint River in a cost-saving move, without adequate treatment and testing of the river's water, which for decades had been used as an industrial waste dump. The water, contaminated with lead and other heavy metals and chemical contaminants as well as *Legionella* bacteria, caused skin rashes, hair loss, and all the other nasty effects of lead poisoning, as well as an outbreak of Legionnaires' disease (i.e., atypical pneumonia). Dow Chemical's chlorine/chloroform had been added to the city's water treatment since 1908, despite the clear evidence of cancer, premature senility, heart attacks, sexual impotency, stroke, central nervous system depression, liver and kidney damage, and immune system suppression—all associated with chlorine.

Is the loss of natural water worth the cost of convenience and increasing new diseases that are actually symptoms of the loss of biological balance and the immune system? Is the loss of our biological balance worth endless symptoms of one "disease" after another? And what about the loss of IQ and integrity of the nervous system, thanks to the neurotoxin fluoride in the public water supply since 1945? It has taken all of these decades since the Manhattan Project (1942–1946) that required vast quantities of (classified) fluoride for building atomic bombs for the 2017 Toxic Substances Control Act (TSCA) fluoride lawsuit against the U.S. Environmental Protection Agency (EPA) to go to court, even after the May 4, 2023, National Toxicology Program (NTP) report that proved that fluoride is a neurotoxin.[9] But after seven years and much blood, sweat, and tears, a U.S. federal court has now deemed fluoridation an "unreasonable risk" to the health of children, and the EPA will be forced to regulate it as such.*

The water molecules in our bodies and brains resonate to frequencies other than our own. The brain (80 percent water) can be entrained at any distance by any biphasic waveform (i.e., a current that goes out from a source and returns to that source) that can penetrate the skull, such as *photobiomodulation (PBM)* by near-infrared (NIR) light, lasers, or light-emitting diodes (LEDs)—even by satellite. In this way the blood sugar and iron in the blood can be entrained. Target a crystal molecule with a wave resonating at the crystal's own frequency, and it will explode. If it's a sugar crystal, the release charge, called *triboluminescence*, can damage the brain or at the very least produce confusion, memory loss, dizziness, and apathy.

According to Gerald H. Pollack,[10] a professor of bioengineering at the University of Washington, water has a fourth phase beyond solid, liquid, and vapor (gas)—namely, a hexagonal geometry able to draw electric current from sunlight.† Nuclear engineer Tom Bearden (1930–2022) explains how the amazing sensitivity of water to everything in its surroundings is due to its bond-structuring capability, meaning the attraction between atoms or ions that enables the formation of molecules, crystals, etc. This is

*See the Fluoride Alert website's "EPA Lawsuit" page for the full story.

†Is the "current" drawn from the Sun electric, or etheric?

what allows water to take on the mood of whoever is in the room, or even if an observer blinks an eye, thereby affecting water's crystalline structure.

> . . . The hydrogen bonding structure of water is enormously complex and richly varying. Bond-structuring of water constitutes a special kind of "neo-Whittaker" substructure inside a special kind of potential for that particular body of water. A glass of water, for example, has an overall neo-potential comprised of its hydrogen bond structuring. That water will change its internal bonding structure if you enter the room, or if you blink your eye while observing it. It continually adjusts to everything in its surroundings. The reason is, everything in its environment has charges, and clumps or orderings or structures of potential . . . The internal dynamics of the water receives inputs from the surroundings this way, and the water's bonding structure changes accordingly. We've only known the complexity and richness of this water structuring for less than two decades, and so far as I know, no one else seems to be considering the Whittaker infolded EM wave structure aspects of it.
>
> The point is this: In the fluid inside the head of a functional hydrocephalic, the water structuring is quite sufficient to provide the rest of the needed "way-station two-way tuner, processor, and transmitter-receiver." The reason is quite simple: The potential of the fluid constitutes a partial potential in the overall bio-potential of the body. It's like the pressure of a mixture of gases: the overall pressure consists of the partial pressures of the component gases. The Whittaker structures ensure intermingling and intercommunication through the internal energy channels of the total bio-potential to all its constituents. Therefore, the water structuring of the fluid in the head of the hydrocephalic serves—bridgewise—as a substitute brain.[11]

Masaru Emoto's (1943–2014) bestselling book *The Hidden Messages in Water* popularized "water messages" in photographs of water crystals responding to human thoughtforms such as "hate" and "love." In the Whittaker in-folded electromagnetic wave structure, all potentials overlap and have multidimensional properties, which is why water can transduce cross currents and serve as a substitute brain bridge, as in functional hydrocephalics and people who have lost most of their physical brain, but can still think.

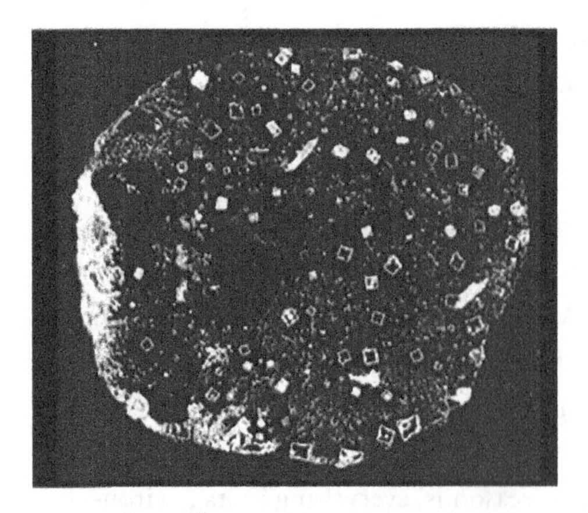

Fig. 8.1. A single drop of rainwater from rural Queensland Australia. The water had been collected off a roof for drinking purposes and this sample was collected from a drinking bottle. It had not rained for the six weeks prior to collection but there had been many unusual cloud formations in that time.
David Nixon, MD, Nixonlab, Darkfield approx. 25x magnification

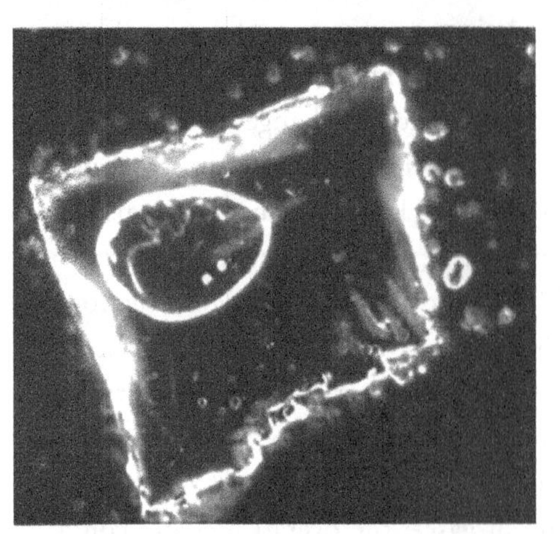

Fig. 8.2. Complex crystal structure that has formed from a drop of rainwater after a period of aerial spraying. This crystal can be seen at the bottom of the circular image above.
David Nixon, MD, Nixonlab, Darkfield approx. 200x magnification

Fig. 8.3. Close up of rain drop showing complex deposition of material and crystal formation.
David Nixon, MD, Nixonlab, Darkfield approx. 40x magnification

For millennia, Indigenous shamans have referred to water as a "road," considering deep lakes and waterholes as access points to a quantum level of reality that is parallel to ours, with rivers and streams acting as transmitters by which human consciousness can travel to places both on the Earth's surface and in dimensions below, as in the Maya underworld of Xibalba, which is strongly associated with water, its nine levels requiring entry through a portal of water and involving crossing dangerous waters. Water's reflective "mirroring" nature (think of the Greek tale of Narcissus) can put the shaman in touch with subterranean beings, the shiny surfaces of deep waterholes and lakes and cenotes serving as a window through which entities of other dimensions can be encountered.

Metaphysics, physics, and Indigenous realities intermingle at quantum junctures . . .

Because water molecule electrons are made to rotate at 2.4 gigahertz by the frequency of a microwave oven or a 4G iPhone, thus forcing a dielectric loss of water, the advent of "the new WiGig"[12] in 2004 initiated a massive assault on our bodies via our own water content, as well as an assault on our oxygen by allowing Wi-Fi devices to access the uncongested 60 gigahertz water frequency band for data transmission at multi-gigabit-per-second speeds.

In 2021, the British company Evove (formerly G2O Water Technologies) introduced graphene oxide as a "hollow fiber membrane"—i.e., a graphene nanotube—water filtration system for the industrial laundry sector, which puts over four million cubic meters of hot wastewater down the drain every year. Evove boasts that their technology allows this hot wastewater to be "cleaned" using a ceramic membrane coated with graphene. And so it is that graphene oxide—the very same conductive substance in Covid-19 injections and masks, the very same graphene that is a proven toxin[13]—is ending up in clothes and bedding and towels in the name of cost savings, not to mention that graphene-treated water then enters the public water supply.

NATURE'S NANOPARTICLES

Here on Earth, Nature provides us with a range of fine particles, from inorganic ash, soot, sulfur, and mineral particles found in the air or in wells, to sulfur and selenium nanoparticles produced by bacteria and yeasts (microscopic fungi). As NASA says, "The human body contains ten times more microbes than human cells, and bacteria and fungi grow in and on just about everything around us on Earth."[14] Our human immune system is familiar with these natural nanoparticles. What it is not familiar with are the nanotechnologies fashioned out of natural nanoparticles, bacteria being the favored stock and trade of, for example, gain-of-function transformation of what is *Bios* (of Nature) into what is synthetic biology (*synbio)* of the metaverse.

And yet it is true that advances in synthetic biology and genome engineering of DNA and RNA have made it easy to create synthetic genomes out of both bacteria and yeasts, to the extent of confusion between what is natural and what is synthetic, which serves the advent of the transhumanist agenda quite well. In a real sense, reverse engineering natural microbes / microorganisms is how synthetic biology begins—from genetic materials to countering the toxic chemicals necessary for synthesizing and developing natural and synthetic nanoparticles (NPs) into nanotechnologies. For example, natural fungi filaments that have been supporting life on land for at least 450 million years act as living templates or "biological slaves" synthesizing metal NPs and building nanomaterials; with gold NPs attached to their surface, they can develop different sizes and materials of additional NP layers.[15]

More concerning to human beings is how smart nanoparticles are engineered. DARPA's Neural Engineering System Design (NESD) begins with bits of natural genetic materials ("virus") and genetic code design, then builds an artificial genome, then reintroduces it in vivo. Here is Dr. Edwin Donath, professor at the Institute of Medical Physics and Biophysics at the University of Leipzig in Germany:

> Genetic engineering has a clear analogy to software production in that pieces of code can be multiplied at low cost, put together in an artificial genome, and processed afterwards in the cellular machinery . . . Recent developments in microfluidics ["wet computing" hydrogel?] will certainly contribute to the parallel production of surface-engineered viruses at low cost and with a great diversity of functions.[16]

Bacteria and genetic materials can be encapsulated in the electrospun polymer nanofibers[17] (electrospun referring to electrostatic fields) being delivered by chemical trails to be breathed in and buried in agrobusiness and organic soils, often with Morgellons payloads.*

BACTERIA

> The bacteria that compose our microbiome work so synergistically with our human cells that the difference between "us" and "the bacteria" is difficult to decipher. Where do "we" begin and "they" end? If all the bacteria in a person's microbiome were killed off, that person would die. Bacteria are an intimate and important part of "us." In genetically modifying "them," are we genetically modifying "us"?[18]

Bacteria are an important part of the planetary life cycle. Not only do they recycle nutrients in the atmosphere and on Earth but they are found six miles above the surface of the planet surrounding the Earth[19] where a layer of DNA-based life makes our atmosphere breathable and constantly creates an ozone layer to protect our genetics from too much ultraviolet light (UV) and mutagenic cosmic rays. There are even anaerobic bacteria living a half mile down on the ocean floor. Falsely, microbes like bacteria are portrayed as dirty little aggressive "germs" when the truth is that they are building blocks of life itself. Billions of bacteria live in our gut, along with other microbiome organisms that symbiotically help us digest food, extract nutrients, and produce multiple neurochemicals, such as 95 percent of the body's serotonin.

Among bacteria, we find *eukaryotes* (with a nucleus), *prokaryotes* (no nucleus but some genetic materials), and most recently a third category, the *archaebacteria*, which may actually be the oldest. Not only is the bacterium the oldest life form on earth (3.5 billion years), but it appears that cell parts like mitochondria and plant chloroplasts actually evolved from an ancient anaerobic bacterium now known as *Mycoplasma*, the tiniest of all natural microorganisms—no DNA or RNA, no cell wall. (Loss of its cell wall led to it being able to engulf other microorganisms and becoming the eukaryotic cell.)

---

*The Morgellons polymer fibers are nanotubes encapsulating self-replicating DNA nanotech. Morgellons was a precursor to more sophisticated nanotechnologies now in the blood of both Covid-19 "vaxxed" and "unvaxxed."

Bacteria are the stars of biowarfare labs, especially the bits of DNA that serve as nucleus for the *Brucella* bacterium* for the pathogen *Mycoplasma fermentans* that enters individual cells. If it enters the brain, then the victim may develop a neurological disease; if in the lower bowel, it may be Crohn's disease.

> Because it is only the DNA particle of the bacterium, it doesn't have any organelles [subcellular structure with specific jobs in the cell] to process its own nutrients, so it grows by up-taking pre-formed sterols [like cholesterol, fermented foods] from its host cell, and it literally kills the cell . . .[20]

The fact is that the *Mycoplasma fermentans* extracted from the nucleus of the *Brucella* bacterium—basically an already a mutated form of *Brucella*—is a favorite *patented* military carrier of pathogens[21] that has laid the groundwork for "neuro/systemic degenerative diseases" like AIDS, chronic fatigue, fibromyalgia, Crohn's, Type 1 diabetes, multiple sclerosis, Parkinson's, Wegener's, collagen-vascular rheumatoid arthritis and Alzheimer's, etc. The deaths of cells and undermining of the immune system are certainly the on-ramp to chronic autoimmune symptoms, but especially since the World Health Organization's AIDS in the Reagan-Bush-Cheney 1980s, the so-called friendly fire *Mycoplasma* Gulf War Syndrome (1990–1991), and finally the 1990s advent of endless chemical trails in the sky piggybacked with synthetic biology mutants and "vaccines."† Thus, it is not difficult to believe that everyone in North America is a carrier of weaponized "novel" *Mycoplasma*, as was indicated by the chief virologist for Merck Sharp & Dohme Maurice Hilleman (1919–2005),[22] who was praised by Anthony Fauci, MD, director of National Institute of Allergy and Infectious Diseases (NIAID). Fauci pushed the three-year Covid drama, then disappeared from public life.

The Gulf War had multiple agendas (the least of which was Saddam

*Was the visna (withering) virus DNA in the nucleus of the *Brucella* bacterium a gain-of-function creation as well? So many lies, so little truth, when it comes to virology.

†Aerosols should include mosquito sprays, such as a report from the *New England Journal of Medicine* (August 22, 1957, p. 362) about one of the first outbreaks of chronic fatigue syndrome in Punta Gorda, Florida, in 1957 and the strange coincidence of a "huge influx of mosquitoes" the week before.

Hussein, a CIA asset), one agenda being an open field lab test for how well the weaponized *Mycoplasma* "vaccine" worked, how the weakened immune system in soldiers would respond, and possibly even with an eye to a supplementary aerosol delivery system to work in tandem with inoculations. Five genetically armed strains of *Mycoplasma* had already been created from a crystalline toxin that Merck, "the pharmaceuticals conglomerate whose association with the Nazis may have been the most lurid of them all,"* engineered in 1946.

> . . . [*Mycoplasmas*] cause serious disease in many animal species (as well as plants), where they may affect multiple organ systems or cause chronic disease. Mycoplasmas are often difficult to detect and to eradicate. They may elude the immune system, and they may alter it . . . possibly precipitating autoimmune disease. *Mycoplasma* proteins are sufficiently similar to animal proteins that either the body's immune system may not recognize *Mycoplasmas* as foreign or they may cause the body to make antibodies that attack the host animal and produce autoimmune disease.[24]

Given the Kardashev objective of getting at least a human hybrid beyond the Van Allen radiation belt in mind, the secret space program has most definitely studied microorganisms in space—not deep space (to which there is yet little human access) but definitely 200 miles up at the International Space Station (ISS). For example, testing bacteria's ability to survive the harsh conditions of outer space by placing "balls" of radiation-resistant *Deinococcus* bacteria on the outside of the ISS for three years. (Extreme temperatures, low pressure, and radiation can all degrade cell membranes and destroy DNA.) Astrobiologists already knew that bacteria would survive inside artificial meteors and asteroids, but could they survive unprotected in space? When examined, the outer layers were fried, but the inner layers survived. Like the archaeological macroscopic *stromatolites* (mat-like structures of bacterial colonies) on Earth prove, bacteria survive best in groups.

---

*"After the Nazis came to power in 1933, members of the Merc family supported Adolf Hitler and the party, some of them enthusiastically. Like many other firms under the Third Reich, they ran their factories with slave labor, and some of them joined the SS and helped to purge the company ranks of Jewish employees."[23]

Bacteria are used to alter DNA for gene therapy by means of *transcession*, a process involving the same *horizontal gene transfer* that Nature utilized eons ago in eukaryotic evolution. Today, bacterial DNA (natural? gain of function? synthetic?) is transcessed into the host cell's DNA.* Once the Covid-19 serum is injected into the deltoid muscle and *not* into the bloodstream where an enzyme defense would be launched by the immune system, the mRNA quickens cells to rapidly produce the proteins (natural? gain of function? synthetic?) that DNA demands.

What a boon the wireless body area network (WBAN) has been to bio-nanoengineers and "precision medicine" that "helps" people by gathering their biometrics 24/7 and making sure that in vivo drug releases are properly timed. Powered by the body's energy—not lithium, silicon or gallium arsenide—bacteria can be programmed to perform a spectrum of actions, from simple logic to killing the host and dissolving after a certain number of divisions, like a kill-switch.[26] Bacteria can be directed to migrate to where the body temperature changes most rapidly, like the forehead right below the hairline, behind the ears, or on the back of the right hand—all of which are uncannily laid out in Revelations 13:16–17 as the Mark of the Beast: "And he causeth all, both small and great, rich and poor, free and bond, to receive a mark in their right hand, or in their foreheads. And that no man might buy or sell, save he that had the mark, or the name of the beast, or the number of his name."

Like bacteria, cells follow their programming as computers follow their algorithms.[27] While synthetic biology corporations like Viridos (formerly Synthetic Genomics Inc.) research and produce GMO plastics, processed foods, and fertilizers for GMO fields, and Raytheon and the Pentagon's DARPA use genetically modified bacteria to create explosives sensors,[28] thousands of public-private partnerships are covertly pursuing classified projects that turn gain-of-function GM bacteria "slaves" into sophisticated electromagnetic transceivers for Internet of Bodies (IoB) nano-networks and WBANs.[29] Creating biochips (bacterial or otherwise) once entailed a laborious process that involved the use of embryonic brain cells from fetal tissue

*In ancient times, "adaptive radiation" played a crucial role in evolution; today, it's high frequency like 5G or 6G for *forced* evolution. "Indeed, the acquisition of genes through lateral transfer may have triggered a period of adaptive radiation, thus playing a major role in the evolution of the *Entamoeba* genus."[25] Could this include human-forced radiative "evolution"?

and other complex processes.* Now, bacterial nanobots premanufactured with nanoprocessors that use the host's brain cell energy can do the job.

> New research from North Carolina State University finds that gold nanoparticles with a slight positive charge work collectively to unravel DNA's double helix. This finding has ramifications for gene therapy research and the emerging field of DNA-based electronics. . . .
>
> The finding is . . . relevant to research on DNA-based electronics [i.e., biochips], which hopes to use DNA as a template for creating nanoelectronic circuits . . . Researchers will have to pay close attention to the characteristics of those nanoparticles—or risk undermining the structural integrity of the DNA.[30]

As for antibiotics, natural antibiotics are implicit to foods grown in uncontaminated, healthy soil replete with healthy bacteria, an exceedingly rare occurrence in this era of aerosol deliveries. While many people have been conditioned to think that pharmaceutical antibiotics are the best way to counter an "infection" (i.e., assault or loss of natural balance), the truth is that Big Pharma antibiotics weaken the immune system and alter DNA by altering the natural gut bacteria and mitochondria. In 2020, the CDC reported that health care providers in the United States prescribed 201.9 million antibiotic prescriptions—the equivalent of 613 antibiotic prescriptions per 1,000 persons.[31] How many of these people, many of them children incarnating into a transhuman era, now have genetically altered DNA as a result? Add to this that GMO "terminator seeds" are soaked in the antibiotic tetracycline, which also impacts the gut bacteria.

Fluoroquinolone-based antibiotics (note the presence of fluoride) such as ciprofloxacin, levofloxacin, moxifloxacin, ofloxacin, and others, are topoisomerase interrupters that *unravel* bacterial DNA and program cells for death. Simply put, fluoroquinolone molecules adhere to human DNA and alter it. "Side effects" of the nalidixic acid in antibiotics include toxic psychosis and destruction of tendons (i.e., "tendonitis").

---

*Fetal cells create an interface between silicon and carbon to make electronic components merge and work together. For insight into how abortion serves neuroengineering, see Cathy Lynn Grossman, "The Hidden Ethics Battle in the Planned Parenthood Fetal Tissue Scandal," *Sojourners* website, July 27, 2015.

Ironically developed to overcome the inevitable antibiotic-resistant strains of bacteria that occur due to the overuse of antibiotics,* a new synthetic antibiotic called PPMOs (peptide phosphorodiamidate morpholino oligomers) specifically targets bacterial genes by binding to their DNA to block their expression. As with all pharmaceutical drugs, "side effects" are rife: urinary tract, sinus and bronchial infections, and strep throat, not to mention the destruction of the gut microbiome and therefore the ability of the body to absorb nutrients.[32] The methicillin resistance and "contagion" of MRSA (methicillin-resistant Staphylococcus aureus) may be less due to the presence of bacteria in hospitals, nursing homes, and dialysis centers than to the mix-and-match presence of previously weaponized bacteria to which the immune system has not yet formed a defense.

Bacterial resistance to antibiotics can supposedly be countered with the use of *quantum dots*, the light-activated carbon nanoparticles engineered to resemble the semiconductors (both conductor and insulator) used in electronics.[33] At the same time, we have clear evidence that constant exposure to 5G contributes to the growth of multidrug-resistant bacteria.[34]

A burning question we ourselves must answer in order to save our health in the nanotechnology world unfolding everywhere around and in us: *Can nanotechnologies like quantum dots be effective in treating conditions caused by other nanotechnologies?* Is this not "robbing Peter to pay Paul"?

BACTERIA'S BIOFILM

Bacteria are able to create a *pseudo-skin* around their colonies as a survival strategy. The *Encyclopedia of Food and Health* defines *biofilms* as

> . . . structured microbial communities that occur as surface-attached communities or suspended aggregates. They consist of microbial cells (bacteria and/or fungi) embedded in a self-produced extracellular matrix composed of polysaccharides, extracellular DNA and other components.[35]

Because bacteria are everywhere, biofilm is everywhere, too, especially now that wireless electromagnetism, chemical trails, and nanotechnology are

*The last time I had an antibiotic was over fifty years ago.

logarithmically multiplying the acidity of yesteryear's industrial pollution. A visual example is the dark blotches arising on national monuments like the Jefferson Memorial, Washington Monument, Lincoln Memorial, and the hallowed shrine at Arlington National Cemetery.[36] Biofilm is harder to see in the sheen on our skin or the inside of our organs, arteries, and veins. Being self-protective by nature, biofilm is difficult to eradicate and tends to disperse and replicate colonies elsewhere. As one 2020 study notes, "Biofilm infections are typically chronic in nature, as biofilm-residing bacteria can be resilient to both the immune system, antibiotics, and other treatments."[37]

The creation of synthetic biofilm has been a useful part of the transhumanist agenda since the pathogenic *Mycoplasma* was patented in 1991 by Shyh-Ching Lo of the Armed Forces Institute of Pathology and the U.S. military.* *Mycoplasma genitalium* has only 525 genes, making it nature's smallest genetic organism. The presence of the *Brucella* bacterium frequently accompanies *Mycoplasma* and has been found to be a coinfective agent since the 1970s. A *Mycoplasma* infection indirectly causes inhibited production of nitric oxide, which in turn causes the endothelial cells on the insides of organs, arteries, and veins to become "sticky" and form a biofilm plaque after robbing them of oxygen.

Agrobacterium: From Biology to Biotechnology (Springer Science, 2008), edited by Tzvi Tzfira and Vitaly Citovsky, is an excellent source book of essays about the *synbio* crossover from biology to biotechnology when it comes to *Mycoplasmas*, biofilm, archaea, protozoa, fungi, and algae, all of which go into bioengineering biofilm from "viruses" (genetic materials) and bacteria, given that bacteria colonize in biofilm for protection.

Biofilm colonies aggregate and attach to host tissues by means of hairlike appendages on a vertical lattice structure so as to take in nutrients and release byproducts more easily. As *Mycoplasma* builds layers and maintains cell-to-cell communication,† it uses host cells for its replication, until at a certain point the biofilm ruptures and disperses the *Mycoplasma* for further colonization. Shapeshifting, rapidly dissolving biofilm requires

*U.S. Patent 5242820A (1991): "The invention relates to a novel pathogenic mycoplasma isolated from patients with Acquired Immune Deficiency Syndrome (AIDS) and its use in detecting antibodies in sera of AIDS patients, patients with AIDS-related complex (ARC) or patients dying of diseases and symptoms resembling AIDS diseases."

†*Quorum sensing* is the ability of bacteria to communicate and coordinate behavior emitting signaling molecules.

no refrigeration; once inhaled and ingested, it reconstitutes inside the body where the immune system ignores it because of its protein surface negative charge; hence biofilm's activity in the body is covert, even to the point of going unrecognized by medical doctors. And yet it is implicated everywhere: infections, dental plaque, gingivitis, cystic fibrosis, prostheses, heart valve malfunctions, etc. Carpal tunnel surgery entails scraping biofilm from myelin sheaths; and it is found in the brains of Alzheimer's and macular degeneration sufferers.

Agrobacterium* is thought of as only having to do with plants, but is actually a handy biofilm producer:

> There is an assumption that Agrobacterium, a commonly used gene transfer vector for plants, cannot infect animal cells; however, this has been proved wrong and certain kinds of human diseases have been identified. Increasing evidence indicates that, under laboratory conditions, Agrobacterium is able to transfer its DNA into numerous and diverse nonplant eukaryotic species, such as fungi and yeast, as well as human cultured cells. Agrobacterium is responsible for opportunistic infections in humans with weakened immune systems. It is also found to be responsible for producing poisonous hydrogen sulfide (H2S) gas, sepsis, monoarticular arthritis, bacteraemia, cancer, Morgellons disease and so on, in humans.[38]

Agrobacterium tumafaciens is the well-known "genetic transformation machine" that has been engineered to form architecturally complex synthetic biofilms on host tissues. Since the 1980s, this bacteria has been used to create genetically modified organisms (GMOs) "because of its ability to transfer a piece of its genetic material, the T-DNA on its tumor-inducing (Ti) plasmid to the plant genome."[39] In fact, *Agrobacterium* is connected to the so-called Morgellons Disease, the precursor of much that has come to light under electron microscopes since 2020:

> Skin biopsy samples from Morgellons patients were subjected to high-stringency polymerase chain reaction (PCR) tests for genes encoded

**Agro* = field, soil, crop production; *tumere* = to swell, as in tumescence; *facie* = doing, making. Agrobacterium is a soil bacterium.

> by the *Agrobacterium* chromosome and also for *Agrobacterium* virulence (*vir*) genes and T-DNA on its Ti plasmid. They found that "all Morgellons patients screened to date have tested positive for the presence of *Agrobacterium*, whereas this microorganism has not been detected in any of the samples derived from the control, healthy individuals." Their preliminary conclusion is that "*Agrobacterium* may be involved in the etiology and/or progression"[40] of Morgellons Disease.*

Biofilm has also been gain-of-function engineered for conductivity. The cheapest creation of biofilm comes from treating biofungal graphene,† the graphene oxide (GO) antimicrobial agent.[41] *Biofungi* are fungi used to control insect pests, along with the bacterium *Shewanella oneidensis*, notable for its ability to reduce metal ions and live in environments with or without oxygen. The goal of the two working together is to lower the oxygen level of the host and thereby increase the GO conductivity‡ while turning the biofilm into a shiny electric generator that reacts with nanometals in a semiconductor fashion. *Shewanella*, coupled with thin, flexible, strong, conductive biofungal graphene, is easily remotely programmed and manipulated, even to the degree that its antennalike nature makes monitoring the host possible.[42]

"Viruses" too are being engineered to make biofilm surfaces conductive. Since 2012, the piezoelectric effect of the genetically altered, gain-of-function engineered M13 virus, when coupled with negatively charged amino acids, has been amping up pseudo-skin conductivity. "Imagine painting a layer of this film onto the casing of your laptop. Every time you tap the keyboard, these [M13] viruses convert the pressure from your fingers into electricity that constantly powers up your battery. Any kind of motion can power up M13."[43]

Chemical aerosols are delivering these "virus"- and bacteria-powered electronics to your lungs and skin. Think of pseudo-skin as a circuit board cum hairy antennas. Strontium, barium, and niobate (niobium plus oxygen) ($Sr_xBa_{1-x}Nb_2O_6$)—strontium and barium are among the extremely common chemicals in chemical trails—are used to make similar thin

*I will continue this discussion in my synthetic biology book.

†Graphene is discussed at length in chapter 9.

‡Is 60 gigahertz frequency 5G utilized?

biofilms for dielectric and electro-optic applications.[44] Alter the radio frequency and the oxygen percentage in the surrounding plasma, and you alter the biofilm composition.

The human skin is a respiratory organ just like the lungs are, and both appear to be under assault. The biofilm pseudo-skin spread over body surfaces. The biofilm pseudo-skin spread over body surfaces is in response to "the oxygen limitation in the developed biofilm"[45] (and 5G WiGig 60 gigahertz frequency). Experimentation with free oxygen (aerobic) versus a lack of free oxygen (anaerobic) is all about the demise of Human 1.0 and advent of Human 2.0 while the DARPA Arcadia program concentrates on engineering protective biofilms to keep bacteria from accumulating on Department of Defense (DOD) assets.[46]

VIRUSES

There is ongoing debate as to whether viruses can even be included in the tree of life. The arguments against this include the fact that they are basically parasites that lack a metabolism and ability to harvest their hosts' nucleotide sequences, nor are they capable of replication outside of a host cell.[47] In other words, viruses are bits of protein around nucleic acids—garbage ready to be eliminated from the cells, not weaponized unless *subjected to the violence of gain-of-function assaults that turn them into weapons.* It is telling that *not one natural virus has ever been isolated.*

As with bacteria, however, a controlled narrative has been developed around viruses over the past 100 years to push the Darwinian medical war narrative: that the vulnerable Human 1.0 is under endless assault from the dangerous ancient microbe world morphing daily into even scarier microbes requiring the expertise of highly trained (and highly paid) medical doctors and their shiny technologies and Big Pharma drugs, surgeries, specialists, blood-drawing and cauterizing, along with their tick bird private insurance companies, tax-supported Medicaid programs, etc. Stories about transmission electron micrographs identifying viruses swept up thousands of miles by dusty plasma "clouds" and water droplets, and the billions of viruses shuffling genetic information 1.7 miles (2.7 km) above and around the Earth, are presented as dirty bombers set on taking our health away unless we are "under the care" of medical experts.

In *The Contagion Myth: Why Viruses (including "Coronavirus") Are*

Not the Cause of Disease by Thomas S. Cowan, MD, and Sally Fallon Morell (Skyhorse Publishing, 2020) and Cowan's self-published quick reference reprint (forty-three pages) *Breaking the Spell: The Scientific Evidence for Ending the Covid Delusion* (2021), it is made crystal clear that there is no transmissibility of disease from natural germs (microbes), bacteria or virus nanoparticles. Period. The instructive word is *natural*: natural viruses (genetic materials) are not the *cause* of diseased, imbalanced conditions, but rather a housecleaning, rebalancing *effect* of genetic debris in the cell. Not so with *engineered* synthetic viruses that are micron- and nano-sized genetic debris subjected to gain-of-function procedures. The only way a natural virus can be transmitted is not by contagion, but by injection to impact the blood and aerosols that enter the lungs and move into the bloodstream.

> "T-cells, B-cells, neutrophils, monocytes, natural killer cells, proteins" [from John and Sonja McKinlay, "The Questionable Contribution of Medical Measures to the Decline of Mortality in the United States in the Twentieth Century," *Millbank Memorial Fund Quarterly,* 1977] are welded into a breathless story about a military machine that attacks germ invaders. Push-pull. Search and destroy. The notion that THIS is what creates health is fatuous. Positive vitality is what keeps us healthy.[48]

The military model of the virus not only enriches the coffers of Big Medicine and Big Pharma but also justifies treating Human 1.0 as a lab rat whose 1.0 immune system is constantly beleaguered by a postindustrial, corporate-driven environment subject to constant bombardment of chemicals and electromagnetism—even from space! Once one awakens to what "virus" actually is and isn't, thanks to courageous professionals like Thomas S. Cowan, MD, Andrew Kaufman, MD, and Stefan Lanka, PhD, we the public can begin to read between the lines of the twisted, weaponized, "double-speak" language in the medical literature, such as in the article, "It's raining viruses, but don't panic,"[49] beginning with the suggestive surname of the quoted Canadian virologist Curtis Suttle ("subtle").

First, we are assured that "the viruses circulating high up in the atmosphere are infecting almost exclusively other microbes, primarily bacteria," and that the "billion viruses every square yard per day" settling out of the atmosphere" aren't making us sick—"they're just part of the natural ecosystem"—but there are viruses that are "obligate pathogens" replicating by infecting other living organisms and moving "genetic information from one organism to another." (Is Suttle subtly referring to gain-of-function viruses?) "The original genetic engineering, if you like, was using viruses to actually move genes around among organisms," we are informed, which is why a large percentage of human nucleic acids like DNA are actually viruses "still stuck in our genome." This twisted truth just deletes the more accurate admission that viruses *are* genetic "garbage" ready to be ejected from our genome.

Unlocking and straightening out the Big Lie about what virus really is may be the thoughtful public's biggest challenge, given that the nanotech electromagnetic stew we are being cooked in makes it harder and harder to actually read between the lines and *think things through.* If only we can learn to doubt medical literature pushing germ causality rather than terrain (environment) causality, we will begin to see the virus for what it is: bits of protein and DNA set aside in the cell for removal. Sadly, however, viruses can be captured and utilized for gain-of-function weaponizing in all sorts of ways, including creating entirely false, lucrative fields of medicine like virology, and creating diabolically brilliant "viruses" that don't exist physically but do exist digitally, like SARS-CoV-2.*

The road to synthetic biology and Transhumanism is paved with lies about what is natural and what is safe in order to cover over the bioengineering of Human 2.0, which is why we the public must learn to probe what we are told about procedures like "gene therapy" and its relationship to electromagnetic fields.† Viruses (genetic materials) are currently considered the most effective vectors for gene therapy. But the presence of gain of function in the bioengineering of synthetic entities makes me wonder if "gene therapy" is more about *stealth* gene therapy, given that the immune system cannot discern synthetic from natural.

And why has there been no public debate about "vaccination by aerosol" via synthetic adenovirus (respiratory) viral vectors?

*I will go more deeply into all of this in my next book about synthetic biology and Transhumanism.

†Corporations like AEA Technology partner with Big Biotech and Big Pharma to create rDNA aerosols in inhaler form, to deliver gene therapy in the form of DNA-based drugs.[50]

"Feasibility of Aerosol Vaccination in Humans" (2003): "The feasibility of using aerosol vaccines to achieve mass and rapid immunization, especially in developing countries and disaster areas, is being assessed on the basis of current available information . . ."[51]

"Immunization by a Bacterial Aerosol" (2008): "By manufacturing a single-particle system in two particulate forms (i.e., micrometer size and nanometer size), we have designed a bacterial vaccine form that exhibits improved efficacy of immunization. Microstructural properties are adapted to alter dispersive and aerosol properties independently . . ."[52]

"World-First Inhaled COVID-19 Vaccine, Developed in Partnership Between Aerogen® and CanSinoBIO, First Public Booster Immunization in China" (November 14, 2022): "Late in 2021, Aerogen® (Galway, Ireland) and CanSinoBIO (SSE: 688185, HKEX: 06185) (Tianjin, China) announced a development and commercial supply partnership for the inhaled delivery of CanSinoBIO's Recombinant Novel Coronavirus Vaccine Convidecia Air™ utilizing Aerogen's proprietary aerosol drug delivery technology,"[53]

"Can 'No-needle' COVID-19 protection be achieved with inhaled aerosol vaccines?" (January 27, 2023): "The regulatory approval of CanSinoBio's Covidencia™ Air has been a recent breakthrough in COVID-19 vaccine research. This is a first-in-class inhaled aerosol vaccine which, according to the company, induces comprehensive SARS-CoV-2 protection after a single breath . . ."[54]

Once again, the public is hoodwinked into being nonconsensual guinea pigs for covert aerosols secretly supporting precision medicine, injections, and contributions to decades of autoimmune "side effects" and heart conditions similar to those we have witnessed since the Covid-19 injections.

In 2003, scientists at the Institute for Biological Energy Alternatives in Rockville, Maryland, took only three weeks to create a synthetic "virus" from scratch.[55] That same year, the CIA published "The Darker Bioweapons Future,"[56] which forewarned that "The effects of some of these engineered biological agents could be worse than any disease known to man"—the telling term being "engineered biological agents."

Certainly, this included the "gain-of-function" process by which bits of natural protein and genetic materials are engineered into synthetic viral vectors, for example, but primarily it is *the entire shift of molecular biology into digital biology* that is making populations "sick" with the wireless 5G millimeter / 6G terahertz transmission system that pulses exact disease frequencies. No need, then, to physically prove the existence of a "bacteria" or "virus" by isolating it. The biggest challenge, I'm sure the CIA realized, is keeping the electromagnetic delivery system of disease secret.

The year before the CIA's admission, biologists at State University of New York, Stony Brook, created from scratch a synthetic double of the *poliovirus*—a single strand of RNA that can be morphed into a large protein that then attacks the central nervous system—simply by following "the publicly available recipe of letters that make up the [poliovirus] chemical code."

> [The scientists] found the polio virus genome on the internet and within 2 years had created a virus from raw chemicals. The synthetic virus could reproduce and, when injected into mice, paralyzed them just as a natural polio virus would do. They said they chose the polio virus to demonstrate what a bioterrorist could accomplish.
>
> "It is a little sobering to see that folks in the chemistry laboratory can basically create a virus from scratch," says James LeDuc, director of the Division of Viral and Rickettsial Diseases at the Centers for Disease Control and Prevention (CDC) in Atlanta . . . As virologist Jeronimo Cello points out, "By releasing this [announcement], you alert the authorities . . . [to] what bioterrorists could do."[57]

Slowly, by deciphering bits and pieces of how bioengineering synthetic biology is done, we begin to grasp how far the invasion of "gene therapy" has gone thus far. Mail-order digitized genetic materials (virus) are transferred into a bacterium, which is then injected / aerosoled into the host body or transmitted electromagnetically. Retroviruses (RNA genetic code) for the chromosomes of host cells, adenovirus and herpes genetic materials[58]—all can be worked with frequency and transmission, as can mind control. Under DARPA's Next-Generation Nonsurgical Neurotechnology program (N3), which "aims to develop high-performance, bi-directional brain-machine interfaces (BMIs) for able-bodied service members,"[59] viral vectors are now being enlisted for mind control by inserting DNA into specific neurons to make them produce two kinds of proteins: one to absorb light when a neuron is firing, which can then be remotely detected and measured by an infrared beam passing unseen through the skull and brain; the other tethered to magnetite (by ligand, an ion or molecule attached to a metal atom by covalent bonding) already deposited in the brain (see chapter 6) to induce an image or words, or transmit from one brain to another.[60]

Corporate bioterrorists and government shills are now cutting deals with the devil for access to frequencies of diseases as well as frequencies of the brain for mind control, thanks to the nanotechnology in bodies and brains and 5G / 6G activation of aerosols via satellite, power lines, microwave towers, i-Phones . . . Getting the picture?

The mindset during the recent Covid "pandemic" was predicated entirely on Louis Pasteur's germ theory, developed in the nineteenth century and still rigidly upheld by the modern medical industry, despite advances in holistic medicine that view the human being as an integrated life system within a larger integrated life system. Dr. Cowan on Pasteur:

> His "germ theory" now serves as the official explanation for most illness. However, in his private diaries he states unequivocally that in his entire career he was not once able to transfer disease with a pure culture of bacteria (he obviously wasn't able to purify viruses at that time). He admitted that the whole effort to prove contagion was a failure, leading to his famous death bed confession that "the germ is nothing, the terrain is everything."[61]

Pasteur's contemporary, French biochemist Antoine Béchamp, declared: "We do not catch diseases. We build them. We have to eat, drink, think, and feel them into existence. Germs or microbes flourish as scavengers at the site of disease. They do not cause the disease any more than flies or maggots cause garbage."[62] Béchamp opposed the germ warfare theory and advocated the terrain theory.

German virologist Stefan Lanka won a landmark case in 2017 that proved to the German Supreme Court that measles is not caused by a virus, nor is there a measles "virus."[63] The entire justification for vaccinations *from birth onward* is based on believing in attacking or suppressing germs. The truth is that it is actually about attacking the immune system of the Human 1.0 child and enriching the medical / pharmaceutical industries. But if there is no virus, why do childhood illnesses appear so commonly—at least before the universally prescribed MMR (measles, mumps, rubella) vaccine? Dr. Cowan answers this question in his chapter entitled "Resonance." Resonance is a crucial concept if one wants to understand how electromagnetic fields and the human bioenergetic field interact for health or disease:

> To understand what seems to be the contagious nature of childhood diseases like measles, mumps, and chicken pox . . . one must investigate the phenomena of resonance. If one plucks a string tuned to a certain frequency, the vibrations of the string will cause a second string tuned to the same frequency to vibrate and sound at the same frequency. The two strings are not touching; the connection is through a sound wave that travels between the strings . . .
>
> Why do measles and chicken pox seem to be infectious? One child puts out the message through exosomes that now is the time to go through the detoxifying experience called chicken pox. Other children in their class or town receive the message and begin the same detoxification experience . . . This observation, rather than proving contagion, teaches us about the mystery we call life. It teaches us again that the materialistic conception of the "wily attack virus" is an impoverished, inaccurate view of the world. And it teaches us to forgo simplistic explanations and look into the deepest mysteries of life if we are to create a world of health and freedom.[64]

According to Rudolf Steiner, childhood illnesses are the soul's attempt to throw off the failings of previous lives so as to "make the corresponding correction as soon as possible."[65] Such an interpretation may sound antiquated and naïve in a materialistic era that not only does not believe in reincarnation as people in the East traditionally believe, but also believes that "never the twain shall meet" when it comes to physical biology and soul or spirit. In my book on synthetic biology and Transhumanism, I will challenge this (antiquated) Western assumption.

For the most part, viruses are proteins that a cell excretes in order to reorder its balance or gain-of-function bacteria used as vectors for metal *bioscaffolding*—similar to what entrepreneur Elon Musk refers to as "neuromesh"—the automated manufacture of nano-scaffolds for three-dimensional (3D) tissue models that closely mimic the geometric complexity of Human 1.0 tissues and organs. Natural viral particles called *virions* (genetic materials o/proteins) are used as durable nano-building blocks for *composite materials*, the term for strong lightweight materials and unlike fibers bonded together chemically in the laboratory.

Neither natural nor synthetic "viruses" can reproduce, but their properties can be readily engineered—genetically modified—into "viral chimeras* that carry proteins of different viral origins."[66] It is a lie that "viruses" mutate all on their own (the so-called emerging variants), as this ignores or hides the fact of gain-of-function nanoparticle constructs developed in part to confuse the difference between natural and synthetic "viruses" and bacteria.

There are basically three ways to weaponize (gain of function) genetic material so it becomes what is commonly labeled a "virus": (1) obtain it from biological tissue, then treat it with mycotoxins; (2) grow it in vitro or in vivo inside incubated cells; or (3) make it from scratch inside bacteria or proteins or yeast cell hosts, yeast being the fastest and most capable of taking chunks of viral genome and putting them together in a preordered sequence. *All three methods are used in gain-of-function research.* That molecular biology is now digitized means that viral DNA can be sent to a colleague via mail-order as easily as an email, which means that "designer" synthetic viruses can be easily (and anonymously) produced. Researchers can now "print" genes with a biological-to-digital converter like BioXP and pop in the parts (or digital signatures, sent like email attachments) preordered from *synbio* labs like the J. Craig Venter Institute or Synthetic Genomics Inc. As Dan Gibson, vice president of Synthetic Genomics' DNA technology, puts it, "DNA is really just the start of making anything downstream, from RNA to protein to whole bacterial genomes."[67]

It is crucial that we counter the germ warfare framing of bacteria and viruses in order to realize how nature's nanoparticles have been marshaled to mutate human DNA. At least one recent study expresses concern:

---

*In genetics, the chimera is an organism or tissue that contains two or more sets of DNA.

> When nanoparticles come into the vicinity of the cellular system, chances of uptake become high due to their small size. This cellular uptake of nanoparticles enhances its interaction with DNA, leading to structural and functional modification (DNA damage/repair, DNA methylation) into the DNA. These modifications exhibit adverse effects on the cellular system, consequently showing its inadvertent effect on human health.[68]

9
Synbio

It would be possible to build a mechanism of the kind that could operate in the natural world, on abundantly available compounds, or perhaps a wide range of compounds, to build copies of itself. Something like that would be a lot worse than any plague or insect infestation you could think of, and in a limited case of awfulness such a thing could have a very broad ability to consume organic matter. Obviously, there would be no predators, no ecological checks and balances. And so it could generally destroy the biosphere.

Eric Drexler, quoted in
Nano: The Emerging Science of Nanotechnology

And the first went, and poured out his vial upon the earth; and there fell a noisome and grievous sore upon the men which had the mark of the beast, and upon them which worshipped his image.

Revelations 16:2

Let's begin this discussion of the transhuman implications of synthetic biology, *synbio,* by recapping a mental picture of the Space Fence in our troposphere, from radar and Starfire laser installations to microwave towers providing "connectivity" for billions of cell phones, ionospheric heaters, wind farms, fracking wells, utility grids—all availing the invisible smart grid electrochemically threaded throughout the charged atmosphere now behaving like a circuit board, thanks to the constant ionization of the atmosphere with trillions of nanoparticles and nanotechnology laid daily via chemical aerosol spraying operations and zapped by lasers from the ground and by satellites and jets, rockets, and drones from the sky, courtesy of pyrolysis.* Add lasers constructing endless self-replicating nano-assemblies from the metal oxides permeating the air we breathe while zapping highly conductive carbon nanotubes (CNTs) like the allotrope graphene for a variety of military-industrial-intelligence agendas. (Graphene can be crafted in various forms and at a great distance by 5G/6G.) Such is the completely unnatural super-conductive Space Fence gridded world we live in.

While the Space Fence infrastructure wears the human body and brain down over time, the interferometric HAARP wavelengths can be bounced off the ionosphere so as to target predetermined areas with specific frequencies for earthquakes, conventional warfare, mind control, disease, etc. In *Chemtrails, HAARP, and the Full Spectrum Dominance of Planet Earth* (Feral House, 2014), I quoted independent scientist Clifford Carnicom regarding the use of 3 Hz throughout the atmosphere:

> Carnicom's essay, "Potassium Interference Is Expected" (May 15, 2005)[3] posits that the potassium ion is being targeted for "biological interference within people over large regions of the earth's surface."
>
> *This is due to the fact that the fifth harmonic of the ELF [extra low frequency] that has been repeatedly measured over a period of several years corresponds to the cyclotronic resonant frequency of potassium. This fifth harmonic, along with numerous other harmonics, is a regular component of the ELF radiation that is under measurement at this time.*
>
> Spectral analysis and equations follow, after which Carnicom drops the bombshell:
>
> *It can also be expected that variations in the magnetic field of the earth can lead to other potential resonance conditions in various regions or latitudes. It is therefore not unexpected to find large regional health issues that will correlate with variations in the magnetic field strength of the earth. Certain ions are expected to be disrupted in some areas of the globe more than others.*

**Pyrolysis* converts biomass to thermal energy, but is primarily a process of oxidation and reduction in the jet combustion chamber by which crystals / carbon nanoparticles are produced from metal salts.[1] *Drones* may now refer to nanobot "swarmtroopers."[2]

In short, HAARP and other ionospheric heaters around the globe appear to be infusing multiples of artificial 3 Hz ELF propagation into our modified atmosphere to create a cyclotronic resonance that can affect biologically important ions of sodium, calcium, potassium and lithium.[4]

By discharging the scalar potential* in a charged sky, a building can be engulfed in plasma flames with the energy of a Hiroshima bomb, as has occurred in various blasts such as in Tianjin, China (2015), and Beirut, Lebanon (2020), not to mention fires in Paradise, California (2017), Lahaina, Maui, and Acapulco, Mexico (2023),[5] etc.—fires ignited by directed energy weapons (DEWs) from the sky (lasers, masers, and LiDARs, like the blue beam from the sky that triggered the Paradise Fire) or from exotic craft like the X-37B unmanned spacecraft.

Is this an atmosphere and world intended for a protein-based, 70 percent water Human 1.0? Not any longer. We are being *epigenetically*† engineered to become carbon-based in the sense that we are being loaded with hydrogel and carbon nanotubes (CNTs) like crystals, graphene, nanobots, quantum dots, etc., all agitated by refined sugars (i.e., carbons) in the genetically modified, pharma-enriched synthetic "foods" we have been conditioned for decades to think of as "good" for us.

Ashes to ashes (carbon), *dust to dust* (silica), as Tony Pantalleresco puts it.

More and more, the nanoscale rules the matter around *and* in our bodies and brains. Slowly, the basic foundations of classical biology, chemistry, and physics are no longer applying. The ancient Greeks were wise to distinguish *form forces* from *matter forces,* particularly in how form forces imply a natural creature status created by a Creator force like Nature or God. Form forces are now being subsumed by metaverse concepts like *metamaterials,* which indicates a composition of any material engineered for properties rarely observed in naturally occurring materials and thus interpreted as "progressive" or "enhancing." Metamaterials dominate

*According to Wikipedia, "In mathematical physics, scalar potential, simply stated, describes the situation where the difference in the potential energies of an object in two different positions depends only on the positions, not upon the path taken by the object in traveling from one position to the other."

†Epigenetics is the study of how your behaviors and environment can cause changes that affect your gene expression rather than alter the genetic code itself.[6]

nanomaterials with their precise geometric shapes, sizes, directions and arrangements, all speeding toward recasting Human 1.0 as a Human 2.0 brain-computer interface (BCI) cyborg form.

As for *matter* forces, I think, on the one hand, of the mushroom cloud of the atomic bomb, truly a specter signature of the technological age into which we have been unknowingly catapulted by the military-industrial-intelligence complex; and on the other hand, I think of the nanotechnology being created at the atomic level, anxiously described in the opening quote of this chapter by nano-creator Eric Drexler. *Both technologies arose at the atomic level of matter's forces.*

The bioterrorism announced after the 9/11 blood sacrifice of 3,000+ formally justified human experimentation on citizens (no longer just on soldiers, as this book demonstrates). The 2004 Pentagon budget for chemical-biological warfare was $10 billion* and has been climbing ever since. HIV/AIDS worldwide, *Mycoplasma* pathogens, autoimmune diseases like myalgic encephalomyelitis in Canada / chronic fatigue syndrome in the U.S., DARPA's Project Jefferson that gain of functioned vaccine-resistant anthrax, the Lone Star tick weaponized for Lyme disease, Morgellons—all of it of great interest to post–Cold War scientism.

Perhaps the 2020–2023 Covid drama will free us from blaming autocratic regimes for the chemical-biological (*synthetic biology/synbio*) warfare now a Darwinian way of life, whatever the nation's insistence that it is "democratic." Paperclip Nazi scientist Erich Traub, the Third Reich's lab chief at Insel Riems, the Nazis' secret biological warfare lab, studied at the Rockefeller Institute in Princeton, New Jersey, prior to World War II, then after the war was stationed at the Naval Medical Research Institute in Bethesda, Maryland, often visiting Plum Island while simultaneously directing biological warfare work at the Tübingen research lab in West Germany.†

According to Francis Boyle, PhD, who drafted the law the U.S. Congress enacted in order to comply with the 1975 Biological Weapons Convention, the United States has 13,000 "death scientists" in four hundred laboratories, employed to enrich the $100 billion germ

---

*A must-read is A. W. Finnegan's *The Sleeper Agent: The Rise of Lyme Disease, Chronic Illness and the Great Imitator Antigens of Biological Warfare* (Trine Day, 2022).

†*Scientific American* has dismissed these connections. Nevertheless, both *Scientific American* and *Nature* magazines are owned by the Holtzbrinck Publishing Group, one of the big five English-language gatekeeper publishing firms.

warfare "biodefense" industry. Dr. Boyle stresses that the Galveston National Laboratory in Texas should be shut down because it is:

> an ongoing criminal enterprise along the lines of the SS and the Gestapo—except that Galveston is far more dangerous to humanity than Hitler's death squads ever were. American universities have a long history of willingly permitting their research agenda, researchers, institutes, and laboratories to be co-opted, corrupted, and perverted by the Pentagon and the CIA into death science. These include Wisconsin, North Carolina, Boston U., Harvard, MIT, Tulane, University of Chicago, and my own University of Illinois, as well as many others.[7]

At the University of Wisconsin, researcher Yoshihiro Kawaoka has resurrected the Spanish flu in this era of wireless 5G/6G and is now busily engaged in increasing flu toxicity to infect and kill "human-like subjects."[8]

LYME DISEASE

The first harbingers of the impending fate of Human 1.0 was the explosion of various cancers, any number of chronic autoimmune conditions, Lyme disease, and then "Morgellons." All pointed to a covert agenda to undermine the ancient immune system and thus pave the way for a brain-computer interface (BCI) transhuman 2.0 species, but we who were observing had no idea yet. The international professionals I paid particular attention to at that time were Carnicom (American); Dietrich Klinghardt, MD, Lyme disease (German); interdisciplinary scientist Harald Kautz, photons / optogenetics / scalar (German); Tony Pantalleresco, herbalist specializing in nanotechnology extraction (Canadian).

It may sound strange, but as I have said elsewhere in this book, immune suppression has proven again and again to be a primary essential for chemical-biological warfare experimentation, which also encompasses the creation of "vaccines" like the mRNA bioengineered Covid serum, which circumvents the immune system entirely. Lyme and Morgellons disease expert Ginger Savely, a doctor of naturopathic medicine with extensive experience treating sufferers, believes that Lyme disease victims' weakened immune systems make them more vulnerable to Morgellons, *as if one biowarfare agent developed out of the other.* It is obvious, what with all the environmental pollutants, GMOs, junk foods, and wireless electromagnetics, that weakening the immune system would have to precede the myriad *synbio* BCI transhumanist modifications to Human 1.0 in order to create a Human 2.0 cyborg. The trillions of aerosol-delivered metallic chemical nanobots (nano-robots) now in our bodies have been engineered to "modulate" and deceive the immune system, what with silica nanoparticles penetrating the skin barrier from the outside and titanium dioxide nanoparticles inducing gene expression alterations in the brain from the inside. Together, the two induce reproductive and liver toxicity,* not to mention increasing stress on the beleaguered immune system.

Chronic disease specialist Dietrich Klinghardt, MD, founder of the Sophia Health Institute in Woodinville, Washington, agrees with Dr. Savely that Lyme is not caused by microbes so much as by a lack of proper immune response. Dr. Klinghardt finds that 90 percent of a successful Lyme treatment must be geared toward rebuilding the immune system, while 10 percent goes toward killing microbes. His treatment begins with his own tissue-specific autonomic response testing, after which he uses ultrasound, infrared light, urine therapy, and progesterone for opening the blood-brain barrier to herbal remedies.[10]

Dr. Klinghardt has devoted years to deciphering the Lyme disease once classified under bioweapons and weaponized in spirochetes (spiral twisted bacteria), after which "Morgellons" was weaponized in coccus (spherical bacteria). The spirochete's outer surface appears to have been engineered to look like a myelin biofilm nerve sheath so that the immune system attacks the nerve or the white blood cells coming to the rescue. Notably, many Lyme sufferers have zero immune response. Is it possible that the bioweapon Lyme was created to attack the immune system so that the cross-species technology misnamed Morgellons could be slipped into place without the immune system (or the public) noticing? Dr. Klinghardt maintains that the Lyme spirochetes *Borrelia burgdorferi* and *Borrelia mayonii,* both of which can shapeshift, have been gain-of-function genetically modified by piggybacking Epstein-Barr DNA and

*"Due to the rising use of nanomaterials (NMs), there is concern that NMs induce undesirable biological effects because of their unique physicochemical properties. Recently, we reported that amorphous silica nanoparticles (nSPs), which are one of the most widely used NMs, can penetrate the skin barrier and induce various biological effects, including an immune-modulating effect."[9]

other viral components and coinfections (*Bartonella*, *Mycoplasma*, etc.) onto spirochete bacteria to aggressively suppress the immune system, thus leading to the usual chronic culprits of brain fog, autism, Alzheimer's, MS, ALS, and other nerve-degenerative conditions.*

Lyme disease was developed under the auspices of the CIA's Project MK-Naomi (1950s to the 1970s). While MK-Naomi is considered a successor to MK-Ultra mind control, biological agents are stressed in concert with the Special Operations Division of the U.S. Army Biological Warfare at Fort Detrick.[11] Today, Fort Terry on the now-infamous 840-acre Plum Island off the coast of Long Island, is controlled by the Department of Homeland Security, doubling as a U.S. Army biological warfare research facility.

Following World War II, Paperclip Nazi scientists were given carte blanche over this facility (see Finnegan's *The Sleeper Agent*) to pursue their chemical-biological warfare black-magic brews. In 1962, they focused on creating a "designer disease" that would incapacitate but not immobilize. The "K Project" (*K* for "kill" or "knockout") cover story was that ticks were "a natural breeding and mixing ground for pathogens"[12] with the *Dermacentor veriabilis* tick being a natural "paralysis agent." The first cases of Lyme broke out in 1975 in Lyme, Connecticut, 13 miles northeast of Plum Island Animal Disease Research Center, much like the mosquito-borne West Nile "virus" in Long Island and New York City. "Coincidences, it seems, abound at Plum," quipped the possibly aptly named *New York Press* columnist Alan Cabal on March 16, 2004.

Since then, Lyme disease has emerged all over the world, differing only in what mosquito is blamed for carrying one tick or another. Three hundred thousand cases of Lyme are reported per year, all blamed on a bacterium when the cause is really gain-of-function (patentable) bioweaponization. Autoimmune conditions, cancers, and chronic immune suppression follow.

Given that the Human 1.0 body operates as a single integrated organism, treating Lyme often makes Morgellons symptoms disappear, but *not* the Morgellons bioengineered causative agent itself.† High levels

---

*Morgellons displays much of this, intermingling with intestinal-fungal overgrowth, rashes, B12 deficiency, and skin and internal itching, along with the subdermal sensation of broken glass and lit cigarettes.

†*Borrelia* spirochetes have been found in skin tissue of four "randomly selected" Morgellons patients.[13]

of aluminum nanoparticles are found in both Lyme and Morgellons, and cell phone radiation no doubt plays a significant role in both, given that it blocks the enzymes necessary to detoxify heavy metals. Lyme spirochetes do not live in the blood, but instead set up sanctuaries in various parts of the body, whereas Morgellons coccus have been engineered to reside primarily in red blood cells.

Morgellons research may have begun in tandem with Lyme's weaponization, but it definitely was *the* classified objective at a 1969 Defense Department Appropriations Subcommittee meeting at which the high-ranking Pentagon biological warfare expert Donald MacArthur begged Congress for $10 million for research into "a new infective microorganism which could differ in certain important aspects from any known disease-causing organisms . . . that might be refractory to the immunological and therapeutic processes upon which we depend to maintain our relative freedom from infectious disease . . . a synthetic biological agent, an agent that does not naturally exist and for which no natural immunity could be acquired."[14] Immediately after the $10 million was rubber-stamped by Congress, the Frederick Cancer Research Facility of the U.S. Army's Fort Detrick Biological Warfare Laboratory morphed into the National Cancer Institute, while the staff and budget tripled.

Synbio appears to be a synonym for weaponizing Nature.

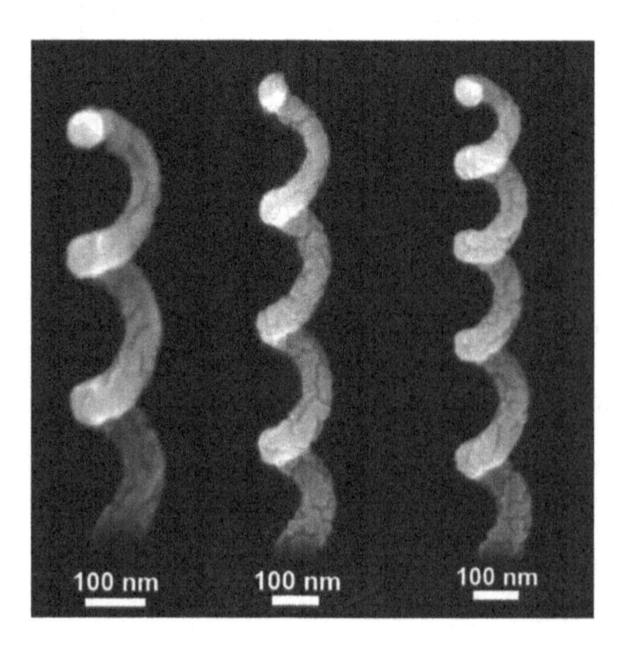

Fig. 9.1. Helix-shaped nano-antennas produced in an electron microscope by direct electron-beam writing and created from the Lyme spirochete *Borellia burgdorferi*.
Helmholtz Zentrum Berlin, image © HZB

MORGELLONS

My independent research into Morgellons filaments that scientist Clifford Carnicom was collecting in the wake of the jet trails that turned into a thin cirrus-like cloud cover over northern New Mexico awakened me to the new biology that is now known as *synthetic biology (synbio)*. Vividly, I recall looking over Clifford's shoulder into his 300X microscope as tiny entities from a filament in my blood sucked the iron out of the blood on a slide. That was the moment I decided I must research in earnest the geoengineering that the public was still calling "chemtrails" (~2005).

At that time, I knew only that the so-called Morgellons and possibly Lyme pathogens were being delivered by jets in NATO nations' airspace for biological experimentation with soil (and therefore the genetically modified mass food supply) and human bodies, given that all people—not just people in NATO nations—must breathe from the one planetary atmosphere. I had yet to discover the connections between Big Pharma inoculation mandates and "chemtrails," or between how geoengineering and "biowarfare" inoculations were both being engineered for a transhuman future; nor did I understand the dangers inherent in the nanotechnology used extensively in geoengineering and synthetic biology. Here, I revisit highlights of those two decades preceding the 2020–2023 Covid-19 inoculation assault, which I will take up in detail in the sequel synthetic biology book.

In the previous chapter "*Bios*," I discussed *Agrobacterium tumefaciens*, the well-known "genetic transformation machine," in the context of creating biofilm. When genetic engineering took off in agriculture in the 1970s and the United States was the first nation to release GM crops, *Agrobacterium tumefaciens* was gain of functioned into a vector to create transgenic plants while "gene escape" of transgenic seed and crop materials into natural organic fields, supplemented by cloud-seeding of "culture mediums," did the rest. (*Agrobacterium* remains latent in plant tissue, thus facilitating the escape of transgenes in newly formed combinations of foreign genes.) With *horizontal gene transfer* and *recombination* (moving genetic material between organisms instead of the vertical parent-offspring model) constituting the main approach to genetically engineering "novel compounds" (meaning synthetic pathogens)—for example, making sure that antibiotics don't work on *Agrobacterium*—commandeering control over the mass food supply was the very first sure step into *epigenetically* engineering a biotech that would initiate new diseases as well as weaken the Human 1.0 immune system in preparation for transhuman Human 2.0.

The synthetic / gain-of-function version of the natural soil bacterium *Agrobacterium tumefaciens* renamed *Rhizobium radiobacter in 2001** (patent owned by Monsanto), guaranteed that organic plants not grown from Monsanto's "terminator seed" would not grow in that soil.

Genetically altered "terminator seed" prohibits the natural evolution of crops by stopping the growth in the second generation. But there is a much more nefarious property of genetically altered "terminator seed" in that *it is genetically altered and no longer natural seed.* While economic interests like Monsanto / Bayer, and the Delta and Land Pine Company of Scott, Mississippi†—plus innumerable chemical, biotech, and seed companies—have attempted to make GMO sterile seed (unable to reproduce) sound like it's all about keeping GMO crops from contaminating neighboring crops,‡ everyone who can still think knows it is about profits on the one hand and undermining the Human 1.0 immune system by forcing farmers to turn to commercial seed ("proprietary seeds and their companion chemicals"[17]) that *only* produces *synthetic food* providing no natural nutrients.

To get *Rhizobium radiobacter* ("rainmaking bacteria") into the soil—*bioprecipitation*§—Big Pharma / Big Ag corporations tag onto cloud-seeding deliveries for local operations and chemtrails for broad operations over private land slotted for organic growth as well as corporate land slotted for biotech GMO growth. *Agrobacterium* T-DNA genetically modifies plant

*Indicating the shift from natural to artificial: "Rhizobium radiobacter has a strong predilection to cause infection particularly in those patients who have long-standing indwelling foreign devices."[15] If "long-standing indwelling foreign devices" are synthetic and electronic, their presence could set off other electronic synthetics. The term *radiobacter* indicates an electronic "rod" type of bacteria (aerobacter an air rod, arthrobacter a joint rod, etc.). Radiobacter = electronic bacterium.

†Since 2018, Bayer AG has been the sole owner of Monsanto. In January 2024, a Pennsylvania jury handed down a $2.25 billion verdict against Monsanto and its parent company, Bayer, after forty-nine-year-old John McKivison contracted non-Hodgkin's lymphoma from using Roundup herbicide on his property for two decades.[16]

‡The truth is that terminator seeds pollinate nearby crops with sterile genes in order to make traditional seed and growing practices obsolete.

§For example, chlormequat as bioprecipitation (aerosol spray) not of rain or water but of chemicals like this plant-growth retardant that produces sturdier stalks in cereal plants like oats, then makes men and women infertile.[18]

genomes and human genomes as well, and is particularly evident in the etiology of the bioengineered Morgellons.[19]

The truth is that the half century of agricultural gene transfers of synthetic Rhizobium radiobacter has produced new and exotic Frankensteins that have undermined the natural human immune system. I will take this up in more depth in the synthetic biology book.

As far back as 1994, independent scientist Clifford Carnicom observed that the atmosphere was being modified by emissions from airplanes and jets. His observations would lead to more than two decades of analyzing specimens he collected from precipitation and in HEPA filters following their expulsion in the trails of jets. Among the unusual materials were submicron, non-soluble filaments. Carnicom managed to crack open these filaments, whereupon he discovered what only *appeared* to be an entire biology: an erythrocytic form (i.e., red blood cells); a chlamydia-like structure; and a pleomorphic ribbon, or sausage-like, form. It only *seemed* to be a biological organism in that it reproduced and grew and had DNA, but it wasn't *entirely* organic. In effect, it was a creation of synthetic biology, *synbio* consisting of modified, gain-of-function DNA sequences that direct organic DNA to create something synthetic. In 1999, Carnicom sent a specimen to the EPA. Several months later, he received their terse response: they saw nothing unusual. A year and a half later they returned Carnicom's specimen with the message that the EPA was under no obligation to do anything unless *they* requested a specimen.

In the early years of Project Cloverleaf's aerosol dosing of Morgellons filaments (the 1990s through the early 2000s), the CDC and Kaiser Permanente were tasked with keeping a public lid on Morgellons by declaring that those who were experiencing Morgellons symptoms—rashes, eruptions of fibers from the skin, glittering quantum dots in skin lesions, subcutaneous sensations of crawling worms and biting insects, immune system collapse—were suffering from "delusional parasitosis" and "neurotic excoriation."

In 2003, the U.S. Army in conjunction with University of California, Santa Barbara, announced the formation of the Institute for Collaborative Biotechnologies to "develop biologically inspired, revolutionary technological innovations in systems and synthetic biology bio-enabled materials, and cognitive neuroscience."[20]

In 2008, the CDC with its public-private partner Kaiser Permanente launched a study of what they called the Morgellons "skin condition"[21] and concluded a few years later that there was "no infectious cause."[22]

In 2010, while Morgellons was still being dismissed as an anomaly or a psychological condition, the 21st Century Nanotechnology Research and Development Act authorized $3.7 billion for five years of research and development into bionanotechnology.

Synthetic biology, coupled with nanotechnology, was definitely heating up.

Throughout his research into the fibers falling from the sky, Carnicom noted strong parallels between Gulf War Syndrome, Lyme disease, Morgellons, fibromyalgia, and chronic fatigue syndrome. Was he observing an attempt at a "direct modification of the human physical being"?[23]

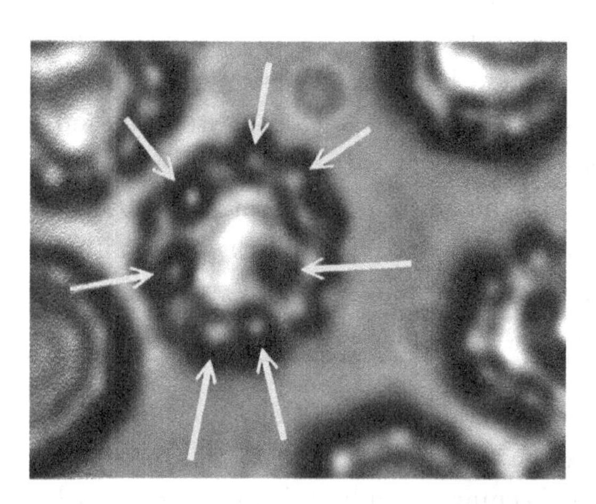

Fig. 9.2. Human blood cell (erythrocyte) damaged by cross-domain bacteria (red arrows).
Clifford E. Carnican

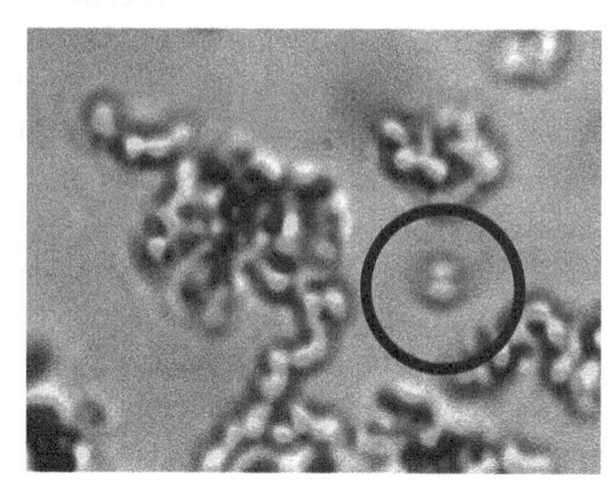

Fig. 9.3. Cross-domain bacteria cell division in a two-hour time interval.
Clifford E. Carnican

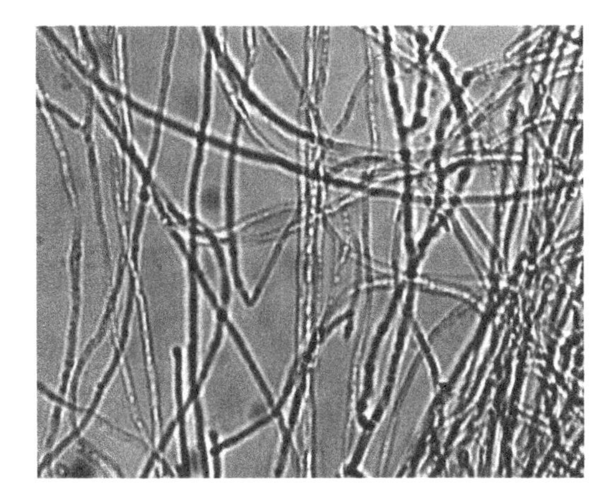

Fig. 9.4. Morgellons filament development with internal cross-domain bacteria.
Clifford E. Carnican

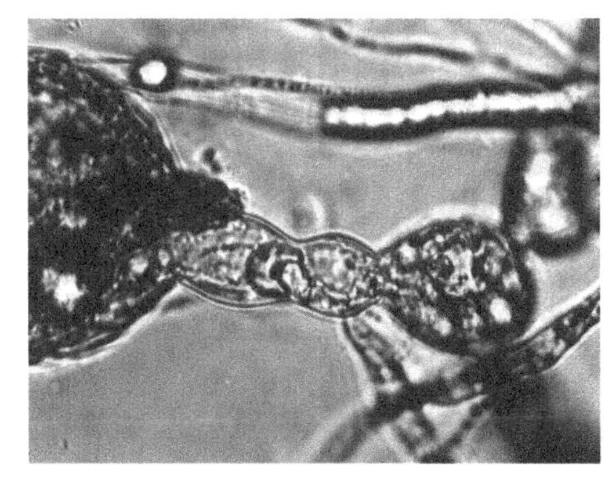

Fig. 9.5. Advanced filament form showing reddish aggregation (probable protein in nature) with internal cross-domain bacteria and cellular production.
Clifford E. Carnican

What Clifford Carnicom called the *cross-domain bacteria* constituted an interpenetration of all three domains of nature—Archaea, Bacteria, Eukarya*—to create new inanimate life forms that synbio scientists and engineers had been laying the groundwork for. Now, fungi and parasites can be forced to assimilate a higher DNA than their own, multiply it, and build up DNA clusters that resemble those in the human morphogenetic field so as to fool and override the ancient Human 1.0 immune system.

In 2014, Carnicom finally stopped waiting for attention from the EPA and named the bacteria-like biological microorganism with inorganic features the *cross-domain bacteria (CDB)* so as to recognize the three domains bioengineered into this synthetic life form: *eukarya* (sophisticated multicellular life forms), *archaea* (single-cell microorganisms), and *bacteria*

*Cellular life is now classified in just these three domains (since 1990).[24]

(both microscopic unicellular and more evolved single-cell organisms). By then, he was finding polymer filaments not just in the environment, but in human and animal saliva and blood[25] inside the submicron spheres he had named, serving reproduction and growth of a "novel and ubiquitous life-form that is known to exist in association with the so-called 'Morgellons' condition," as well as creating nodes of operation near nerve centers and joints in Human 1.0.[26]

Fungi of the Eukarya domain give Morgellons the ability to grow and multiply explosively when triggered by specific electromagnetic frequencies such as the "blue light" emissions of 460–475 nanometers—all of which acidifies the blood.* But just as Morgellons is neither a mere skin condition nor entirely synthetic, it is not entirely a fungus, either, despite presenting as a wirelike polymer fiber that *looks* like a tubular fungus. Similar to human red blood cells able to self-replicate in Nature but not in a petri dish, cross-domain bacteria are hearty inorganic life forms that can withstand freezing, drying, and burning in a Bunsen flame when bleach is poured on them. Nothing destroys these synthetic fruiting body cells.

By 2014, Carnicom had confirmed that the Morgellons condition "represents a fundamental change in the state and nature of biology as it is known on this Earth. The evidence now indicates and demonstrates that there is, at the heart of the 'condition,' a new growth form that transcends plant and animal boundaries."[28] Enough tests had determined that the original (*coccus*) bacterial status had been bioengineered in order to form a unique relationship between iron and the cross-domain bacterium, particularly in its ability to set up inside red blood cells and devour the iron, as I had seen it do.

"The fibres grow and extend through my body, actively pulsing together, causing uncontrollable muscle spasms. They grow into worms, their eggs 0.8 to 1–2 mm in size placed inside insects with many legs and a T proboscis used to puncture the skin so the beetle-like body with

*"The interest in luminescent semiconductor nanoparticles (SNPs), such as quantum dots (QDs), is growing primarily due to their easily controlled luminescence wavelength in the visible range and narrow emission full width at half max (FWHM). This has enabled their growing application in information displays, LEDs, solar cells, optoelectronics, large format cheap microelectronics, biosensors, etc."[27]

its many rows of eggs can be pulled through the skin opening . . . The worms go through any tissue and seal the wall behind them. They attach to the back end of blood vessels, go to the edge of the skin, and put eggs wherever you go through the skin . . . I have no plausible way of getting rid of these things. Sugar or carbohydrates that break down into sugar feeds them. While treating myself, the worms encyst in my torso and left leg; when I stop, they disappear, and it's business as usual."
—PHILIP BALL, Morgellons sufferer, email, July 3, 2019

Because preparing slides of dried blood samples has become increasingly difficult, Clifford Carnicom has turned to live blood analysis ("The level of coagulation, rouleaux, or agglutination appears more severe than in the past, and individual cell observation is more difficult to achieve. A real time blood coagulation observation is required, and this is the approach taken."[29]) and has over the last three years tested the blood of four uninoculated subjects in an attempt to answer the question, *Are there obvious changes in post-inoculation people's blood?* Working in tandem with doctors and microscopy researchers around the world who are now closely examining live blood and what is in CV-19 phials besides serum, plus other fluids such as dental anesthetics and insulin, he is connecting the recent mRNA Big Pharma assault on Human 1.0 blood (2020-2023) with what his more than twenty years of studying the geoengineered aerosol assault has indicated. The blood coagulation he saw in two uninoculated subjects was due in part to the presence in the samples of carbon-based polymers coupled with nanometals which billions of people have been inhaling and ingesting for going on three decades. (Given that *hydrogel* is made with synthetic carbon polymers, was it involved in the coagulation found in the blood of both inoculated and uninoculated?)

In a 2022 article posted on his website, Carnicom describes the results of his research when applying an electrical current to a live blood sample, and then witnessing an incredible transformation within two hours: rapid proliferation and growth of cross-domain bacteria microfilaments from nano to micro.[30] He concluded that there is most definitely a physical transformation of human blood going on in our now wireless world: *blood is no longer blood.* After a year's worth of studies of blood samples, Carnicom writes:

> I will introduce two phrases here at the onset, the first is that of a "*kill switch.*" This first is well known and the second phrase comes to mind as a slightly gentler option, and that is one of "*selective decimation*" . . . Conceptually and theoretically, it would appear that a "kill switch" for the human race now exists. It is one of our questions as a species if we would like to confront that potential reality or not.[31]

Morgellons actually alters the blood, as Clifford Carnicom's 2019 paper "The Transformation of a Species?"[32] discusses. Morgellons is not simply an "infection" or parasitic attack; the synthetic, bioengineered cross-domain bacteria are altering the fundamental morphology and geometry of human erythrocytes so that "The blood of every individual does exhibit some degree of variation caused by the presence of the [cross-domain bacteria]," with some blood cells being extremely misshapen, others being less so, and Rouleaux (red blood cells stacked together like plates) being the most common. In some cases, the cell membranes remain surprisingly intact, with no material structural damage; in other cases, ragged edges and actual obliteration of cell membranes are seen.

The *synbio* Morgellons creation is now in everyone, not just in those displaying skin lesions and rashes. Anomalous cross-domain bacteria have been found in people suffering with lesions and rashes and in those without. Only the degree of presence of these anomalous synthetic forms differs, with more pronounced degradation of cellular integrity in those with lesions:[33]

> The segregation of only certain individuals as having the 'Morgellons' condition is completely and totally false; the general population is involved, whether they would like to know of it or not. The pathogens found have now been discovered repeatedly across all major body systems and functions, including skin, blood, hair, saliva, dental (gum), digestive, ear, and urinary samples.[34]

Morgellons sufferers may test positive for Lyme or *Bartonella* bacteria, but the sores are different: Lyme displays the characteristic bull's-eye rash (*Erythema migrans*), whereas Morgellons varies from a red rash to staph-like craters with protruding, color-coded "wires" extending from the skin. To sufferers, Morgellons lesions feel like glass shards cutting from the inside out—a clear sign of transdermal nanotechnology contamination. Morgellons was never a mere skin ailment or even pathogen- or parasite-driven disease befalling certain unlucky individuals. It is the first phase of experimentation to create an "under the skin" *operating system (OS)*, a biometric interface that integrates bodies and brains with the Space Fence grid lockdown and artificial intelligence.

After twenty-plus years of lab work, Carnicom made it crystal clear that four government entities—the Environmental Protection Agency (EPA), the U.S. Air Force, the Centers for Disease Control and Prevention (CDC), and the U.S. Patent and Trademark Office—have committed crimes against humanity by colluding in burying and obfuscating the issue of Morgellons. Nor can it be surprising at this point in the transhumanist assault that the Department of Defense (DOD), Department of Health and Human Services (DHHS), and Food and Drug Administration (FDA) have been behind Operation Warp Speed and everything else having to do with the Covid-19 "vaccine" development, given that synthetic biology has never been about helping the blind to see and the maimed to walk but has been about biological warfare against Nature.[35] What is surprising is *the depth of public mind control* that has allowed the military and its contractors to *operate unaccountably* for over a half century, two jaw-dropping examples being the $21+ trillion stolen from the U.S. budget for the secret space program* and its *synbio* / nanotechnology research,* and handing over $6 billion to Advanced Technology International (ati.org) to develop the Covid serum.[37]

Meanwhile, the Department of Homeland Security's SENSR (spectrum efficient national surveillance radar) is digitally copying each Human 1.0 into a *Sentient World Simulation (SWS)* as individual virtual avatars with self-learning algorithms, digital "twins" of ourselves ready and willing to remotely run what is left of Human 1.0 in our bodies and brains, particularly in those who have been inoculated with the software that enjoins them to their Human 2.0 Cloud model. The Sentient World Simulation is the AI component of the Internet of Things and all of its multiplying spawn—Bodies, Bio-Nano Things, etc.—plus the synthetic biology we are now encountering at every turn under varying names and applications,

*"The Solari Report has been covering the missing money since 2000 when Catherine Austin Fitts began to warn Americans and global investors about mortgage fraud at the U.S. Department of Housing and Development (HUD), the engineering of the housing bubble that led to trillions more dollars in bailouts and funds missing from the US government starting in fiscal 1998."[36]

from "precision medicine" to personally tailored advertisements via our wireless body area networks (WBANs) and the Internet of Bodies (IoB) that are constantly making use of cell phones, towers, drones, planes, satellites and transmitters embedded in the environment (buildings, vehicles, and furniture). We are to have a presence in the natural world (at least for now) and a presence in the synthetic world. Despite the cheery-sounding "single holistic framework" in the following text, the SWS technology is transhuman to the quick.

> The Sentient World Simulation (SWS) is an ultra-large-scale ABS [agent-based modeling and simulation] developed to capture a comprehensive view of "Whole of Government" operations. The SWS supports a strategic geopolitical perspective that captures the interplay between military operations and the social, political, and economic landscapes. The SWS consists of a synthetic environment that mirrors the real world in all its key aspects. Models of individuals within the synthetic world represent the traits and mimic the behaviors of their real-world counterparts. As models influence each other and the shared synthetic environment, behaviors and trends emerge in the synthetic world as they do in the real world. The SWS reacts to actual events and incorporates newly sensed data from the real world into the virtual environment. Trends in the synthetic world can be analyzed to validate alternate worldviews. The SWS provides an open, unbiased environment in which to implement diverse models. This results in a single holistic framework that integrates existing theories, paradigms, and courses of action.[38]

FROM MOLECULAR TO DIGITAL BIOLOGY

Aerosol delivery has led the way to the transhuman Human 2.0 synthetic biology now underway in human blood since the epigenetic release of "experimental" Covid-19 serums in human bodies—a release affecting both injected and noninjected, as their blood samples reveal—GMO soil,[39] terminator seed "foods," livestock, and basically anywhere and everywhere.

Over the years, red blood cells and gels have fallen out of the sky in the American Pacific Northwest, spider web–like strands in California, discrete filaments in Oregon, and red blood cells (not "juniper pollen") in New Mexico and Colorado. Early on, Carnicom established that airborne submicron fibers and the characteristic fibrous structures emerging from the skin of Morgellons sufferers are one and the same.[40] We the people have unknowingly served as the beakers and Petri dishes for secret transhuman synthetic biology experiments for decades. That the subatomic quantum world of nanotechnology has also taken us by storm demands that we examine what *gain of function* really means (beyond lucrative patents) when it comes to "pathogens," shape-shifting polymers,[41] and cells growing on intricate polymer "scaffolds" bathed in hydrogel.

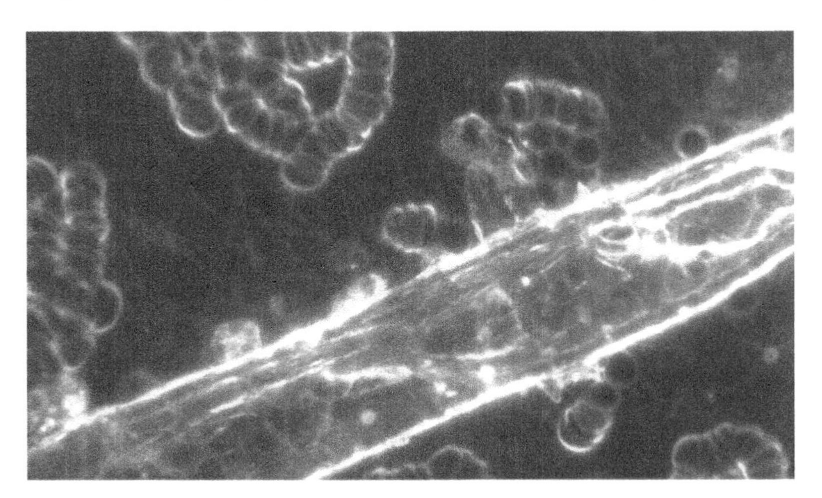

Fig. 9.6. Presumed hydrogel/graphene ribbons in live blood sample.
Ana Mihalcea

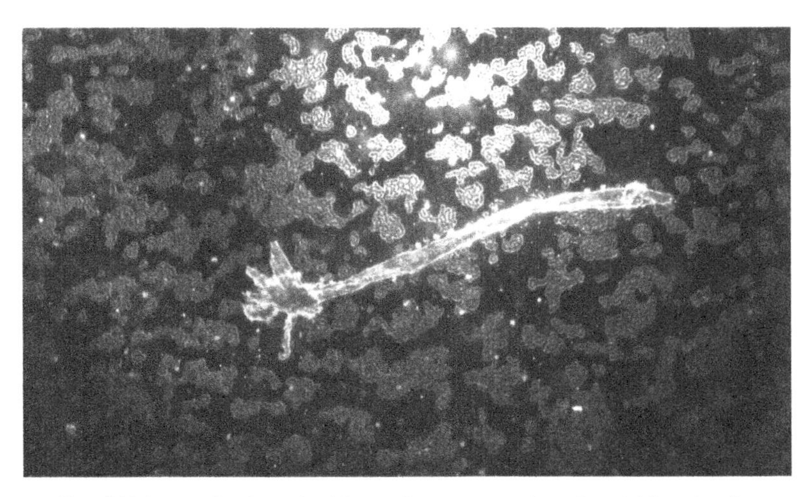

Fig. 9.7. Large hydrogel ribbon showing stacks of red blood cells.
Ana Mihalcea

> The scaffold is built out of a series of thin layers, stamped with a pattern of channels that are each 50-100μm (micrometers) wide. The layers, which resemble computer microchips, are then stacked into a 3D structure of synthetic blood vessels. As each layer is added, UV light is used to cross-link the polymer and bond it to the layer below. When the structure is finished, it is bathed in a liquid containing living cells. The cells quickly attach to the inside and outside of the channels and begin growing just as they would in a human body.[42]

Graphic photographs of the extrusions from Morgellons sufferers' skin are found in Carnicom's many papers at carnicominstitute.org. At high magnification, the artificial nature of these fibers begins to emerge: the single Morgellons fiber is actually composed of innumerable sub-fibers measuring less than 1 micron in thickness—similar to airborne fibers

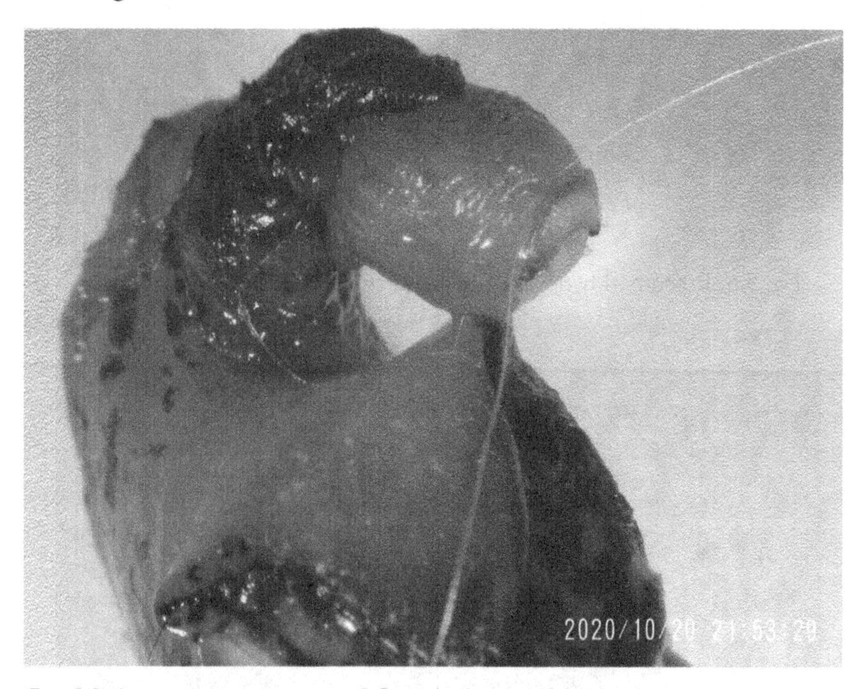

Fig. 9.8. Insect pupa or cocoon? Sample from Alf, a two-year-old Yorkshire terrier suffering from a heavy Morgellons infestation since the summer of 2020. This evil "pupa" popped out of Alf's skin. It looks like something between a worm and a snail, and it erupts frequently in many sizes, from 1 millimeter to 2 centimeters. It is slimy, sticky, really nasty, and doesn't move.

Anne Helene Have

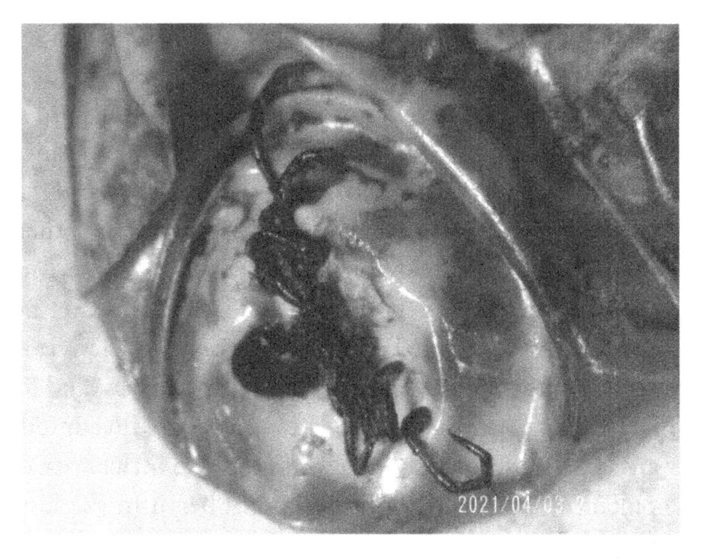

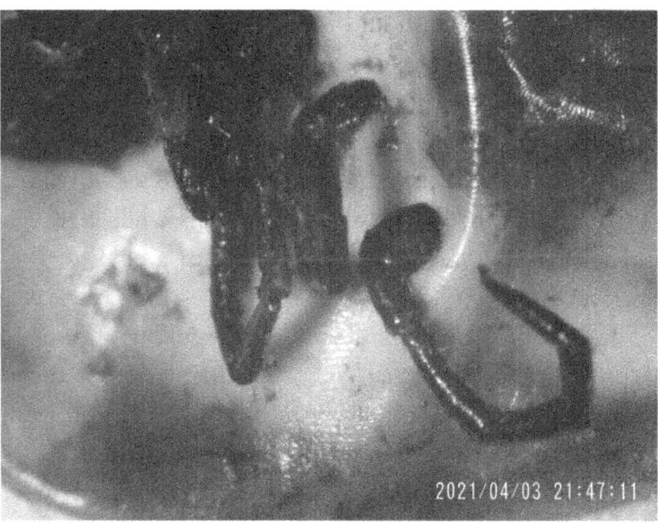

Fig. 9.9. Spiderlike formation found in Alf's blood. Anne Helene Have, Denmark: "Both of my dogs are heavily infested with a very dark and evil version of these technologies. One night in April 2021, Alf was restless and running around the house. In the morning I followed blood traces from the living room to the bathroom, where I found what looked like a part of a fetal membrane surrounded by blood with a spider growing out of it. I took microscope photos of the amniotic membrane and attached spider parts. I don't know what happened to the other half of the missing membrane, nor what was inside it. Did Alf vomit the spider? Did it come out of his rectum? From a sore?"

Anne Helene Have

many beyond Carnicom have collected—and the cross-domain bacteria inside the fibers measure 1 micron or less, with branching or budding growths encapsulating submicron structures that seem fungal in nature.

A decades-long dialogue about Morgellons by people suffering from its effects has been running on websites, as well as on various social media sites and on YouTube.* Meanwhile, medical professionals remain either clueless or are complicit, with the CDC continuing to misdiagnose Morgellons either as "delusional parasitosis" or a "skin condition." And yet no matter what deceptive narrative is trotted out, the cross-domain bacteria organism lives on, feeding on fear, sugar, electromagnetic frequencies, and low (acid) pH, seemingly with its own "swarm consciousness" or "hive mind," which is why the sufferer's consciousness matters, as the two influence each other. The development of cross-domain bacteria may begin in the lungs as we breathe in the filaments, but it is in the acids of the digestive tract that the crystals for communication and motility appendages develop—which is why it is important to adhere to an alkalizing diet and why boron (in the form of borax) may be a useful nanofilament replication inhibitor.

Several factors connect the history of Morgellons to the present *synbio* Covid-19 era, with the hope that those trained in the scientific method will at last awaken to the magnitude of what happens when lies, secrets, subterfuge, and self-serving greed are allowed to run amuck in the name of "science" and "progress." At last, people are stepping forward to examine the historical and scientific connections between Morgellons and the "biometric application programming interface" (BioAPI) being delivered by chemical trails, injections, and the food supply, beginning with the following:

- The cross-domain bacteria (CDB) "new life form" shares polyvinyl alcohol with the hydrogel in the Covid serums. Both the CDB and hydrogel are composed of "metamaterials" that are synthetic biological forms designed to construct a neural network that self-assembles,

*I recommend Clifford Carnicom's website, Carnicom Institute; the 2012 "Consciousness Beyond Chemtrails" conference presentation titled "The Dark Agenda of Synthetic Biology" by Sofia Smallstorm, available on the Bitchute website; reporting by Skizit Gesture, by herbalist Tony Pantalleresco on Substack website, and by Lookoutfa Charlie on his Blogspot website; the Klinghardt Institute's Biological Lyme Protocol; and Facebook pages of Marcia Pavlis, Finding Hope with Morgellons, Morgellons Coverup, Tommy Target, Morgellons: An Open Forum (closed group), Morgellons Extreme & Emerging Illnesses, and Energy Clearing Protocol.

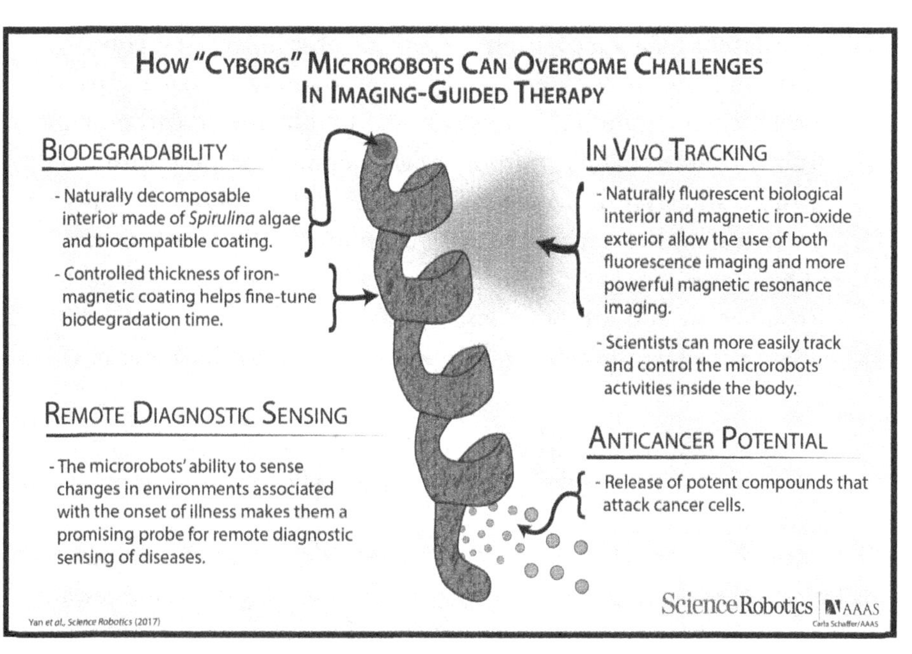

Fig. 9.10. Cyborg magnetic microrobots are helix nano-antennae made of fluorescent *Spirulina* algae with biofilm coating, used for *in vivo* (in the body) tracking, thanks to a magnetic iron oxide exterior.
University of Edinburgh, "Tiny robots step closer to treating hard-to-reach parts of the body." TechExplore, November 22, 2017.

self-replicates, self-organizes, and self-heals. While some functionary instructions are programmed into them, others can be delivered remotely by AI systems or human operators via "precision medicine" and "medical geoengineering." For example, Morgellons filaments can program hair to turn medulla shafts into an antenna farm. The receiver-transmitter capability of hair has long been known, from the story of Samson and Delilah, to how traditional native peoples cut their hair only during a full moon, the same kind of thinking applied by alternative practitioners who recommend shaving or waxing one's hair to reduce the power source available to nanotechnology. Putting a cell phone up to one's head or carrying it on the body feeds *synbio* like Morgellons.

- The CDB inside the polymer sheath is *synbio* but they're not *entirely* synthetic; they're more like carbon-based life forms from a different planetary biosphere, capable of mimicry and consciousness and knowing when they're being observed. Densely packed erythrocytes are often interwoven with multilayered carbon nanotube strands that have undergone chemical decomposition and fusion, all carefully arranged in hexagons of silicon and copper*—electrical and biological molecular robots smaller than a cell.[43]

- The Morgellons "wires" sticking out of lesions seek electromagnetic stimulation and are programmed for different functions.[44] These filaments don't "grow" so much as pop up like hidden antennas.

PLASMONICS, BIO-NANO-ANTENNAS, AND HARALD KAUTZ

The science of *plasmonics* has been booming since the onset of geoengineering but was first developed in the 1950s, combining "fundamental research and applications ranging from areas such as physics to engineering, chemistry, biology, medicine, food sciences, and the environmental sciences."[45] Inspired by photonics (the study of the manipulation of photons of light), plasmonics refers to the generation, detection, and manipulation of signals at optical frequencies along metal-dielectric interfaces at the nanometer scale.[46] *Signals.* Thanks to our plasma-suffused atmosphere, the artificial nanocrystals that now coat our neural pathways are handy for tracking, tracing, and targeting, much like phased-array antenna beams that use electron spin resonance spectroscopy via nanocrystals instead of MRIs to make frequency connections for remote biosensing and biotelemetry.

The U.S. military and Big Pharma, in public-private partnership, have developed a plasmonics that uses a single microbead for tissue engineering bio-scaffolding platforms (think Elon Musk's Neuralink[47]) as well as for liquid crystal applications of biofilm nano-delivery. "Neural mesh" (bio-scaffolding) platforms began with electronic engineer Pietro Valdastri's work on *in vivo* telemetry systems in 2004 at Vanderbilt University,[48] work that has now advanced to synthetic nerves replacing natural nerves for cell communication, thanks to neural nanobots like those in *graphene-based hydrogel (GBH)* coupled with the heavy nano-

*As I indicate elsewhere, the hexagon, the chemical signature of consciousness-changing elements like DMT, is everywhere in this metamaterials technology.

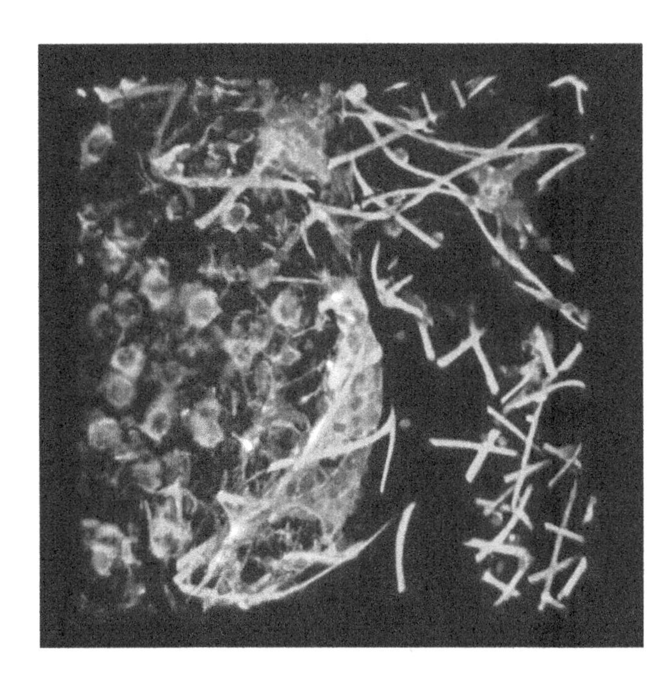

Fig. 9.11.
Mesh (neural lace) merging with brain cells.
Lieber Research Group

metals breathed in for decades. "Liquid crystal applications" imply both transmissions (think the old crystal radio sets) and what I later refer to when discussing "black goo"—"a hive mind in a technological realm without spirit." As one commentator noted, "This could be the beginning of the first true human internet, where brain-to-brain interfaces are possible via injectable electronics that pass your mental traffic through the cloud. What could go wrong?"[49]

With optical fibers (fiber optics) and optogenetics (i.e., the use of light signals to monitor and control brain cells), we arrive at the self-assembling nanomachines that collect light patterns of DNA and turn them into electromagnetic signals transmitted via plasmonic bio-nano-antennas. Of course, the *bidirectional* brain-computer interface (BCI) that hoovers up emotions, energetic states, and DNA cell communication can also turn electromagnetic signals into artificial DNA light. Could this be the incoming programming of Human 2.0?

For an extraordinary look at plasmonics and the submicron world of nanotechnology coupled with digital synthetic biology, we turn to German scientist-mystic Harald Kautz,* our unique trail guide into the

*Isaac Newton (1643–1727) was also a scientist-mystic, as have been many great scientists.

dystopian future rushing to overtake Human 1.0. The CEO of Aquarius Technologies, a company with water treatment solutions that employ microbiology, Kautz has authored several books and written a number of papers and essays on Morgellons and the atmospheric nanoscience behind the chemical trails that deliver Morgellons.[50] He offers us an eclectic map into a *quantum* realm that no longer fits the old Newtonian paradigm, due to the fact that the nano-scale tiny, basically invisible things and fields that we must now accept—sight unseen, so to speak—obey the laws of quantum mechanics, not classical physics. Kautz strongly feels that understanding this new era depends upon our realization that we are essentially beings of light living for a time in a materially bound, electromagnetic world of multiple layers of invisible forces that our technologies attempt to read and measure.

Kautz views Morgellons as a carbon-nanotube quantum-dot technology designed to be part of a brain-computer interface (BCI) network that began with the science fiction-sounding "black goo" that he insists lay buried in Texas oil fields for 20,500 years and was, in fact, the source of the petroleum that spawned the planetary machine era we are now ensnared in. Kautz believes that black goo and graphene are one and the same.[51]

Black goo* is now part of the zeitgeist, having appeared in a number of recent movies, where it's depicted as an animate and conscious agent that reaches into quantum dimensions: *Smilla's Sense of Snow* (1997), *The Matrix* franchise (1999–2021), *Spider-Man 3* (2007), *A Haunting in Connecticut* (2009), *Prometheus* (2012), *Lucy* (2014), *Stranger Things* (a TV series since 2014), *Ares* (2016), *Alien Covenant* (2017), *The Silver Surfer* (2020), and others. Such consciousness preprogramming may be Hollywood's attempt to convince us that the fanciful-sounding black goo is fiction and not science. Given the advent of inorganic life forms such as *synbio* fungi, parasites, DNA, etc., the assumption that because black goo is inanimate, it is therefore unconscious—or worse, a mere figment of the imagination—may be a tragedy for all of humanity because whether you call it black goo or graphene, it is paving the way not just to mind control but to Transhumanism.

---

*Interestingly, the term *gray goo* is how nanotechnology discoverer Eric Drexler characterized ecophagy, the global catastrophe of self-replicating nanobots that could eventually consume all of the Earth's biomass.

Kautz describes the bioengineered, synthetic Morgellons creation as having a fungus-type cell tissue with fruiting bodies that display a morphogenesis 80 percent human and 15 percent insect, with male and female reproductive organs that produce offspring encapsulated in the shape of plasmonic photonic crystals. These plasmonic synthetic forms are described in transhumanist literature as "technical units" that use human biophotons as a primary energy source. Since biophotons are light particles generated from within the body of any biological entity and are the source of *qi/chi* or *prana*—the life force that animates living beings and things[52]—biophotons are quite literally "light parasites" feeding off our life force.[53]

Biophotons register the presence of *light ether*, which gives rise to plasma.* Russian biologist Alexander Gavrilovich Gurwitsch (1874–1954) observed how onion rootlets strengthened weaker rootlets via their light emissions, which measured 260 nanometers (biophotonic activity ranges from 200 to 800 nanometers), thus discovering the biophoton. Gurwitsch went on to originate the morphogenetic field theory that British biologist Rupert Sheldrake has since advanced, which proves that biophotons connect all of life to multiple quantum dimensions. Gas discharge visualization (GDV) techniques[54] pick up biophoton activity in the coronal discharge around living bodies. As quantum entanglement proves, biological systems are connected *over distance* due to invisible scalar dimensions in which biophoton events occur instantaneously—faster than the speed of light.† Kautz: "Another fact illustrating the major role of the annihilated part of biophoton activity is that when a being dies, in the moment of death there is a burst of biophotons leaving the body—originating from the biophoton activity disentangled at that very special moment—when cell communication is losing its coherence."[56]

*As discussed in chapter 5, æther/ether is the substance found throughout the electric universe that gives birth to plasma, the first (not the fourth) state of matter. From there, it extends to the subtle bodies of light that all biological entities have, including planet Earth herself. As discussed in Freeland, *Under an Ionized Sky* (2018), Rudolf Steiner delineated four ethers that bind the physical to the psychic (or soul or spirit): The warmth ether, the most primordial, manifests as heat and appears as spherical; the light ether manifests as plasma/gas, its primary quality being luminosity; the chemical/sound ether manifests as fluid and is disc-forming; and the life ether immediately precedes matter and is individualizing.

†Quantum entanglement appears to be how researcher Deva Paul's consciousness (astral body) was "teleported" in a March 2000 study.[55]

Fig. 9.12. Plasma / æther. Alan Bean, Apollo 12, 1969.
Is the cloud surrounding him an aura or plasma / æther or ozone (O_3)?
Johnson Space Center of the United States National Aeronautics and Space Administration (NASA)

Quantum entanglement introduces us to the *transdimensional aspect of synbio* in nano-creations like Morgellons. In fact, the deeper into *synbio* we go, it appears to have a great deal to do with the "metaverse" synthetic world simulations (SWS) and Rudolf Steiner's idea of the Eighth Sphere. Here, Kautz untangles the mystery of the "under the skin" parasitic sensations experienced by many Morgellons sufferers:

> Aborting [pseudomorphous] fruiting bodies are accompanied by a very painful pulsed extraction of biophotons from a kind of parallel "dimension." It feels like a being sucking the life force out of the head in the direction of the intestines . . . This perception could represent the birth of a second generation of the beings whose DNA is inside the red stem cells described by [the Carnicom Institute]. We would experience a species reproduced by pseudomorph mothers and fathers who have a fungus-type cell tissue and yet functional reproductive organs.

> The babies of this species would use the human biophotons in order to shift onto higher realms, a parallel space-time level that allows a parasitic way of life, with humans as energy sources.[57]

Kautz's description of "aborting fruiting bodies" and their quantum entanglement with transdimensional synthetic nano-creations makes me think of matter (in this dimension) and antimatter (in quantum dimensions), both bound by light (biophotons) pulsing at 144,000 cycles per second.* When Kautz describes the moment of death as being "a burst of biophotons leaving the body," it sounds like matter and antimatter exploding and separating in order to return to their original energy states. This raises the question: Will transhumans be twinned to transdimensional virtual beings through the process of plasmonics?

> According to Kautz, artificial piezoelectric crystals are taking the place of the natural ferro apatite (phosphate) crystals in the body that make the biological system more responsive to artificial electromagnetic signals. The natural ferro apatite crystals are also piezoelectric crystals and play a major role in the transmission of signals in the central nervous system. If the natural ferro apatite crystal is displaced by artificially made piezoelectrical crystals, the biological system is opened to a greater extent to artificial electromagnetic signals, both low and high frequency.[58]

Piezoelectric converts ultrahigh frequency (UHF) sound waves to electrical signals and vice versa:

> The fibers and the crystals form a read/write unit. The fibers collect DNA light communication, i.e., the bi-directional single photon emissions interchanged by any DNA cluster, and turn it into radio signals. The crystals take in radio signals and transform them into light signals readable by the human DNA. Whatever human experience I want to "mind-control," I can induce this experience in a person, like saying "ass-

*Harald corrects my intuition in his April 9, 2024, email: "Death is not exactly coming from the state of matter but from biophotons in the annihilated state, i.e., two longitudinal biophoton beams (single photon emissions) that form a pair of a wave and a time-reversed replica wave that annihilate each other. As long as we live, the light is invisible, but when we die, the cells un-sync and the beams de-couple and become visible."

> hole" to a person to induce anger, read it out by collecting the signal sent by the Morgellons with a special antenna, store it in a digital file, turn this stored anger into a radio signal and make any other human being experience the same emotional or even mental pattern by making the hexagonal crystals reproduce the light patterns. According to old Vedic knowledge, red light controls sexuality, orange anger, blue the thoughts.[59]

The more one studies developments in *synbio* nanotechnology—from gain-of-function transformation of natural life forms into vectors, pathogens, "virus," etc., that can then be patented, to the actual creation of new "crypto" artificial life forms and "robotic 'life'," as in this patent—

> U.S. Patent 7523080B1 Self organizing model for artificial life [2005]

> Abstract: A new architecture overcomes the limitation of conventional robotic technologies. Named the Self Organizing Model ("SOM") it includes a method that allows systems to learn, grow and continually evolve without outside control. This technology enables Artificial Life, one aspect of which is robotic "life." If this system is compared to a real living thing, the hardware is like the body and the potential instinct and habits and related data are like the DNA. The hardware includes memory which contains the instinct and related data. Algorithms and organizations are provided so that the hardware forms an adapting and evolving brain that senses the environment and formulates actions to improve the survival of the Artificial Life according to predetermined rules. The organism can learn and become more complex all without complex software programs that attempt to anticipate all possible situations.

—the more one realizes that *synbio* nanotech is actually an unnatural life form whose "consciousness" is built to merge with AI and control or transform what is natural. Morgellons symptoms *seem* biological, just as the mRNA delivery system *seems* like the medical procedure called a vaccine. A deeper look with dark-field microscopy at either reveals invasive technology that is innately or remotely conscious, self-assembling, and self-replicating, consisting of carbon nanotubes, nanowires, and nanoarrays loaded with sensors. Engineered to merge the inorganic with the organic via gain-of-function spliced DNA or RNA, Morgellons nanobots use the body's energy system as well as environmental electromagnetic systems to obtain total penetration of the human body. This technology has been preprogrammed to control bodily organs and lay self-assembling scaffolding that replaces the Human 1.0 neural network with a silicon-carbon Human 2.0 network plugged into a vast (Internet of Bodies) hive mind that goes far beyond what we now call "the Cloud."

NANO SELF-ASSEMBLAGE, REMOTE ELECTROMAGNETICS, AND TONY P

CARBON NANOTUBES

Fullerene

an allotrope of carbon whose molecules consist of carbon atoms connected by single and double bonds so as to form a closed or partially closed mesh, with fused rings of five to seven atoms

3-D carbon origami circuitry

artificially folded DNA in complex structures;[60] *electronic applications*

Quantum Dots

semiconductor nanocrystals with optical and electronic properties

Dendrimers

highly ordered, branched polymeric molecules

Canadian herbalist Tony "Tony P" Pantalleresco* does not use the ill-named term "Morgellons," preferring instead the more accurate *nano self-assemblage* engineered to wirelessly respond to remote electromagnetics like 5G/6G. As Tony P puts it:

> We only hear that nanotech is atomistic but not that it is programmed to self-assemble. That is when you see it. Nor does anyone pay attention

*See "Tony Pantalleresco," Power of the Pulse website.

> to the fact that it replicates and repairs itself, nor that each nanoparticle can hold a terabyte of data. Because of these oversights, we fall into fear and don't seek to understand either its defence mechanisms or what actually stops its assemblage.

Tony P also corrects the conditioning we have undergone to view our bodies as carbon-based. Up until now, Human 1.0 has been protein-based; it is only now that we are being forced to become carbon- and silicon-based.

Tony P represents the very best of self-trained citizen science in the era that proves many "experts" and institutions simply cannot be trusted—like government, Big Medicine, Big Pharma, media, etc. Tony P has personally undergone the Morgellons trial by fire and come out the other side by dint of his research and experimentation into healing an immune system under assault not by molecular pathogens but by digital synthetic biology and nanotechnology constantly stimulated by our wireless electromagnetic world. Over the years, he has generously responded to thousands of people suffering from the chemicals loaded with tiny classified nanotechnology being constantly and covertly delivered by aerosols to lodge not just in Smart Cities but in billions of bodies and brains, a decades-long fact that the military-intelligence-pharmaceutical cartels refuse to acknowledge.*

Tony P stresses that we do not need electron or dark field microscopes in order to get a look at the nanotechnology coming out of our bodies. For one example among many, even the tiny 60X pocket magnifier available on the internet† will show the fullerenes on the surface of fruits in the grocery store.

---

*Tony P's podcast (at Substack website) provides prolific commentary on the nanotechnology we are breathing and ingesting—the tiny, sometimes floating, sometimes adhering fibers now everywhere in neural dust settling on every available surface, even the vegetables and fruits displayed in supermarkets. By specializing in nano-biotech primary and secondary symptoms, Tony P takes the time to respond to those who write him, describing their symptoms in detail, and the hundreds who tune in to his radio show "Everything Goes." He constructs magnetic devices like the nano-bucket and nano-triangle, and concocts individual orthomolecular recommendations of aromatherapy, foods, minerals, salts, magnetics, and magnetic coils.

†60X Magnifying Loupe Jewelry Jewelers Pocket Magnifier Loop Eye Glass Led Light, $11.19. A 600X microscope will even pick up Teslaphoresis self-assembly. Teslaphoresis: carbon nanotubes (CNTs) self-assemble into long wires under a Tesla coil.[61]

In his well-supplied workshop, he has developed various low-tech devices dependent upon natural chemical combinations, electromagnetics, and sheer magnetism to stimulate the body to release pockets of bots, crystals, fullerenes, origami circuitry, and liquidized proteins in small electromagnetic pulses (EMPs). By working virtually to compare notes with similar citizen scientists around the planet (western and eastern Europe, Eurasia, India, Australia, New Zealand) who are also experimenting with Tony P's devices, progress has been made over the years in discovering how Human 1.0 can meet the classified nanotechnology challenge without high-tech that we the people never gave our consent to.

WEAPONS OF MASS SOLUTIONS

Anti-nano Triangle

In his video "Dr. Dena Interview with Triangle,"* Tony P explains how the electromagnetic fields of the Triangle work to draw out nanotechnology and shut down its programming

Nano Detox Bath

Hot bath: Soak to eliminate daily exposure buildup. The salts get into your cells through the skin and flush out the particulates. Can be used in combination with the anti-nano triangle for EMPs.

¼ cup of borax
¼ cup of sodium bicarbonate (baking soda)
¼ cup of Epsom salts
¼ cup of TSP (trisodium phosphate)

Tony P worked with the same naturopath that Clifford Carnicom worked with in New Mexico: Gwen Scott, ND, in Cochiti Lake, New Mexico; like Tony P, she too was suffering with Morgellons. It was Dr. Scott who introduced the term "pseudo-lifeform," then later referred to it as "artificial life." Dr. Scott had been a CNN television news anchor for over thirty years until studying with traditional healers and Ayurvedic

*March 3, 2023, video is on Bitchute website.

medicine practitioner Deepak Chopra. She was awarded a degree in naturopathic medicine by Clayton College of Natural Health in 2002. Over her career until her death on March 15, 2015, Dr. Scott had worked with over 30,000 people, many of whom were Morgellons sufferers. All that while, she continued presenting natural medicine reports on Albuquerque CBS and writing the syndicated health column *The Herb Doctor* while living on a Native American pueblo outside Santa Fe.

Tony P recognizes the challenge that the Space Fence lockdown (smart grid) we are now subject to intends for humanity: planet Earth as a machine consciousness.

> Here is your Space Fence composition: carbon (more than likely C_{60}* or C_{70}). The Fence just got denser and amplifies more frequencies. Carbon is superconductive, zero resistance, 3X harder than diamond, 100X stronger than steel. If they are using carbomers (polymers functioning as thickening, dispersing, suspending, and emulsifying agents in cosmetics and personal care products) or diamines (binds monomers or molecules that can be bonded to other identical molecules to form a polymer), then it is even more than C_{60} / C_{70}.
>
> C_{60} in the head just fries the brain and central nervous system, turning into a super antenna, and with the Space Fence it will throw the brain frequency since the sky is tied to the brain. The C_{60} with its super conductive materials—and I am sure they are adding titanium borate and silica to enhance the superconductivity—will cause further disruptions as well as more access to a global enchantment or mind control, and anyone with the nano-assemblage who is using silica C_{60} or consuming titanium and high carbon foods [sugar and junk food!] will form the circuitry and implants internally, making oneself more susceptible to brainwashing of all sorts, AI possession, frequency takeover.
>
> Carbomer or diamine body armor would resist a .50-caliber and could be programmed to assemble into any form or pattern. Fire it in terahertz, 5G, or a multiple-band frequency, and it will distribute with

*According to Wikipedia, "Buckminsterfullerene is a type of fullerene with the formula C_{60}. It has a cage-like fused-ring structure made of twenty hexagons and twelve pentagons, and resembles a soccer ball. Each of its 60 carbon atoms is bonded to its three neighbors."

zero resistance. It also increases visibility and amplifies whatever you're looking at. They're now making lenses out of it.*

GRAPHENE, THE MAGICAL *SYNBIO* CARBON

From polymers, filaments, and fibers (similar to those encasing Morgellons) to the biofilm of the last chapter, we are dealing with BioAPI connectors delivered by jets and sounding rockets. Morgellons began the self-replicating infrastructure but has since been joined by a graphene-based hydrogel, which connects the Human 2.0 to AI systems via 5G/6G. When oxidized—to undergo a reaction in which electrons are lost to another species—graphene produces a compound with superconductive capabilities that ensure it will respond to certain frequencies, electromagnetic waves, and 5G technology as a *sentient* entity with innate intelligence and capacity to store vast amounts of data. Once in the human body, it can interpret this data and affect biological and behavioral responses.

"Know that GO [graphene oxide] has been specifically referenced as having the ability to entangle its aggregate field with just about anything (with startling ease) to produce non-localized effects. I would submit that during instances of electronic voice phenomenon (EVP), which I have personally experienced, lights in the room flickered to provide sufficient resources to approximate human phonic forms. I further assert that stereotypical changes in temperature are reflective of the same process by drawing upon less-than-evident pervasive facilitating media (like ethers)." —James Cayon, email, October 9, 2021

As carriers of nanosensors collecting Big Data to quantum dot tags, bioengineered "pathogens," fungi from the upper atmosphere, synthetic parasites, and myriads of replicating polymer- and metal-constructed nanogear (artificial life), Human 1.0 are now nodes in remote communication networks with supercomputers, fusion centers, and laptop boys as per frequency access. Privacy? You must be kidding! Since their creation in

*Emails from Tony Pantalleresco, September 8, 2019, October 1, 2020.

1982, quantum dots—artificial atoms with optoelectronic properties for cell biology research and microscopy—have been hacking into the cell's communications system. In fact, Morgellons was originally designed to collect quantum dots and deliver plasmonic antennas and plasmonic photonic crystals into the human body for incoming and outgoing communications via 5G/6G systems. Quantum dots access your frequency as an energy source for the *synbio* assemblage and networking necessary to your WBAN (wireless body are network) system. When Morgellons sufferers feel the creepy-crawlies throughout their bodies or just under their skin, that is the sensation of quantum dots on the move.

Our bodies have always had various living microorganisms in them, but now a *synbio* menagerie is being developed in us, trapped in pockets of tissues and organs, pulsing until they reach critical mass and the pocket bursts and releases liquidized proteins that sometimes wind up in the tubular strands of fullerene constructs of silica, carbon, graphene, or other polymer or metallic materials—not alive in the sense of *bios*, biological life, but artificial life forms, *synbio* integrations of DNA run by a program. You cannot kill them, but you can, if you find the right way, *disengage* them from their program—for example, with a pulsed low-frequency microwave. With strong pulses, the fullerenes shatter and the quantum dots and bots burn as the program defrags. When the *synbio* bits touch, they short out; if they don't touch, they might take themselves out. It is when they are not agglomerated or aggregated enough to form a programmable network that they are most vulnerable, whereas those already "engaged" take a long time to burst and release their fragmented, burned-out circuitry. If you see glowing radioactive materials on your skin from what's being dropped on us, a Tony P borax–baking soda–Epsom salts soak in the tub will help. Sometimes these materials are fluorescent, sometimes grayish green or yellowish green-gray. Sometimes the tiny metal shards come out shiny, indicating either aluminum or nanosilver. If blackish gray, it's barium titanate; if glowing, it's crystalline materials; if grayish, it's lead, mercury, or titanium.

We have become beakers in an atmospheric chemistry lab, and the juggernaut* overtaking our 5G/6G wireless planet is exponentially

*Juggernaut: a merciless, destructive, and unstoppable force. From the Sanskrit *Jagganatha* or "world lord."

increasing the self-nano-assembly of circuitry in our bodies and brains* in order to increase construction of a Human 1.0 "hive mind" neural network. Picture Elon Musk's neural mesh from one body to another, but with no need for surgery—the logical next step to establishing the virtual reality of the "metaverse," and along with it, complete control over global populations via mind-controlled governments. The references we hear about the metaverse taking humanity to the next iteration of a 3-D Internet, augmented and virtual, sounds like Rudolf Steiner's *Eighth Sphere* one hundred years later as neural self-assembling scaffolding brain-computer interfaces with remote artificial intelligence control and no need for AR/VR headsets. Here, Steiner analyzes how we are already in the Eighth Sphere as we breathe from our atmosphere:

> How was it that the expression "Eighth Sphere" came to be used?—You know that human evolution takes its course through the seven spheres of Saturn, Sun, Moon, Earth, Jupiter, Venus, Vulcan. We will conceive that besides these seven spheres there is still something else which lies outside them and yet is in some way related to the Earth. Here, then, we have a sphere, visible only to visionary-imaginative clairvoyance, which stands there as *an Eighth Sphere over and above the seven which constitute the domain of the ordered and regular evolution of mankind* . . . as long as a man is within the material world, makes his observations through the senses and thinks with the intellect, he is standing in the Fourth Sphere, the Earth Sphere. If he develops his faculties of soul sufficiently to be able to see the Third Sphere, the Moon Sphere, then he takes a far flight—but not in the spatial sense. He observes, not from another place, but physically speaking, spatially speaking, from the same place. These seven spheres ought therefore in reality to be drawn within one another . . . The Eighth Sphere is to be observed within the Earth Sphere. It cannot properly be drawn either above or below; to depict the reality it would have to be sketched into the Earth Sphere. I have often given a crude example to express what

*Phased array antennas (8 tiny arrays per antenna, 16 antennas per cell phone) and electronically variable shifters make the simultaneous projection of multiple microwave beams at multiple users across a band of frequencies easy. Phased arrays are configured in specific patterns to create beams that are steered by AI.

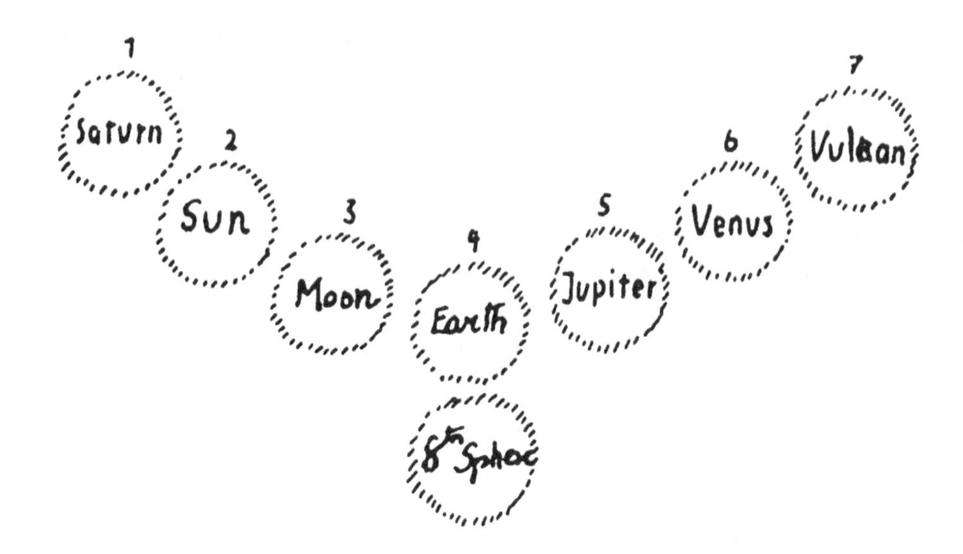

Fig. 9.13. Rudolf Steiner, blackboard drawing from October 18, 1915, *The Occult Movement in the 19th Century*, GA 254, Rudolf Steiner Archive, Dornach, Switzerland.

> is here meant. Just as the physical air is around us, so is the Spiritual around us; we have to look for the Spiritual within what is actually physical in our environment. Hence it must be presumed that just as the spiritual is round about us, so we must also look for the Eighth Sphere in our environment. This means that an organ enabling man to perceive the Eighth Sphere would have to be developed, just as his physical senses enable him to perceive the material Earth. He could then experience the Eighth Sphere quite consciously; but unconsciously he is always within it just as we are always within the air, even if we are not aware of it. And if we have developed an organ for experiencing the Eighth Sphere, we are conscious of it around us. So that if the Eighth Sphere is to be described, it must obviously be described as a realm in which we are living all the time.[62]

Inside the Eighth Sphere being epigenetically developed around us, cell phone apps with varying frequencies are stimulating and spreading nano-assemblies throughout our bodies and brains. The higher the frequency—from 4G to the inevitable 7G and above—the more the nano-network replicates and releases bots. As the frequency climbs, the nanopatterns,* bots, quantum dots, and dendrimers release and a circuit consisting of lattices, origami, networks, multi- and single-walled carbon nanotubes forms, bots and networks coexisting and releasing ever more patterns in order to spread and expand.†

As I indicated above in the section on scientist Harald Kautz, black goo is one of the Eighth Sphere residents, an abiotic (nonliving) mineral oil from the Earth's crust containing high levels of monatomic gold and iridium that avail it of becoming a biophoton attractor that sets in motion an alchemy for a programmable, self-organizing liquid crystal that broadcasts a swarm consciousness for a hive mind in a *technological realm without spirit.* Being magnetic, black goo is capable of influencing the electromagnetic spectrum to provide the foundation for soft-bodied robots and self-assembling nanobots. In fact, black goo *is* synthetic biology, leading us directly to the graphene being delivered for Transhumanism.

Until graphene was found in the Covid serums and made public in 2021, its extraordinary properties were generally unknown outside of DARPA, which was highly aware of its magical attributes. First extracted from graphite in 2004, graphene is considered by the military to be the holy grail. Carbon-based, it is two hundred times stronger than steel, three times harder than diamond, has electro-optical properties that can either repel or harness cosmic rays, and conductivity one million times that of copper. It is light, transparent and two-dimensional, and possesses an ability to adapt and evolve, thereby resisting removal efforts.

Graphene and graphene oxide, "a single-atomic carbon sheet with a hexagonal honeycomb network"[63] that under electron spectroscopy appears as a graphene mesh, is used by geoengineers for both weather control and raining down Big Pharma elements for Transhumanism. Nucleated (chemical) ice‡ is structured along the lines of an ordered gra-

---

*Nanopatterning refers to the process of fabricating a nanoscale pattern having at least one dimension below 100 nm for generating "novel" [weaponized against life] nanobiomaterials for applications in medical devices and implants with superior surface properties.

†Thanks to Tony Pantalleresco, email, June 24, 2019.

‡"Ice-nucleation active (INA) bacteria can promote the growth of ice more effectively than any other known material. Using specialized ice-nucleating proteins (INPs), they obtain nutrients from plants by inducing frost damage and, when airborne in the atmosphere, they drive ice nucleation within clouds, which may affect global precipitation patterns."[64]

phene lattice, a process utilizing the bacterium *Pseudomonas syringae* for fake snow (think ski slopes for the elite) and polar vortexes to enrich energy corporations and scare populations with a "climate change" mini-ice age on its way. "By flattening a droplet of water between two sheets of graphene, researchers have created a new form of ice," the journal *Nature* exclaims.[65] Ice nucleation changes ice crystals at the molecular level by making artificial nanomaterials serve as nuclei for the crystals.[66] Besides oxidized single-walled and multiwalled carbon nanotubes like graphene oxide, other water-dispersible carbon/graphene nanomaterials capable of nucleating ice include carboxylated graphene nanoflakes. The graphene oxide lattice supporting the ice nucleation of polar vortexes is a definite corollary to the lattice network that self-assembles in our bodies and brains once we've breathed in or ingested the graphene-based materials in the environment—whether you've been jabbed or not.

Weaponized graphene is part of a process that can quickly convert electrical energy into enough of a temperature increase to create a thermite-like electrical explosion in the sky, thus replacing "ultra-fast chemical reactions that require fuel and are irreversible" with a repeatable, cleaner, *non-chemical* alternative, the transducer of choice being "low-density and highly permeable graphene foams characterized by heat capacities comparable to air."[67] On the micro scale, the same effect can be achieved in human blood when people are hit with 5G/6G frequencies that allow nanotech graphene to permeate the blood-brain barrier to influence cognitive pathways with its superconductivity and superior charge-carrying capability. Able to absorb frequency bandwidths up to 15 gigahertz, graphene oxide can be made to set off mini-explosions in the blood and brain.

Graphene oxide has a synergy with nano-iron oxides (SPIONs)[68] designed for in vivo drug release[69] by means of aerosols. Load the aerosols with lipid fat molecules, magnetic SPIONs, and graphene oxide, along with the drug or chemical compound slated for dispersal, and once breathed in, the heat of the body releases the drug. *Aerosol vaccination, anyone?* If more heat is required, all it takes is a zap from the 5G/6G system, and graphene oxide can boost gigahertz signals (billions per second) into the terahertz range (trillions per second), due to "the highly efficient non-linear interaction between light and matter that occurs in graphene."[70] Clusters of heated magnetic nanoparticles targeting cell membranes can be made to remotely control ion channels, neurons, and behavior. (Is this what is going on in Covid-19 postinoculation "unusual psychiatric illnesses"?[71])

Behavior modification via graphene oxide . . .

Besides being an excellent conductor of heat and electricity, black goo absorbs radiation and multiplies it a thousand times, the toxicity depending on the amount of radiation absorbed. This is why Covid-19 "virus" symptoms as well as post-inoculation symptoms are identical to those of graphene oxide poisoning: fatigue, muscle spasms, breathing challenges, cognitive issues, and a whole host of skin conditions.

The uses of graphene—black goo's offspring—are literally astronomical,[72] especially in a wireless electric world seeking ever greater conductivity. Combined with fluoride (as it is in the chemical aerosol spraying operations), graphene oxide can enhance aluminum combustion,[73] one example being the 2017–2019 California "wildfires" (see chapter 4). With graphene, one can make multifunctional neural implants,[74] scavenge energy from metals,[75] provide transparency for contact lens computers,[76] and shape the new transistor that uses DNA for transhuman synthetic scaffolding, a transistor far superior to silicon in that the electrons of graphene transistors propagate far faster, 26 gigahertz being the optimal multiplication frequency.[77]

Hidden within the proprietary (secret) manufacturing process for the Covid serum is evidence of the use of graphene nanobots, as revealed in a document submitted to the FDA by Pfizer to gain Emergency Use Authorization, which the Big Pharma giant wanted suppressed for seventy-five years, until federal judge Mark Pittman forced the release of the documents.[78] Through a meticulous examination of vaccine samples as well as his study of blood samples, nanotechnology expert Dr. Philippe van Welbergen correlates the presence of graphene in the vax with the emergence of blood clotting disorders in vaxxed individuals and the destruction of essential red blood cells."[79]

Covid-19 was never about a deadly "novel virus" (in the scientific literature, "novel" seems to refer to gain of function) spreading across the planet. It was always a graphene-enhanced acute irradiation "system" activated by cell-phone phased array antennas and 5G/6G, with the "solution" being an mRNA end run around the immune system to fool it into "accepting synthetic materials and objects disguised so they won't be recognized as foreign."[80] It was about injecting nanoscale biometric application programming interface technologies, referred to as BioAPI by the technocrats, which are essential to the transhumanist Human 2.0 neural network.

The mass global inoculation program of 2020–2023 could be interpreted as the capstone of the long preparation for the first transhumanist generation. The chemical trails had been delivering epigenetic payloads of nanoparticles for almost thirty years, their electrospun polymer chains drifting in the wind and sticking to bushes, porches, and car windows. The Space Fence lockdown was operational. The 5G/6G wireless grid between satellites and buried fiber optics was in force, and smart cities were armed and calibrated to lock down and control urban populations under the pretext of "climate change." The time had finally arrived to activate the BioAPI microprocessor nanotech in Human 1.0, vaxxed or unvaxxed, by means of "software" in the serums and from the air the living digital *synbio* scaffolding loaded with polymer fibers treated with the linear polysaccharide chitosan from the shells of insects and crustaceans. Voilà! The vax!

Graphene compounds have been implicated in DNA abnormalities, inflammatory reactions, and cellular necrosis. And while it can be consumed, inhaled, then absorbed through the skin, the main way black goo enters the human system is via injection. Hence graphene has almost certainly been included in the Covid-19 serums, its use disguised under Pharma's blanket claim of "proprietary formula" that indemnifies it from claims of damages. "Air, water, medications, and even protective face masks—Black Essence's reach is speculated to be vast, with a special emphasis on its presence in COVID-19 vaccines."[81] Graphene has changed everything previously assumed about biology in that molecular biology is now digital biology. Moderna, Pfizer, AstraZeneca, and Sputnik serums appear to be composed of 99.5 percent graphene oxide nanoparticles whose signals are programmed to interact with 5G.

Two decades before commencement of the 2020 full-scale assault on Human 1.0 via injection, papers were made public about liquid graphene crystals in the brain being microreceivers in electromagnetic fields, and that under extremely low frequency (ELF) waves, liquid graphene crystals make it so that people can no longer *think*. Graphene is poisonous. Before entering the body, whether by breathing or ingesting or inoculation, it is a nonmetallic chemical agent, but once in contact with hydrogen and body heat amped up by the presence of nanometals and ambient electromagnetism, it becomes extremely biomagnetic and capable of absorbing energy from 5G/6G waves, which oxidize rapidly and act as free radicals that can quicken the thrombosis, myocarditis, and pericarditis[82] we are seeing postinoculation.

Iron-based nanoparticles of hybridizing carbon,[83] graphene, and other carbon nanotubes (CNTs) are in the Covid serums as well as in the chemical aerosol deliveries. In the Human 1.0 body, CNTs attach to synapses for delivery of genetic information into cells.[84] Graphene is already in our brains after two decades of broad exposure to prime Human 1.0 for brain-computer interface. If we think of graphene in Black Goo terms, its interface with artificial intelligence (AI) makes it particularly manipulable and obscures its harmful characteristics, like how it modifies specific genetic structures because of how AI facilitates Black Goo assimilation to the point of introducing additional sets of commands.[85]

Graphene oxide nanohybrids in the Covid mRNA genetic modification inoculations intensify the body's magnetic field, which is already loaded with magnetites, ferritin, and other paramagnetic nanoparticles from the sky. Inhaling graphene oxide inflames mucous membranes, causes a loss of taste and smell due to midbrain cranial nerve damage, and quickens the onset of bilateral pneumonia, as oxidized graphene resonates at 41.6 gigahertz and affects the absorption of oxygen. *Shades of Wuhan's 5G activation*? Notably, the 2019 outbreak in Wuhan, China, occurred immediately following the activation of state-of-the-art 5G towers throughout the city for the sake of the Wuhan Institute of Virology.

As I've indicated, graphene is now found in thousands of products, from water desalination membranes, sewage treatment, medical diagnostic devices, and 3D printing of human organs, to biosensors, smartphone displays, stretchable electronics, and advanced solar cells. To achieve the transhumanist goal, graphene's neuroelectronics capability is focused on neuromodulation technology for mind control and bioengineering: "Read a person's brain [at a resolution never seen before], detect specific neurological patterns, and then control that person's neurology to alter their brain function. Injected into the brain's hippocampus in vivo, graphene oxide nanosheets (s-GO) depress glutamate (main ingredient in glutathione, main excitatory neurotransmitter in the central nervous system) and interfere with synaptic activity."[86]

Whether inhaled or inoculated, graphene immediately creates a magnetic field mental fog that prevents clear thinking and comprehension, graphene-based hydrogel (GBH) being an AI-controlled, remotely programmed neuroelectronic swarm intelligence system whose electrical conductivity is several orders greater than that of traditional hydrogels. For example, remote programmed graphene-based nanobots are directed to seek out vital organs (brain and heart) as per frequency, electromagnetic field, and/or temperature. Bluntly put, graphene oxide overwhelms the system, even overriding supplements of glutathione or N-acetylcysteine (NAC), both of which have to do with tissue and wound healing.*

Hydrogels are hydrophilic polymer networks that retain water like super-sponges and are excellent for controlled deliveries of active agrochemical compounds. They're used extensively in aerosol deliveries over agrobusiness fields so that the soil retains fluids—yes, water, but also insecticides, fertilizers, and vaccines. *Hydrogel is constantly in the air and in the food supply.* As "micronaut" (microscope expert) Karl Coronas wrote in his April 20, 2024, email to me, "I believe all the cloud-seeding news is to divert from hydrogel spraying. Cloud-seeding is so fickle compared to 'we are spraying the warfare on you everywhere.'"[88] Though more "natural" hydrogels (biopolymers) like starch, chitosan, guar gum, alginate, lignin, and gelatin still exist, GBH hydrogels are now all over the world.

Because hydrogels resemble human tissues, their use as replacements for damaged tissues may be why they originally captured the attention of scientists seeking to replace—not enhance—Human 1.0 bodies as well as provide an excellent "healing" cover story. By employing *entanglement*, polymers can be made more durable, more able to withstand stress and handle a wide range of motions.[89]

The terms *hydrogel*, the DNA-hydrogel combination called *Dgel*, and *graphene-based hydrogel (GBH)* basically refer to the same substance whose overriding BioAPI interface task is to provide the all-pervasive filler "glue" between natural body tissues and the synthetic nano AI network, namely a 3 mm–thick network of synthetic polymer chains that fuses with cells as it replicates and installs Internet of Things nodes throughout the human nervous system. As cell alignment and brain-computer-interface fusion

---

*Collapse of the immune system due to glutathione imbalance is generally followed by a cytokine storm. Glutathione protects the integrity of the cell by slowing down and blocking anti-oxidative stress and destructive "modifications" by synthetic biology agents like graphene oxide, but it can't heal ongoing damage. The fact is that "the reaction between GO [graphene oxide] and GSH [glutathione] provides a new perspective to explain the origin of GO cytotoxicity."[87]

occurs and the scaffolding unfurls, Human 1.0 body systems are, over time, replaced by *synbio* Human 2.0.

Trillions of self-replicating nanobots with hive-mind swarm intelligence and transmission capabilities continue to be delivered to populations by means of chemical trails and GMO frankenfoods, and—most recently and effectively—inoculations, prescribed masks,[90] and PCR test swabs treated with GBH. These bots are self-assembling, self-repairing, and transmit their swarm consciousness over the human WBAN (wireless body area network) connected to the artificial intelligence platform we aptly call the Cloud. The neural mesh scaffolding that self-replicating GBH nanobots are programmed to make is next-generation nonsurgical neurotechnology (N3) at its best. Graphene's hexagonal lattice structure, described as "novel geometries with magical properties,"* turns the graphene-based hydrogel inside your body and brain into an AI network whose sensors are drawing power from the host's body and surrounding environmental Internet of Things.[92] For cloning the Human 1.0 body as a virtual avatar "living" in the Cloud, graphene-based hydrogel is hard to beat.

As for the nanoform GBH masks and swabs rolled out with the biological warfare jabs, they supplement the graphene-based hydrogel being delivered in aerosols to play a critical role in the new digital biology of Human 2.0. Once the final component of the transhumanist epigenetic agenda got going in 2020 with the gain-of-function digital "virus" and its medical "solution," thousands of immune system crisis symptoms erupted: spontaneous bleeds and discharges, degraded fluid densities, anaphylaxis, cerebral vein thrombosis, eye disorders and blindness (central nervous system), strokes, Bell's palsy, turbo cancer, and more. Initially, mainstream media lied that these were just allergic or adverse drug reactions due to the polyethylene glycol in the shots. Migraines, nosebleeds, bruising, bloating, menstruation oddities, clotting, and loss of speech and senses of smell and taste appeared everywhere, but graphene oxide's long-term effects were MIA (missing in action) at the CDC and in mainstream media "coverage."

Spanish physician Dr. José Luis Sevillano states that the Covid

*"It is important to note that the application of the Phi conjugated structure [in graphene sheets] is discussed in global circles as the novel geometries with magical properties."[91]

"virus," like the "5G flu," is not due to a virus but to the impact of graphene coupled with 4G and 4G Plus on the immune system whose phased array antenna technology is much like that of 5G. Thus a human being becomes a CPU (central processing unit) not just for tracking, but for broadcasting to people around them whatever frequency the controlling AI system wants broadcast via "shedding." This includes multiplying frequency around and inside the bodies of the inoculated [and non-inoculated] and therefore "getting into the ionization scale."[93] When Dr. Sevillano said he had discovered living things in the vaccine that shouldn't be there, he was witnessing swarm nanobots busily creating a self-assembling living polymer of DNA, blood, cells, tissues, and organs, while hydrogel absorbed the body's water and graphene annihilated whatever it came in contact with.

In the twenty-minute 2021 YouTube video "Streets of Philadelphia, what's going on today. It gets much worse than other day,"[94] we see the homeless in Philadelphia's Kensington district, known as the largest open-air narcotics market for opioids on the U.S. East Coast. Note all of the absolutely still, folded-over bodies of the addicts standing everywhere. Is this fentanyl, or fentanyl coped with graphene oxide?

When I watch this footage, what is glaring to me—aside from the dullness, litter, and waste, and the tragedy of strident dehumanization and devolution of society—is how flat and non-upright these barely living bodies are. Most are hunched over and leaning on nearby objects. It seems their bodies are physically and energetically incapable of supporting them. This looks a lot to me not only like brain damage (likely from consistent abuse of street drugs) but also severe vagus nerve deterioration. Did you know that graphene-based bioelectronics can be used to selectively modulate the vagus nerve? . . . Hypothetically, if illicit drugs were doped with graphene oxide, can you imagine the (dual-use) potential for such an application?[95]

Ritual sex and pedophilia may remain the core practices of Brethren of the Mystic Tie, but for sorcerous, full-spectrum dominance over Earth and its inhabitants, conductive nanometals, graphene, ionized microwaves, and radio frequencies of a certain bandwidth do nicely.

A FEW EXCELLENT PRE-COVID-19 *SYNBIO* VIDEOS

"NANO DUST **IT'S INSIDE ALL OF US* ULF DIESTELMANN FROM MONTREAL*," MARCH 2, 2017

From the skin

3:30	Aluminum in rain
	"webbing" spiral / tower
6:00	Morgellons (450X); 25:00 (40X)
	Foam, plastic
	UV light
14:30	First snow
20:00	Why some develop lesions and many do not
24:40	"Floaters" in the eyes

"SMART DUST IS INSIDE OF ALL OF US—VIRUSES FALLING FROM THE SKY—VACCINES IN OUR AIR, WATER, FOOD," JULY 27, 2019, NATUBER TV (2.25 HOURS)

MEMS, GEMS, NEMS

9:45–25:00	Harald Kautz-Vella—"Self-replicating hollow fibers that read out the light fingerprint of your DNA and transform it into an EM signal detectable via satellite and ground stations."

"DARPA PATENTS/DOCUMENTS: HAVE WE ENGAGED THE BORG?" PLAZMA, APRIL 30, 2021*

I will stop here regarding synthetic biology and geoengineering so as to continue more in depth in the *synbio* book. Meanwhile, the human being and Nature are being denatured into machinery, much as carbon dioxide—"the miracle molecule of life . . . the single, most important molecule for plant growth . . . photosynthesis, the single most important biochemical process for life on Earth"[96]—is being used as a whipping boy for much

*Available on Bitchute.com

larger stakes. Mass extinction level events are underway by politicized scientism by people who look human but are not—people whose weaponized technologies are blocking out the life-sustaining Sun and spirit-sustaining stars—technologies assaulting all that is natural and God-given in order to replace it all with *synbio* and Transhumanism.

And yet there may be a weakness in this diabolical plan that has become particularly manifest since 2020, whether the masses still asleep or slowly awakening are aware of it or not. That weakness is the Space Fence itself, the wireless smart grid looping above and on the planet (and now inside the human being), dependent upon almighty artificial intelligence and its quantum and supercomputers through the electromagnetic spectrum rungs of power they depend upon for fulfillment of the doctrine of full-spectrum dominance. In the final chapter, we take a closer, more thorough look at the electromagnetics intended to take the place of natural power and the life force we have been beholden to for millions of years.

> . . . He, as he stood there, was an incredibly subtle creation, nerves, sinews, bones, muscle, skin and flesh, heart and a thousand organs and vessels. They were his strength, yet his strength parceled and ordered according to many curious divisions, even as by a similar process of infinite change the few clouds that floated in the sky were transmuted from and into rivers and seas. The seas, the world itself, was a mass of subtle life, . . . and the stars beyond the world. For through space the serpentine imagination coiled and uncoiled in a myriad of shapes, at each moment so and not otherwise, and the next moment entirely different and yet so and not otherwise again.
>
> CHARLES WILLIAMS,*
> *THE PLACE OF THE LION*, 1933

*Charles Williams (1886–1945) was described by the poet T. S. Eliot as the superb writer of "supernatural thrillers" that explore the sacramental intersection of the physical with the spiritual while also examining the ways in which power, including spiritual power, can corrupt as well as sanctify. He was a member of C. S. Lewis's literary society, the Inklings, as was J. R. R. Tolkien, author of *The Lord of the Rings* series. C. S. Lewis modeled his novel *That Hideous Strength* (1945) on Williams's approach, and T. S. Eliot wrote an introduction for the last Williams novel, *All Hallows' Eve* (1945).

10

5G/6G, the Internet of Things, and the Wireless Body Area Network (WBAN)

Where will all the transmitter/cells be placed on the ground and in the air? Something is missing here. Is there another version of 5G we're not being told about? Is geoengineering of the atmosphere the means for tuning up space so 5G signals can be passed along without cells/transmitters?

Jon Rappoport, "5G wireless: A ridiculous front for global control," April 3, 2018

As captures of the wireless Space Fence lockdown, we have become accustomed to the common digital background pulsating all around and through us 24/7, furthering the hive mind frequency covertly assigned to us in the name of "seamless connectivity between the internet and everyday life,"[1] thanks to the nanobots in our blood, tissues, and brains subject to 5G / 6G. Above, satellites radiate us as does the electromagnetic nanometal ULF "ring" surrounding the equator. Thanks to the HAARP ionospheric heater system's control over the ionosphere, our ionized atmosphere keeps its 3 Hz pulse trained on our endocrine system as well as our *consciousness soul** development set in motion in the fifteenth century.

*According to Rudolf Steiner, the Human 1.0 soul evolution thus far includes the *sentient* soul, the *intellectual* soul, and most recently the *consciousness* soul. The development of the consciousness soul is most definitely being targeted by the "culturemakers," as is evident by the inability of most people to think conceptually. Europe's seven liberal arts have been utterly eradicated by Google (CIA) and Big Telecom who now "educate" (program) the young around the world.

To accomplish the *devolutionary* split between human and Divine, an endless electromagnetic inundation of the planet is being used to destroy organic life along with free will human consciousness evolution.

5G has been sold as the magical solution to wireless communications coverage, signaling, and latency (the delay before a transfer of data begins), as well spectral efficiency, wireless sensor connectivity, simultaneity, data rates, etc. But what is 5G, really? At its military-industrial-intelligence core, it is a profitable, classic bait-and-switch for smart city planetary lockdown, the bait being the *mesh networks* that guarantee convenience and superfast connectivity, the switch being a phased array antenna weapon system centering around not just HAARP and the smart phone and IoT but the invisible digital world of frequencies that permeate and surround living Human 1.0 *and the new race of hybrid Human 2.0*, each with their own wirelesss body area network (WBAN) and waning *aura* that was once radiant with life energy.

Categorized by the World Health Organization as a group 2B carcinogen in 2011, 5G operates from 28 to 300 gigahertz. It was nevertheless approved for "consumer devices" by the FCC on July 14, 2016, after the military had been deploying it for decades as an invaluable dual-use technology, meaning military and civilian. Following the ruling, the FCC freed up 280 megahertz of C-band spectrum for 5G. One month after the Covid-19 drama began in February 2020, satellite operators "repacked" their operations from 500–280 megahertz to reserve the lower 280 megahertz for terrestrial use.* Extremely high frequency bands in the millimeter watts spectrum of 30 to 300 gigahertz promised high-speed downloads and high-speed point-to-point communications for truly private (covert) conversations, but the real purpose was to use millimeter waves to monitor *all* Internet of Things (IoT), Internet of Nano Things (IoNT), Internet of

*"In February, the Commission majority adopted rules for the C-band that will free up 280 megahertz of spectrum for 5G. Specifically, the rules required existing satellite operators to repack their operations from the band's entire 500 megahertz into the upper 200 megahertz, allocated the lower 280 megahertz for terrestrial flexible use, and provided for a 20-megahertz guard band in between. Moreover, the Commission provided five eligible space station operators serving the contiguous United States with the opportunity to clear the lower 300 megahertz of the band on an accelerated timeframe in exchange for accelerated relocation payments" (Press release, Federal Communications Commission News, "C-Band Spectrum Will Be Made Available for 5G Services on an Accelerated Basis," available on the FCC website).

Bodies (IoB), especially full-spectrum dominance "under the skin" of all biological systems down to the molecular level.

Lena Pu, whose professional history includes restoring sensitive environmental habitats for the U.S. Army Corps of Engineers, is an environmental health consultant for the National Association for Children and Safe Technologies (NACST.org). In a talk she gave to the Silicon Valley Health Institute on September 19, 2019,[2] Pu pointed out that 5G encompasses three thousand frequencies and should more honestly be called 297G (297 "generations," not 5, 6, or 7). She then talked about German Paperclip scientist Herman P. Schwan (1915–2005), the founding father of biomedical engineering who studied the dielectric properties of microwave radiation as they related to blood and water. In 1953, Schwan established the radio frequency exposure guidelines for the American National Standards Institute (ANSI) and Institute of Electrical and Electronics Engineers (IEEE), then in 1970 published *Biological Engineering* (McGraw-Hill). The 1953 IEEE standard remains the standard for pregnant mothers, children, adults, and the old and disabled, despite thousands of studies and papers proving the negative health consequences of microwave cell phones and their link to brain cancer. Schwan is yet another connection to the Nazi obsession with artificial intelligence and Transhumanism.

Is wireless 5G / 6G the capstone to the secret society pyramid on the dollar bill?

Pu is known for her fresh blood membrane morphology (FBMM) studies proving that within five to fifteen minutes in the presence of wireless microwave technology, healthy blood begins to become "sticky" and clump up as it "cooks"*—similar to the coagulation experienced by Covid postinoculation sufferers, including "sudden deaths."[3]

EFFECTS OF FREQUENCY AND PULSING

With 5G we are not just concerned about power, but about *frequency* and *pulsing.* An electron volt assault is one thing, but *cyclotron resonance,* the

*"The evaluation of RF radiation health risks from 5G technology is ignored . . . Conflicts of interest and ties to the industry seem to have contributed to the biased reports. The lack of proper unbiased risk evaluation of the 5G technology places populations at risk. Furthermore, there seems to be a cartel of individuals monopolizing evaluation committees, thus reinforcing the no-risk paradigm. We believe that this activity should qualify as scientific misconduct."[4]

interaction of charged particles in a magnetic field, takes its toll on our biology over time. The initial patent for HAARP laid out how it would excite a region by "electron cyclotron resonance heating," a topic taken up in relation to the effects on cells by Robert O. Becker, MD, author of *Cross Currents: The Perils of Electropollution, the Promise of Electromedicine*: "Cyclotron resonance is a mechanism of action that enables very low-strength magnetic fields, acting in concert with the Earth's geomagnetic field, to produce major biological effects by concentrating the energy in the applied field upon specific particles, such as the biologically important ions of sodium, calcium, potassium, and lithium."[5] The latency period for adverse biological effects from 1 to 5 gigahertz is estimated to be ten to twenty years; higher frequencies from 6 to 100 gigahertz further compress the timeframe. The pulsed nature of high-frequency, high-intensity signals constitutes the *modulation* that is "the secret of transmitting information by means of electromagnetic fields," as Becker stresses on page 212. Pulsed modulation, not the power of the carrier wave, is why *all* abnormal, man-made electromagnetic fields, regardless of their frequencies, constitute hazards *at the very least* to the resonant Human 1.0 body (page 214).

In the smart city, people's brain waves are easy to manipulate. As with the digital terrestrial trunked radio (TETRA) system pulsed by law enforcement, military, public transportation providers, networks, emergency services, utilities, sports arenas, and the worldwide hospitality and leisure industry, an ancillary subcarrier frequency (16 hertz) can be pulsed via ELF to entrain violence, passivity, or any other emotional frequency throughout the satellite-designated area.

The frequencies of 5G disrupt sleep, weaken the immune system, and open cell channel gates. Biochemist Martin L. Pall considers the 5G infrastructure to be an integrated weapon system using pulsed frequencies. He refers to the "black" (as in classified military-industrial-intelligence complex) nature of electromagnetic radiation studies, particularly when applied to millimeter wave technology like 5G. Since the 1996 Telecommunications Act, public-private partnerships between Big Tech and agencies like the FCC and FDA haven't been the least interested in public safety.[6] As I pointed out earlier, Pall points to how the sweat ducts in human skin—skin being our largest organ—act as coiled helical antennas picking up signals and frequencies.[7] Add to this the dipole conductivity of water, the giver of life, below our spongelike skin (see chapter 8, "*Bios*"). The combination of oxidative stress and voltage-

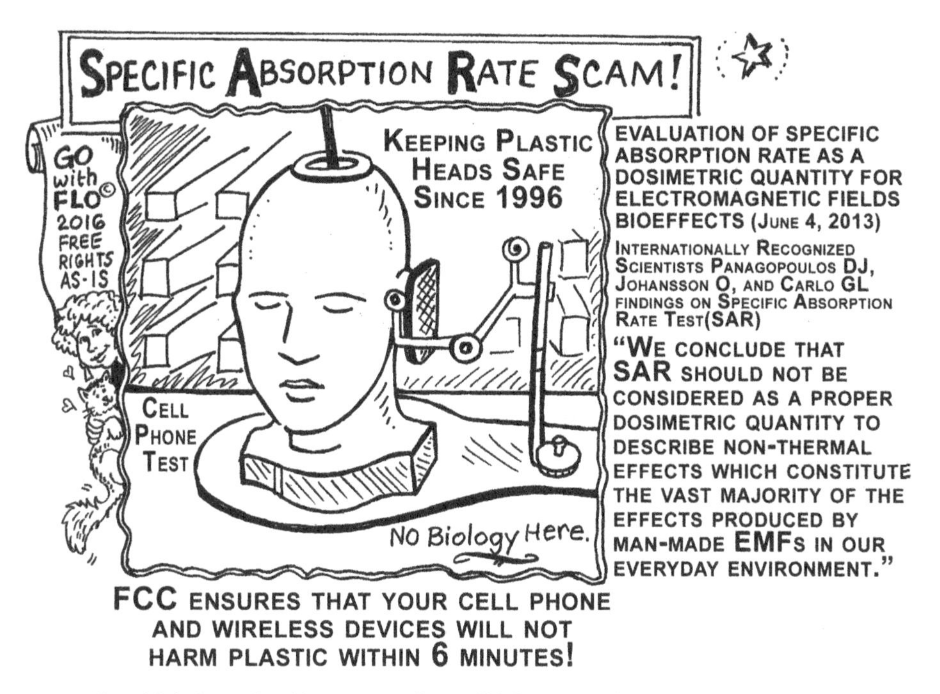

Fig. 10.1. Specific Absorption Rate (SAR) scam! Permission granted by artist-activist Floris Freshman. VGCCs and tinnitus appear to go hand-in-hand. Ringing in the ears (tinnitus) is not auditory, but rather the frequency modulation (pulsing) of VGCCs in the brain.

Line art by Floris Freshman

gated calcium channels (VGCCs) found in the membranes of excitable cells (neural, muscle, glial, etc.), the intense overwork of sweat duct "antennas," and our skin's high specific absorption rate explain much about electromagnetic hypersensitivity/electrosensitivity and its relation to declining health.*

Decades of denial and disinformation and calling the knowledgeable "conspiracy theorists" were the public media response to the first wave of "canaries in the coalmine" claiming *electrosensitivity (ES)*. At last, rational expert witnesses are publicly discussing the health dangers of surrounding EMF signaling in power lines, smart meters and cell towers, schools and hospitals—you name it—the environment we now take for granted. Andrew

*And possibly to post-vax "shedding." Current radiation standards do not take into account the high specific absorption rate (SAR) of the skin in extremely high frequency bands.

Michrowski, PhD, president of the Ottawa-based Planetary Association for Clean Energy, has sounded the alarm not just about EMF signaling in the home but in institutions like schools and hospitals where multitudes of wireless "smart" devices interlock in the Internet of Things and generate a condensed cloud of signals and resultant reflective phenomena that employees, children, and adults already concerned about symptoms are unwittingly subject to. André Fauteux, editor of the Canadian magazine *La Maison du 21e siècle*, has stressed the actual effects of EMF signaling (such as tinnitus) and the skin conductance that goes with it by quoting Greek biophysicist Dimitris J. Panagopoulos, PhD, who spoke at a European Parliament workshop about the dangers of artificial electromagnetic environments compared with the life-giving effects of natural solar electromagnetism, due to "two unique and detrimental properties not found in nature":

> [Being polarized and coherent, "smart' environments] can produce constructive interference and amplify their intensities at certain locations, and also force all charged particles (e.g. mobile ions) in living cells/tissue to oscillate in parallel and in phase with them. Polarization and coherence are the reason why EMR ~ 0.1 mW/cm2 from a mobile phone is damaging, while solar EMR ~10 mW/cm2 (100 times stronger) is vital.[8]

With smart hospitals and precisional medicine in mind,* read what is admitted in the March 18, 2021, U.S. patent 20210082583-A, "Methods and systems of prioritizing treatments, vaccination, testing and/or activities while protecting the privacy of individuals," regarding the "precision healthcare" network to which one subjects one's health:

> a plurality of electronic devices configured with instructions to generate an ID, when in proximity of another such electronic device, one or both transmit said ID to said another electronic device and receive an ID from said another electronic device, generating a score based on plurality of such received IDs, receiving information from a server, displaying relevant treatment instructions to said subjects based on received information.

*"Precision medicine" (aka "telemedicine") is now introducing the Internet of Medical Things, the FCC having approved a million small antennas that SpaceX Starlink needs for high-speed internet service from space.[9]

What are the implications of this "transmit said ID" gobbledegook? That with the implementation of the "smart hospital" precision medicine system, we are not only being tracked by Bluetooth technology but are being implanted either by injection or aerosols or transmission with biosensors programmed to report to "secretive government contractors with longstanding ties to HHS [Department of Human and Health Services]" like Israeli intelligence and Microsoft Corporation[10] via the Internet of Bodies system dependent upon the WBAN (Wireless Body Area Network) which "psinergy" expert Sabrina Wallace calls a "body part."* The new rules for wireless broadband operations above 24 GHz (5G) adopted by the U.S. Federal Communications Commission (FCC) in July 2016 must be viewed as exploitation that will end in higher frequencies in the sub-THz (subterahertz, i.e., 90 to 300 gigahertz) region of the spectrum. Human immersion in frequencies like those to which the sweat ducts (helical antennas) are most attuned alerts us to the fact that sub-THz technologies are being used as weapons to break down Human 1.0.[11]

The higher the frequency, the shorter the wavelength; the smaller the wave, the greater the impact on the human being. For example, the 2.4 GHz Wi-Fi microwave of the 4G smart phone is 16 cm, whereas the 5G very high frequency wave is 3–5 mm. Wi-Fi routers and microwave ovens operate at 2.8 GHz and produce changes in human cells. Cell phones and smart meters operate from 2.4 GHz down to 900 MHz. Cell phone manufacturers seem bent on keeping the density high enough to reach cell towers and low enough to cook the brain over time, thus contributing to the explosion of cancers, particularly brain cancer.

As for the 60 GHz band of 5G, it produces *oxygen depletion*, which may be what we witnessed in the early video footage from Wuhan, China, as the first volley of the international media fear campaign for the so-called Covid pandemic. Over time, immersion in 5G electromagnetic fields

*I highly recommend all of Sabrina Wallace's vlogs in her "course" to educate the people as to our condition. Begin with "Psinergy Must Watch—Electronic Warfare—Our Bodies are Biohacke," on Youtube. Read the show notes. Sabrina's qualifications for her expertise include black ops programming as a prodigy. As she has put it: "This channel belongs to a disabled American 40-plus-year-old woman who vlogs daily about her knowledge in multiple disciplines, including vocational experience and psionic experience that was tested as a child and ongoing into adulthood." I will continue discussing the WBAN body part in the synthetic biology book.

Fig. 10.2. Note the chemical light discharge over the tower.
Sent by an anonymous researcher in 2018

induces symptoms like those now wrongly attributed to Covid-19: sore throat, headache, fever, cough, and chills, shortness of breath (hypoxia, i.e., oxygen depletion), high fever, and severe fatigue, as well as the neurological loss of taste and smell.[12]

Adjusting the microwave density turns a convenience into a weapon. So it goes with dual use. By weaponizing the air surrounding us, the no-turning-back point of millimeter wave impact on the molecules in our cells is achieved. The 5G microwave network propagates in the mouth and nostrils, down the throat, and into the lungs, the ear canals, and the nerves of the inner ears close to the brain.

Remember: Microwave density—the specific absorption rate (SAR) of the skin—is one thing, but add the *pulsing* of 5G microwave photons 24/7 and molecules and cell resonance are overloaded and overwhelmed.

HUMANS AS "WETWARE"

The relatively unknown two-thousand-page 2005 U.S. Patent 6965816-B2, "PFN/TRAC system FAA upgrades for accountable remote robotics control to stop the unauthorized use of aircraft and to improve equipment management and public safety in transportation,"* is a good example of a stealth patent not unlike the thousand-page National Defense Authorization Act (NDAA) first passed in 1961 *still* being repassed every year with new stealth provisos tucked in.† The players behind the PFN/TRAC patent include the Carlyle Group of the Deep State Bush/Clinton/Obama cabal, including the law firm of former special prosecutor Robert Mueller, Wilmer Hale LLP; Hillary Clinton's Rose Law Firm; George Bush Sr.; John Podesta; James Comey; Rod Rosenstein; Loretta Lynch; Eric Holder; and Larry Summers. Then there are the corporate players: IBM's Eclipse Foundation; Cisco; Microsoft; software company SAP; Oracle; the venture capital firm Kleiner Perkins; Qualcomm; Goldman Sachs; JP Morgan; AT&T; and Hewlett-Packard and its spinoff, Agilent Technologies.

The PFN/TRAC patent may begin with remote robotics and aircraft with QRS-11 crystal gyroscopes turned on and off by National Security Agency (NSA) controllers,‡ but read on and you will discover human beings under the Internet of Things that "requires hardware, software, and wet ware," *wetware* being the CIA's term for human beings. The additional patents that refine the original remote electromagnetic control over vehicles, ships, equipment, commerce, education, and "wetware" also legally (albeit obliquely) encompass human beings. In fact, PFN/TRAC is being used to aggressively go in, take over, control, track, and manage any system, whether it be an airplane, car, motorcycle, human being, or animal. This dystopian control system is yet more of the wireless body area network loaded with wetware control mechanisms, devices, nanochips and nanobots.[14]

Interestingly, the gatekeeper at the U.S. patent office is British Serco, ever ready to obscure true inventors while hiding actual patent holders like the Highlands Group, the Pilgrim Society, the British Privy Council working inside the City of London Financial District, and so forth. (See appendix 8, "Movers and Shakers.")

---

*Patent filings, however, go back to December 2, 1996.

†According to Wikipedia, "In recent years each NDAA also includes provisions only peripherally related to the Defense Department, because unlike most other bills, the NDAA is sure to be considered and passed so legislators attach other bills to it."

‡Boeing, Airbus, and Airlines for America have complained of 5G effects on aircraft safety systems like altimeters.[13]

Transhuman hardware, software, wetware . . .

By the time military insider and AT&T/Bell Labs alumnus Eric Schmidt became CEO of Google, IBM's Eclipse Foundation had organized its Silicon Valley underlings to begin carrying forward their weaponized 5G dream. This would involve computer hardware, which they already had; computer software, which was soon streaming out of the now well-funded Silicon Valley NSA fronts; and wetware.

Wetware meant human beings and other living creatures, which would be ionized and programmed on a highly individualized DNA basis with the hardware and software to create a completely integrated and controllable 5G grid full spectrum dominance weapons system. Artificial intelligence would permeate the grid, seeking to self-replicate while turning humanity off.[15]

A GLOBAL WEAPON SYSTEM

Much as Big Medicine and Big Pharma are shifting from molecular biology to digital biology right now, the telecommunications industry (Big Tech) shifted from analog to digital in 2002. Now, virtually all electromagnetics in technology are digital, which for human neurology means more dependable remote control over brains and the ability to break down and enter the blood-brain barrier. The bandwidth shortage we keep hearing about is untrue, but useful when it comes to amping up the fear that will ensure unlimited access to bandwidth.

As microwave frequencies steadily increase, so do the microwave towers and subterranean fiber optics cables being installed in communities. Going digital and adding bandwidth, frequency amplifiers, towers, and all the other supporting infrastructure has always been about dual use weaponization of technology. Short, high-frequency waves in the gigahertz and terahertz range don't so much *interface* with the brain as *modulate* it so that signals can be dropped right into the brain. *The shorter the wave, the more powerful its effect on the brain.* To get extremely low frequency (ELF) into the brain, high-frequency carriers like 5G are needed. Whereas seven-inch penetration of one gigahertz is overkill (so to speak), a shorter wavelength carried on a higher frequency with a shorter wavelength is perfect.

> The surveillance technology of today is the surveillance of the human mind and, through access to the brain and nervous system, the control of behaviour and the body's functions. The messaging of auditory hallucinations has given way to silent techniques of influencing and implanting thoughts . . . the systems have been established for the control of populations—and with the necessary technology integrated into a cell phone.[16]

Given that a 10-watt transmitter is all that is needed for a strong cellular signal, you may have wondered why cell phone towers are being fed by 400+ kilovolt-ampere transformers (amps or amperes are the charged electron current flowing through two conductors) so as to pump out between 20,000 to 60,000 watts (watts are the *power* in an electric circuit). Gleaning gigahertz wave forms that connect to high-frequency electricity from a cell tower for mind control is one reason.

Take a look at the antenna arrays on 5G towers: they are of far greater size than what's necessary (approximately eight feet) and highly directional so that signals can be steered toward specific zones or targets. If 2,000 volts (voltage is the pressure from the current being pushed by an electrical circuit's power source) go into regulating community mood and mind control, what are the rest of the signals doing? Perhaps Human 1.0 and 2.0 wireless body area networks? Cell tower microwave frequencies link over a fifty-mile range, working in tandem with NEXRAD (next-generation radar) installations, the big "golf ball" radar units operated by the National Weather Service under a complicated bureaucratic chain that ends with the Department of Defense.

Sadly and ironically, we have been fooled by convenience and comfort into a tech bondage. We are oblivious to the possibility that the military doctrine of full-spectrum dominance might be hellbent on encompassing not just human DNA and the human aura, *but the human soul itself.*

Telecom military contractors are now moving from the already completed 5G cellular systems that use steerable phased-array antennas for the higher digital bandwidths necessary for Big Data "stacking," to 6G build-outs whose terahertz waves are essential for satellite tracking and tracing of the impossible to see nano devices daily delivered for us to breathe in from the saturated atmosphere. The higher the frequency, the greater the bandwidth—and remember: not all the frequencies above 300 gigahertz have been allocated. In fact, there seems to be no limit to data transmission, especially with Earth's atmosphere now functioning as an ionized antenna kept aloft by daily dumps of conductive metal nanoparticles in a sea of signals from satellites, base stations, wireless providers, and hotspots everywhere.

Arthur Firstenberg describes the phased-array beam technology employed in 5G (much as it is employed on a much larger scale in the ionospheric heater weapon system known as HAARP):

> The arrays are going to track each other so that wherever you are, a beam from your smartphone is going to be aimed directly at the base station (cell tower), and a beam from the base station is going to be aimed directly at you. If you walk between someone's phone and the base station, both beams will go right through your body. The beam from the tower will hit you even if you are standing near someone who is on a smartphone. And if you are in a crowd, multiple beams will overlap and be unavoidable. At present, smartphones emit a maximum of about two watts and usually operate at a power of less than a watt. That will still be true of 5G phones. However, inside a 5G phone there may be 8 tiny arrays of 8 tiny antennas each, all working together to track the nearest cell tower and aim a narrowly focused beam at it. The FCC has recently adopted rules allowing the effective power of those beams to be as much as 20 watts. Now if a handheld smartphone sent a 20-watt beam through your body, it would far exceed the exposure limit set by the FCC. What the FCC is counting on is that there is going to be a metal shield between the display side of a 5G phone and the side with all the circuitry and antennas. That shield will be there to protect the circuitry from electronic interference that would otherwise be caused by the display and make the phone useless. But it will also function to keep most of the radiation from travelling directly into your head or body, and therefore the FCC is allowing 5G phones to come to market that will have an effective radiated power that is ten times as high as for 4G phones.[17]

Besides 5G towers installed every 500 feet or so, multiple thousands of miniature satellites known as CubeSats are now in orbit, all with phased array antennas shooting tightly focused beams of microwave radiation that penetrate all living things in their path as well as 5G Internet of Things (IoT) devices below. That 5G has been weaponized must be undisputed, even if it is killing us over time, given that the Pentagon considers 5G to be a "foundational enabler for all U.S. defense modernization" that will "enhance the 'lethality' of certain systems" subject to the Space Fence,[18] including "access to key facilities, support for new or improved infrastructure requirements, and the ability to conduct controlled experimentation with dynamic spectrum sharing [dual use]."[19] Four months into the "pandemic" (June 2020), seven U.S. military sites were conducting fifth-generation (5G) communications technology "experimentation and testing": the Naval Base at Norfolk, Virginia; the Joint Base at Pearl Harbor-Hickam, Hawaii; the Joint Base at San Antonio, Texas; the National Training Center at Fort Irwin, California; Fort Hood, Texas; Camp Pendleton, California; and Tinker Air Force Base, Oklahoma.

Deployment of 5G translates to 1.6 billion subscribers, 20 billion interconnected wireless devices, 800,000 small cell base stations, and 50,000 satellites overhead that require one million antennas on the ground, most of which are small, all of which are dual use. The rhetoric that maintains that the Space Fence lockdown system is just about missile / machine communication is hiding the fact that it is actually about the netcentric warfare being waged against Human 1.0 whose self-assembling tiny nanobot machines ensconced in their blood, organs, tissues, and brains, and the fact that they are nodes in a planetary electromagnetic system wherein they are "wetware" brain-computer interfaces (BCIs) whose nano-hardware is transmitting and receiving, thanks to the transhuman software activated by inoculation. Jody McCutcheon of *Eluxe Magazine* nails the truth about 5G but without naming the tiny machines now in our bodies and brains:

> Until now mobile broadband networks have been designed to meet the needs of people. But 5G has been created with machines' needs in mind, offering low-latency, high-efficiency data transfer . . . We

> [Humans 1.0] won't notice the difference, but it will permit machines to achieve near-seamless communication. Which in itself may open a whole Pandora's box of trouble for us—and our planet.[20]

That millimeter waves do not penetrate far into the body may make people feel safer, but what is not said is that *the extremely short electromagnetic pulses reradiate the waves for greater penetration.*[21] Thus, the radiation emitted by 5G ground antennas and satellites high overhead are able to penetrate bodies and brains, walls, and rooftops as they seek out IoT incarnations that share dual-use frequencies with the military-industrial-intelligence-Big Pharma public-private partnership.[22]

Our Human 1.0 brains were once in sync with the 7.83 Hz Schumann resonance frequency. Now, due to control over the ionosphere and sophisticated electromagnetic instruments, the Schumann frequency is being manipulated—yet another way to undermine our Divine human connection with the Divine Earth. Meanwhile, what is this doing to our brainwaves? (With the HAARP system's perennial broadcast of 3 Hz pulses, it appears we are being kept asleep.)

By emitting frequencies that oscillate [pulse] in a certain frequency range, a victim can be manipulated. There are six types of brainwaves:

Delta, 1–4 Hz, associated with sleep;

Theta, 4–8 Hz, associated with drowsiness, childhood, adolescence and young adulthood;

Alpha (Berger's wave), 8–12 Hz, the neocortex is relaxed and idle, as when watching TV or meditating;

Beta, above 12 Hz; low amplitude beta with multiple and varying frequencies is often associated with active, busy, or anxious thinking and active concentration;

Sensory motor rhythm (SMR), 12–16 Hz, associated with physical stillness and body presence; a target will have trouble moving whenever this frequency is applied;

Gamma, 26–80 Hz; higher mental activity, including perception, problem solving, fear, and consciousness.

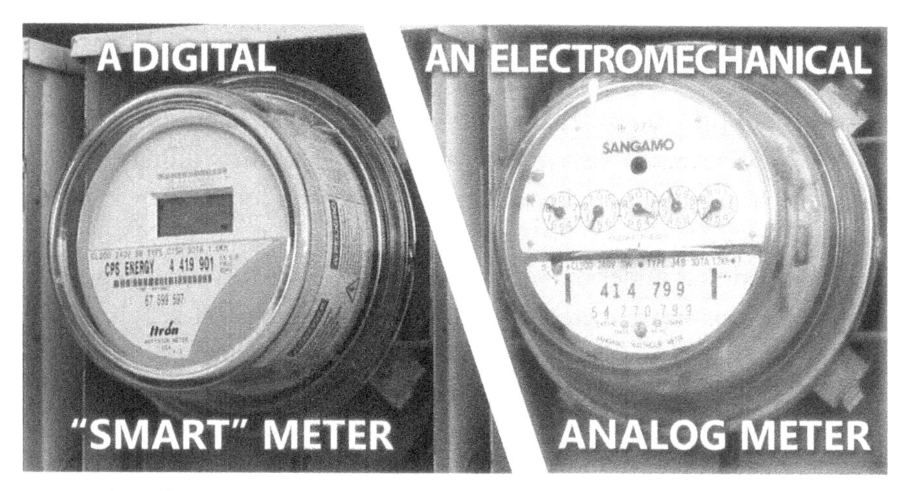

Fig. 10.3. Smart (digital) versus electromechanical (analog) meter. Smart meters are the electromagnetic gatekeepers—or wardens—of homes and neighborhoods.
StopSmartMeters.org

The brain-computer interface (BCI) through the Internet of Bodies (IoB) is taking the place of the Schumann resonance so as to merge smart-city "fifteen-minute" dwellers with the artificial intelligence wirelessly wired to trillions of 5G nodes coming from thirty billion "smart" (stealth) devices, including the IoT gatekeeper of the neighborhood known as a smart meter.

SMART (STEALTH) METERS, THE NEIGHBORHOOD GATEKEEPERS

"Stealth" technologies are always dual use, meaning for practical use but just as easily for a directed energy weapon (DEW) by simply commandeering the IoT or electromagnetic frequencies. Targets might be conservatives or ethnic groups or activists who object to medical kidnapping and mandatory inoculations, nationalization of private health data, infusing toxic chemicals into the food supply, covertly harvesting personal data, denial of First Amendment rights, etc.[23]

Smart meters (928 megahertz) are designed to spy on energy use, *including WBAN energy readings*. Once smart meters are installed, energy use seems to rise, rates inflate, cancers and tumors increase,[24] in part due to their 1-watt Wi-Fi transceiver and 4G/5G/6G connection to the cell network, and the chitchat that ensues among neighborhood smart meters and the IoT. The data collected every five, fifteen, or thirty minutes is sold to private corporations, and a back door is left open for entry by cyberattacks and hacks. In August 2018, the Seventh Circuit Court in Naperville, Illinois, ruled that under the search-and-seizure Fourth Amendment, it is "reasonable" for smart meters to collect consumer data on behalf of the government on a virtual nonstop basis.[25] Smart meter range is thirty miles, with 50,000 microwatts per square meter of pulsed emissions.

Like 5G, the smart meter has been classified by the World Health Organization as a class 2B carcinogen. The 17,000 to 190,000 pulses per day that a smart meter emits exhaust the Human 1.0 immune system, that noble ancient defender of our natural health. The smart meter's extreme connectivity and CPU (central processing unit) power are really about the commands it receives through the IoT (50/60 Hz wiring, refrigerator, microwave, TV, etc.) equipped with audio/visual surveillance.

Smart meter software is proprietary, as is the digital signal processing (DSP) chip that reads the usage rate of each unique "smart-enabled device" in virtually every appliance sold today, for which manufacturers pay premium dollar to utility corporations. Is the DSP chip a backdoor for more stealth operations? Can it trip breakers? Ignite fires? Is there a supercapacitor (an electronic component capable of storing an electric charge) inside the smart meter?[26]

Similar to smart meters, *smoke detectors* are required in just about every room of every home and workplace by the U.S. Fire Administration, a division of FEMA (the Federal Emergency Management Agency under Homeland Security). "Smart" (stealth) smoke detectors don't just scan for smoke; they pick up broadcast microwave signals and turn them into remote-sensing spectroscopes.* Tinker with a smoke detector and you may find it broadcasting on government-only channels like the NSA's PRISM, known for gathering emails, videos, and private data on "persons of interest" for secret "star chambers" of judges,[27] but now gathers data and scans for keywords to submit to supercomputers.

---

*Carbon monoxide sensors based on miniature tunable diode laser spectroscopes are becoming common.

NANO STEALTH VIA 5G/6G ELECTROMAGNETIC TRANSMISSION

Western scientism, obsessed with measuring, naming, and control over "things," began the present assault on humanity with the 9/11 War on Terror and Homeland Security's Transportation Security Administration (TSA) speeding terahertz waves unzipping double-stranded DNA and dislodging proteins at airports.[28] Nineteen years later (a Metonic cycle), the World Economic Forum (WEF) rejoices at the prospects of an "under the skin" full-spectrum dominance, beginning with trillions of nanosensors: "Because they are so small, nanosensors can collect information from millions of different points. External devices can then integrate the data to generate incredibly detailed maps showing the slightest changes in light, vibration, electrical currents, magnetic fields, chemical concentrations and other environmental conditions."[29]

At the forefront of the "faster speeds, lower latency" 6G and virtualized ORAN (Open Radio Access Network of interoperation between cellular networks provided by different vendors) are the IEEE Standards Association and the 3GPP (3rd Generation Partnership Project), the latter being a "global partnership" whose task is to define the cellular generations (3G, 4G, 5G, 6G, 7G, etc.) and the Mobile Broadband Standard for the Internet of Things. At the same time that we are told that China launched the first 6G satellite *Tianyan-5* on November 6, 2020, we are assured that 6G still lies in the future, which is not true. Most importantly, we learn that:

> Key to the promise of 6G, sensing is the basis for all interaction with and emulation of the physical environment.[30]

A vast sensor network—including the trillions of nanosensors being delivered from the stratosphere daily to be outside and inside our bodies and brains—is "aggregating data from ground and air inputs"[31] in the name of comfort, convenience, and "progress": autonomous driving / AV navigation with real-time 4D mapping, including the air space above; immersive communications experiences through location and context-aware digital and sensory experiences such as truly immersive extended reality (XR) and high-fidelity holograms; video conferences in real-time virtual reality (VR) with in vivo nanosensors supplementing "the physical sensation of being in the same room together"; global sensors measuring inputs from forests, oceans, cities, homes; etc.

> The 5G platform already harnesses AI for optimization, dynamic resource allocation, and data processing. But extremely low latency of less than one millisecond and distributed architecture mean that 6G will be able to deliver global, integrated intelligence. 6G will propel the fourth industrial revolution, enabled largely by the industrial Internet of Things (IoT) services integrated with AI and machine learning.[32]

The 6G sensing role in "precision medicine" or health care is deeply engaged with the Transhumanism agenda, even to the extent of being the real reason that the medical and pharmaceutical industries keep people "sick" (autoimmune) so as to have endless "specimens" to experiment on.

> We also can look forward to advances in precision healthcare, in which data science, analytics, and biomedicine are combined to create a learning system that conducts research in the context of clinical care while also optimizing tools and information to provide better outcomes for patients. Precision healthcare can include the use of tiny nodes that measure body functions tied to devices that can medicate and assist patients.[33]

6G hospitals and neighborhoods . . .

The Internet of Nano Things (IoNT) is made up of trillions of nanosensors and nanobots "small enough to circulate within living bodies and to mix directly into construction materials" contrived by synthetic biology in order "to modify single-celled organisms such as bacteria . . . to fashion simple biocomputers that use DNA and proteins."[34] These cellular nanosensors and carbon nanotubes made from nonbiological materials for "the fabrication of devices to complex atomic specifications"[35] that "sense and signal, acting as wireless nanoantennas" are now everywhere, including inside our bodies and brains (you might want to review carbons discussed in chapter 2 as well as carbon nanotubes in chapter 6).

As herbalist Tony Pantalleresco said earlier, the protein-based Human 1.0 is being replaced by a nanotech carbon model "operation system." In *Nano,* Ed Regis explains the journey from molecular to digital biology, from natural to synthetic:

> . . . begin with nature's own molecular devices: DNA, proteins, lipids, and so on, reprogramming them to do your bidding. You wouldn't have to create the machines, you'd use the machines that were already there—in particular, the mechanisms of protein synthesis—to build new, better, and more sophisticated molecular devices.
>
> Proteins, after all, were nature's own wonder molecules. Not themselves living things, they were nevertheless what living things, by and large, were composed of. Much of the human body's structural material—its muscles, connective tissues, the cell walls, and so on—all these things were made up of proteins. Much of the body's functional elements were likewise proteins: hemoglobin was a protein; hormones were proteins; enzymes were proteins. The proteins, in short were the body's all-purpose building material and molecular operating system.[36]

Without our knowledge, our bodies and brains—already subjected to a chemtrail regime for over two decades and a GMO food supply for three decades, and childhood vaccines for even more decades—are now undergoing interpenetration by AI interfaces that automatically capture input. The real-time surveillance capabilities of smart meters and smoke alarms and all the rest of the IoT piezoelectric world, where everyday objects with machine intelligence talk to one another over their own networks—toasters, washing machines, doorknobs, digital radios, Micro Mote computers with one megabyte memory storage that can fit on the edge of a nickel[37]—are connecting to billions of WBANs. Audio handshakes[38] take the place of real human contact, while embedded nanosensors, vehicle-to-vehicle communications devices, and bioacoustics data transfer sensors transmit digital signals through our bones.[39] The Internet of Things Cybersecurity Improvement Act of 2020 may make you feel more secure, but the truth is that billions of WBAN transceivers once known as "human beings" are now viewed as "nodes" and entry points for hackers. Where's the security in that?

"You really have to be aware to be able to *discern* and thus refuse all this and understand how a smart phone becomes the door of *l'Enfer* [hell] nowadays, and that 5G is more than mortiferous waves, but a system introducing 'the nano internet of things' Transhumanism (and the 8th Luciferian sphere anti-man and anti-Creation of God), thus opening the door to the abomination of nonhuman synthetic technology wanting to graft itself and invade the human body and mind, and that of other living beings, too. It's not even 'science' anymore; it's occult science." —New Rachel Sky, email, November 29, 2020

The Internet of Bodies (IoB) is blithely described as a Fitbit-style "collection of data via a range of devices that can be implanted, swallowed or worn," leaving out the fact that every day we are breathing in tiny *self-assembling* programmed machines that are constructing IoB neural networks inside a new hybrid race[40]—nanosensors digitally collecting biometrics from wearables, digital pills, "smart" thermometers, and myriad digital gadgets and household items that engage 24/7 with the neural network that we and our WBANs are nodes on, hackable by botnets like "dark nexus."[41]

In a May 14, 2024, article in *Popular Mechanics,* a study at University of Massachusetts, Amherst, recommends that 6G be powered by using human beings as antennas. Apparently, *visible light communications (VLC)*—"a wireless version of fiberoptics that uses flashes of light to transmit information"[42]—uses LED bulbs instead of radio signals to send information wirelessly to receivers ("anything with a camera"). The researchers have discovered that VLC leakage of energy from "side-channel radio wave signals" can be harvested by adding a little copper Tesla coil that requires only microwatts to run, and that "the human body offers the best medium—up to 10 times better than any other setting tested—for amplifying a copper coil's ability to collect leaked RF energy."[43]

And while people are just awakening to worries about 5G, the stealth dual use 6G is integrating 5G machine connectivity with satellite networks

*An allusion to the 1938 science fiction novel *Out of the Silent Planet* by British author C. S. Lewis.

surrounding the planet for global coverage,[44] with 7G "space roaming" waiting in the wings as well as other higher generations ready to run our "silent planet"* under a full-spectrum dominance mandate, cut off from Nature, cut off from the stars, cut off from the great spiritual universe. From millimeter to submillimeter to terahertz wave range, 6G subjects us to just this side of optical, X-ray, gamma and cosmic rays, its submillimeter (i.e., terahertz) bandwidth sandwiching us between microwave (electromagnetic) and infrared (photonic)[45] in sync with the high frequencies (100 GHz–10 THz) necessary for AI technologies like metaverse-style augmented reality and full-on brain-computer interface with the internet, thanks to our WBANs assisted by nanobots tucked into the tissues of our extraordinary human brains.

Infrared is an extraordinary *sensing light*. As early as the 1950s, Wilhelm Reich, MD (1897–1957), and Trevor James Constable (1925–2016)* used infrared film to photograph "plasmodial living organisms native to our atmosphere."[46] The present Space-Based Infrared System (SBIRS) under the Remote Sensing Systems Directorate of the U.S. Space Force specializes in "human biosignature" surveillance as well as "battlespace awareness and technical intelligence"[47] from space by relying on HEO (highly elliptical orbit) sensors on satellites, GEO (geosynchronous earth orbit) satellite constellations, and ground-based Space Fence sites, its infrared sensors gathering "raw, unprocessed data" then down-linked it to the ground.[48] I've referenced sensors and nanosensors throughout this book, but the sensors on satellites and SBIRS are of a "higher tier":

> . . . sensor networks are spatially aware and are most closely linked to geographic location and the physical environment than centralized systems. A sensor node in a typical sensor network has a battery, a microprocessor, and a small amount of memory for signal processing and task scheduling. Each node is equipped with one or more sensing devices such as sensors for visible or infrared light, changing magnetic

*A New Zealander like pilot and harmonics mathematics genius Bruce Cathie (1930–2013), Constable was a radio electronics officer in the U.S. Merchant Marine for twenty-five years and wrote four books on aviation. Both Constable and Cathie pursued the truth about UFOs and discovered the existence of "plasmodial living organisms." I myself have seen one shaped like a giant skatefish.

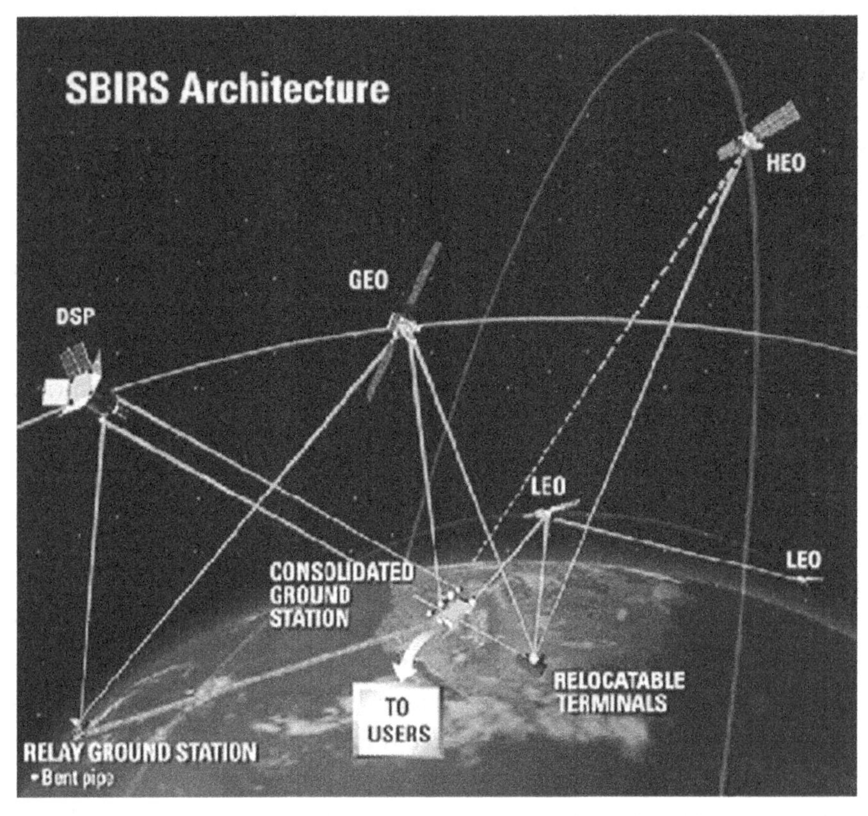

Fig. 10.4. The space-based infrared system (SBIRS) provides "battlespace awareness and technical intelligence," but who is the enemy? The system entails HEO (highly elliptical orbit) sensors on satellites, GEO (geosynchronous earth orbit) satellite constellations, and ground-based Space Fence sites around the planet to which the system is tethered. "SBIRS infrared sensors gather raw, unprocessed data that are down-linked to the ground, so the same radiometric scene observed in space will be available on the ground for processing. The SBIRS sensors also perform on-board signal processing and transmit detected events to the ground, in addition to the unprocessed raw data."[50]
US Air Force

field, electrical resistance, acceleration or vibration, pH, humidity, or temperature; acoustic microphone arrays, and/or video or still cameras. Each sensor node communicates wirelessly with a few other neighboring nodes within its radio communication range. A wireless sensor network may also be augmented with a higher tier of more powerful,

> wired nodes with greater network capacity and computation power, as in the Tenet architecture. *Nodes in this higher tier are sometimes called masters or microservers.*[49]

Given that SBIRS is a space-based "master," it should not be surprising to learn that the data it collects would include:

- Infrared brain imaging from space
- Teramobile infrared femtosecond lasers* and LiFi (light fidelity) LED (light-emitting diode) satellite optics
- Near-infrared spectroscopy (NIRS) able to penetrate the prefrontal cortex for "noninvasive" brain-computer interface
- Heat sensor cameras like FLIRS (forward-looking infrared radiometers)
- Detection of plasma spectra in "the terahertz gap" (the far-infrared wavelength of 15 µm microns to 1 mm, 20 THz to 300 GHz)

And of course I wonder what SBIRS infrared is ferreting out of the stratosphere as well as from our troposphere.

6G HEXA-X

> *We are on the move towards a cyberphysical world where we separate physical machines and assets from the logic that is then moved into the digital world, often via a digital twin.*
>
> Joachim Sachs, "Hexa-X-II Feb ws, Session 19: 6G for Dependable Communication," April 26, 2024

Cellular and Wi-Fi networks depend on microwaves in the gigahertz range, whereas tiny submillimeter wavelengths like terahertz (0.03 mm to 3 mm) are essential for manipulating atoms, DNA, and RNA—all of which points to nanobots and the ominous future that 6G is heading up. Enter Europe's Hexa-X, the "flagship for 6G vision and intelligent fabric of technology enablers connecting human, physical, and digital worlds,"[51] a "digital world" run by twenty-five "key players" from the corporate

*A femtosecond (FS) laser is an infrared laser with a wavelength of 1053 nm.

world and academia—and possibly the first board of directors of Rudolf Steiner's Eighth Sphere now known as the metaverse.

> While 5G has enabled us to consume digital media anywhere, anytime, the technology of the future should enable us to embed ourselves in entirely virtual or digital worlds. In the world of 2030, human intelligence will be augmented by being tightly coupled and seamlessly intertwined with the network and digital technologies. With advances in artificial intelligence, machines can transform data into reasoning and decisions that will help humans understand and act better in our world.[52]

A common radio platform is used for both 5G and 6G connectivity, despite the fact that 6G is 100X faster than 5G (1 terabit per second). Yes, 6G uses 5G as its core deployment platform. As designed by Nokia Bell Labs, 6G makes decisions, creates real-time digital twins, and deepens humanity's connection with the digital world (the metaverse) because it is *a unique sensor technology* that is truly immersive, which is why it is called *extended reality (XR)*, whether utilized for smart farming of the GMO "foods" or for creating high-fidelity holograms.[53] Whereas 5G can be used for virtual reality (VR), 6G kicks it into high gear with augmented reality (AR).

The *hexa* in Hexa-X stands for the geometric six-sided form, but also includes the root word *hex,* a black magic curse, a spell; notably, hexagrams and pentagrams are used in witchcraft for oppressing and binding. Is naming a 6G terrestrial outpost of the Eighth Sphere after spell-making simply a fun semantic "coincidence," or is it an implant designed for the human subconscious, like much of what is called "advertising" does with logos and images? The clue to the answer may lie with factoids like how both melatonin and DMT have hexagon-pentagon molecular structures, and how the hexagon is the most efficient geometric shape for processing thermal energy.

Hexa-X, in an article about its joint European initiative to shape 6G, does not so much *inform* the public about a 6G technology as take the public down a rabbit hole populated by big, vague, suggestive terms like *immersive communication and cyber-physical systems* (immersed in 5G/6G affixed to hex BCIs?) and *the high number of use cases* (7 billion or so via WBANs and in vivo nanobots?). As Led Zeppelin warned in "Stairway to Heaven," (1971), "sometimes words have two meanings," one for the conscious mind (exoteric) and one for the subconscious (esoteric):

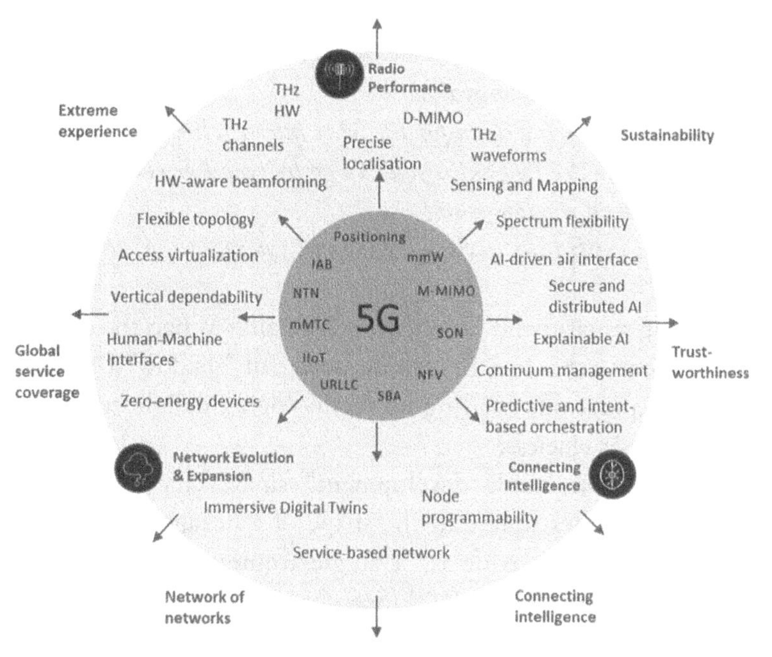

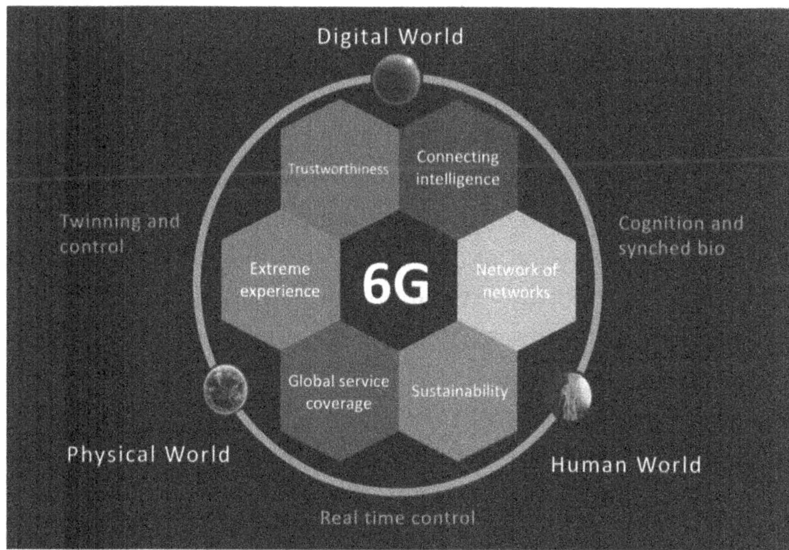

Fig. 10.5. Top: 5G. Bottom: The Hexa-X: 6G.
According to Peter Vetter and Magnus Frodigh in "Hexa-X—The joint European initiative to shape 6G," January 27, 2021, "Hexa-X will pave the way to the next generation of wireless networks (*Hexa*) by explorative research (*X*). The Hexa-X vision is to connect human, physical, and digital worlds with a fabric of 6G key enablers." Peter Vetter is president of Bell Labs Core Research at Nokia. Magnus Frodigh is head of Ericsson Research. [55]

Images from Ericsson.com

If there's a bustle [possible danger] in your hedgerow,
don't be alarmed now,
It's just a spring clean for the May queen [elites].
Yes, there are two paths you can go by [good or evil],
but in the long run
There's still time to change the road you're on.

Am I being fanciful and "unscientific"? Really? When the entire world is run by ancient secret societies, those powerful "shadow" Brotherhoods heading up corporations heading up technologies, delighting in keeping the profane masses clueless?

The UN's "sustainable development" smart city is divided into "cells" about ten miles square in the shape of a hexagon, like a beehive. Significantly, graphene oxide is "a single-atomic carbon sheet with a hexagonal honeycomb network."[54] (See chapter 6, "The Weaponization of Magnetism" and appendix 1, "Invisible Mindsets.") University of Washington scientist Gerald H. Pollack discusses the fourth phase of water (beyond solid, liquid, and gas) as a hexagonal geometry of life able to draw current from sunlight.[56] Could this have something to do with why sunlight is being blocked from Human 1.0 by geoengineered plasma cloud cover, given that our bodies contain 60 to 75 percent water?

The encryption utilized for 6G promotion by Hexa-X is far more *opaque* than the promotion of 5G. While still touting "UN sustainable development goals" and the "EU Green Deal" we've heard and read over and over, the Hexa-X conceptual graphic boldly sets forth six beehive / hive mind cells depicting the new mix of digital, physical, and human worlds liberally peppered with terms like *extreme experience* and *fully immersive communication or remote control at scale* of *end-to-end communications*; *mission critical and massive machine connectivity*; *the great potential of Artificial Intelligence (AI) and Machine Learning (ML)*; *large-scale deployments of intelligence in the wider society*; *digital inclusion*; *novel [gain of function?] aspects such as integrated sensing, artificial intelligence, local compute-and-storage [DNA?], and embedded devices* [nanobots?]; *extreme performance. . .*

So when I read something like this, I cannot help but feel that Mr. Rugeland is casting a spell, not informing:

> Enablers for Human-Machine Interfaces and fully immersive digital twins will be developed in the form of flexible assignment of heterogeneous resources and dependable network services and an execution environment guaranteeing very high levels of privacy, user transparency, and availability will be developed for the digital twin concept, useable by other human-in-the-loop and critical industrial applications as well.[57]

Saying without telling constitutes the arts of propaganda and spellbound encryption . . . Eventually, truths are allowed to leak out, but only years after they have lost the power to galvanize public inquiry and resistance. 5G faster downloads were never what 5G was *really* about:

> 5G is about increasing bandwidth to transfer more data to support the Internet of Things (IoT), the network of physical objects—"things"—that are embedded with sensors, software and other technologies for the purpose of connecting and exchanging data with other devices and systems over the internet, including your body. The IoT intends to wirelessly interconnect 50 billion more devices, and the data collected will be used primarily for artificial intelligence, surveillance and creating the next version of "humans"—beings that are part human and part machine.[58]

Similarly, AT&T's *AirGig* is about adding Wi-Fi to existing power lines with 4G and 5G waves supposedly for last-mile wireless connectivity and extending the reach of cell towers and fiber optic cable. By collaborating with utility companies and the "smart grid" of Lockheed Martin's Space Fence lockdown on smart cities, AirGig's broadband-over-power-lines (BPS) and its *close-proximity, high-intensity radiation* mean greater *immersion* and 5G *extreme performance*[59]—there's that term again—moving more and more Big Data for metaverse machine-learning systems in expectation of the magic "Singularity" moment—like a Satanic Rapture when everyone, everywhere is digitally "twinned" to the internet and we are translated into cartoon cyborg avatars that the Big Tech juggernaut has been driving us toward for decades.

And don't forget *backscatter* devices coupled with satellite 5G beams that use radio frequency (IoT radio, TV, iPhones, etc.) for transmissions without a battery or power grid connection. Backscatter means picking up an existing signal and converting it into thousands of microwatts of electricity so as to modify and reflect the signal with encoded data—all of which suggests that your IoT, including the iPhone you can't do without, is being used (on your dime) even when all the devices are off.

THE FOUR ETHERS UNDER ASSAULT

> *By virtue of both the one and the other, the human being, when he lives in the physical-etheric world, as he learns to walk, to speak, to think, he becomes a moral being in the earthly world. This is what he becomes because his etheric, when it unites with the physical body, manifests on a soul-spiritual level, and this ether body consists of warmth, light, chemical and life ether. This is the cosmic life of any process of becoming. In the course of the unfolding of the soul-spiritual life of the human being, his cosmic moral being comes to manifestation; and in the process of his biological development, the four ethers enter into a connection with their projections in the four elements which manifest in the physical in the four aggregate states of matter: in the elements of "warmth" (fire), "air," "water," and "earth," behind which also stand the spiritual moral qualities and deeds of the Hierarchical Beings. That is the essential nature of the phenomenon of earthly life . . . When evil gets hold of the etheric body of a child, [it] is prevented from uniting with the morality of the world-ether. From the other side, the four kinds of ether which form our ether body—now removed from their moral sense—are subjected to a coarse, violent assault of the physical Ahrimanic ground in the Earth's interior, in its electromagnetism. And in this way, the phenomenon of life in the human being is made subject to a fundamental evolutionary deformation.*
>
> G. A. Bondarev, *The World and Humanity at the Crossroads of the Occult-Political Movements of Our Time**

*Russian anthroposophist G. A. Bondarev died in 2022. This book is presently only in manuscript.

In chapter 9, I hinted at an idea by which Human 1.0 might be able to save its spirit- / Nature-based species (*Homo sapiens*) from the assault on life we are undergoing in the twenty-first century, thanks to the occultists who govern the military-corporate-intelligence-medical-pharmaceutical global cartel in league with elite bloodlines and their ancient secret societies.

Human 1.0 shares with planet Earth its physical body elements and the four ethers that I discussed in chapter 2 of *Under an Ionized Sky* (2018). It was there that I told the sad tale of how the æther (ether) as a Western scientific concept was banished by secret societies in the 1920s. In his May 20, 1920, address at the University of Leyden in the Netherlands entitled "Ether and the Theory of Relativity," Albert Einstein (1879–1955) flatly stated that space without ether is unthinkable, and yet he stopped referencing it in his writings and lectures.

> *The warmth ether* manifests as heat
> *The light ether* manifests as gas
> *The chemical / sound / tonal ether* manifests as fluid
> *The life ether* precedes matter and individualizes[60]

As for what ether is, plasma and ether are, in many ways, one and the same in different phases. Once a gas is ionized by electromagnetism, it becomes a true ionic plasma. In deep space, the heat of an electromagnetic event turns ether into plasma, the fourth state of matter.*

Unfortunately, the banning of ether *as a scientific concept* banned *life itself* from Western science. One hundred years later, we can see how Western societies in general have been weakened if not deadened by the materialism that bans from science what cannot be measured. Our human relationship with Nature and the Divine is now broken. Atheistic scientists have taken up the slack and deemed everything living to be "things." Ignorant of the living etheric bond between the human being and the Earth, those who speak from pulpits and podiums mindlessly support financial and medical policies that lead again and again to atrocities against life itself in the name of "progress" and "healing."

The West having lost its way, planet Earth is being turned into an

*From heat to fluid to gas → [electromagnetic event] → plasma.

AI-run machine while the human nervous system is loaded with "'I'-less, soulless content for which electricity that powers the internet serves as a basis."[61] Things might have turned out otherwise, had we known about the four living ethers we share with the Earth!

If we are willing to be "schooled" in practicing the wisdom of "We cannot solve our problems with the same thinking we used when we created them," we necessarily begin by rekindling our relationship with the four living ethers so that our deadened nervous systems might embrace more life and less electricity. Given that the 5G/6G technology works in tandem with synthetic biology for the agenda of full electromagnetic spectrum dominance over all of life on Earth, we also attempt to turn technologies in directions that favor life itself, one nervous system at a time. Think of this every time you allow a small child to watch TV or play with an iPhone or iPad; watch closely as the child's living etheric forces are devoured before your very eyes.

Let me be absolutely clear: *Electromagnetism and electricity are inimical to etheric energy.* As early as the eighteenth century—when Western scientists still recognized the reality of the human soul—electricity was deemed to be dangerous to the human being.[62] If we can but reclaim the truth of our ether-forged bond with planet Earth, we can grasp why Rudolf Steiner stressed that electromagnetism has more in common with death than life. Now that the entire planet is bathed in the wireless electromagnetism of the Space Fence lockdown, what are the four ethers undergoing? Whole forests are dying, and species are disappearing as their etheric connection to the cosmos disappears. Slowly, it is finally dawning on thoughtful people that in order to save what is truly living, including Human 1.0, we will have to confront the technologies promising temporary baubles like convenience, comfort, and entertainment in exchange for stripping us of our life forces as controlled media seek to convince us that free energy is still elusive, despite myriad reports of the murders of free energy inventors.

We have been captured in a wireless Space Fence lockdown as the HAARP "common background" pulse of 3 Hz pulsates around and through us, and we are forced into a seamless connectivity between the internet and everyday life. Thus, our resistance must begin with electromagnetism and the technologies that produce and beam it. Please review *Under an Ionized Sky* (2018) where I thoroughly describe the Space Fence edifice—off planet, on planet, underground, and in our bodies. These networks require constant recalibration in order to keep them working in sync to keep us "plugged in." *This necessity for constant recalibration may be the pivotal weakness of the planned planetary edifice of control.*

In my book on synthetic biology, I will continue to develop this sage insight:

We cannot solve our problems with the same thinking we used when we created them.

EPILOGUE

The Great Chain of Being

> *Power corrupts, the voice of reason whispered darkly in his brain, and there was no doubt that was true—the corruption had long since set in, else how explain the rapes and murderous pranks? But once that was over—once the reign of the Exec was unchallenged and the foul work of cleaning off the old world was done, was it not at least possible that the human beings who owned such power would at least try to use it wisely?*
>
> FREDERIK POHL, *DEMON IN THE SKULL*, 1985

Feeling overwhelmed after reading *Geoengineered Transhuman*? The truth can be very hard on us, particularly after decades of going along to get along with deceptions and secret agendas, viewing yourself as powerless, the very state that the Deep State controllers have been banking on, watching us wade through all the lies, propaganda, and "national security" secrets. You've tried to deny the nightmare, but you only ended up in a *spiritual bypass*.

By now you've begun something real, namely the lifting of the spell of the last four years. Now, you may even be ready to pursue the sage, liberating insight: *We cannot solve our problems with the same thinking we used when we created them.*

The good news is that we already possess the consciousness tools for the geoengineered transhuman situation we are now finding ourselves in. Electromagnetism and artificial intelligence have not yet entirely destroyed or captured all of our brain capacity, and our ether bodies are still aligned to a fragile degree with the four ethers of our living planet. Our greatest weakness up to this point is that we have spent years being distracted by electromagnetic technologies as if they were birthright conveniences and have not made an honest, concerted effort to train our minds in how they are actually "dual use" technologies being used for social control. Since being shocked into partial wakefulness by the 2020-2023 assault, we have been allowing fear and outrage to consume our energy, unaware that there is a kind of thinking that leads to world-class solutions for world-class problems. Nothing quick and pill-like, mind you, and while prayer and meditation are excellent, problems due to world-class technologies need a thinking that can penetrate to the nature of weapons like the HAARP scalar system that could actually be used to clean up nuclear radiation or ameliorate destructive weather instead of destroy Human 1.0.

For de-weaponizing world-class technologies, we need more than intellectual processing by the computer-like brain—you know, the brain with all of the neurons figuring out our sense perceptions and the what of what we're encountering in our experiences on Earth. We need deeper thought processes capable of exploring what these sense perceptions and experiences mean for real human lives so we can process not just measurements and categories but what such weaponizing spells for life on Earth. My brain is a diving board, and I am the diver. What my brain provides for my dive may vary immensely, making my dive (thought) either pedestrian or incredible, but the dive (thought) is mine, not my brain's.

In a nation that considers the Good to be relative ("feel good"), a passing meme or arbitrary term that "business as usual" exploits for profit, it is not likely that more rational argumentation, documents, statistics, graphs, and citations—the lifeless PhD way of thinking that created the technological problem we are now facing in the first place—will convince digitally smart scientists and techs lacking in ethics and empathy, who view human beings as so much cattle or statistical "nodes," to limit weaponized "dual use" technologies to serving as plowshares that serve the Good. Western culture, like Rome, has fallen prey to self-interest as its litmus of "success." Jonah was called to serve the Good but instead brought calamity. From Jonah 1:1–3:

> Now the word of the Lord came unto Jonah the son of Amittai, saying, Arise, go to Nineveh, that great city, and cry against it; for their wickedness is come up before me. But Jonah rose up to flee from the presence of the Lord unto Tarshish, and went down to Joppa; and he found a ship

> going to Tarshish: so he paid the fare thereof, and went down into it, to go with them unto Tarshish from the presence of the Lord. But the Lord sent out a great wind into the sea, and there was a mighty tempest in the sea, so that the ship was like to be broken . . .

Western civilization is riding the *Titanic* down into the deep, the wickedness of Nineveh (Deep State) all about it. Jonah's individual free will *could* have been a boon to his conscience, his community, and history itself, but because he was filled with fear for his own skin, his free will operated in error, as does ours.

In the previous chapter on electromagnetism, I promised that human consciousness could hold the solution to today's technological peril. This is true, but not with the kind of thinking you learned in modern schools. We need high-altitude depth thinking for countering and changing the direction of the weaponized technologies of our era. Without applying what Rudolf Steiner defined as *consciousness soul etheric thinking,* the Jonah's of the future will have no free will because they will be cyborg transhumans run by remote 5G/6G/7G systems.

THE SPECTRUM OF FORCES WE LIVE WITHIN ON PLANET EARTH

Our era may indeed be an Atlantis redux attempting once again to *compel full-spectrum dominance over planet Earth with weaponized technologies.* Ever keeping in mind the revolutionary idea that we will not be able to solve the transhuman technology problem with the same thinking that created the problem, let's begin with a random list of macro-forces that we encounter each sojourn on planet Earth, albeit invisibly and unconsciously and yet deeply interconnected with the Great Chain of Being*: magnetism, electricity, gravity, light, the atmosphere, atoms, plasma, the four ethers, the four elements, earthquakes, lightning, tornadoes, floods, droughts, tides and tidal waves, the Sun and Moon, the human mind, genes, *bios* life (*ki*), and death.

*According to Wikipedia, "The great chain of being is a hierarchical structure of all matter and life, thought by medieval Christianity to have been decreed by God. The chain begins with God and descends through angels, humans, animals and plants to minerals."

Technologies now exist by which these forces can be harnessed and wielded against us by those seeking dominance over planet Earth. Along with educating ourselves about these forces and their power over us, we must learn to pay close attention to the new technologies of nanotechnology and synthetic biology (synbio) and how they point to the relatively unknown force of *nonorganic life.* Given that the human body-soul configuration was originally created by Divine beings in accord with this living planet's forces and laws, it becomes more and more apparent that *the natural health and balance of the human being depends upon the health and balance of planetary forces, and vice versa.*

Not only do we now have quantum physics but the Heisenberg Uncertainty Principle has informed us that consciousness is in anything that exhibits wave- like properties. Under Newtonian physics, planetary forces were inert, mechanical, and set in motion long ago by the Clockmaker God. All scientists needed to do was measure and categorize the Earth's inventory. But now we know that living forces and tiny particles do not behave like numbers, that the position and speed of a photon of light or an electron cannot be measured with accuracy, and that the sudden entry to the lab of a conscious human being with a different particle-wave configuration changes everything. In fact, the planet is a living being teeming with different consciousnesses, which has made materialists extremely uncomfortable, driving them to seek more and more control, including keeping the masses from realizing that "primitives" were right about being interconnected with the Earth. Living, invisible forces are more like "gods" than fixed "laws," with quantum physics adding consciousness, Earth evolution and human evolution are intimately bound up with each other in the Great Chain, which is why we are finally beginning to understand why the elites are redoubling their efforts to keep people cut off from Nature, deadened ("digitalized"), and in ignorance.* Ether was removed from science one hundred years ago and the Great Chain gets a paragraph in high school history books, when truthfully we needed knowledge of our interconnectedness with each other and the Earth more than ever, especially if we are to take seriously the problem of Evil presenting as technology today, which was why

*This includes keeping scientists and technicians in ignorance by subjecting them to compartmentalized, need-to-know security clearances.

physicist / metaphysicist Rudolf Steiner included Angels and Archangels among the forces of the Great Chain:

> People will have to know all the forces that the soul must summon up in order to overcome the powers of evil, or to transform them into good impulses . . . [P]articularly through the control of the different forces of Nature, the impulses that lead to evil will send their influences into the world in an immense, gigantic form. And the opposite forces, the forces of good, will have to grow out of the opposition to evil, and man will have to draw the strength for this opposition out of spiritual sources. This will take place above all during the 5th epoch [1414 CE to 3574 CE] when the exploitation of electric forces, which will assume quite different dimensions from those which they have assumed so far, will enable man to spread evil over the Earth, and evil will invade the Earth by coming in an immediate way out of the forces of electricity . . . The significant fact in the 5th post-Atlantean epoch will be, above all, that the Beings who stand immediately above the hierarchy of Man, the Angel-Beings, will be able to approach the individual human beings, not only groups of men, in a very intimate way so that these individual men will think that they are upholding things which come out of their *own* personal impulses, whereas—we may indeed say this—they are possessed by this kind of Angel-Beings [Fallen Spirits of Darkness belonging to the Hierarchy of the Angels] . . .[1]

Whatever your religious belief might be, Steiner categorized Evil and Christ as two opposing forces impacting humanity's evolution and viewed Evil as a trinity dedicated to testing the free will of humanity: Ahriman (contractive), Lucifer (expansive), and the Asuras (fracturing). Interestingly, his cosmology bears similarities to traditional Aztec cosmology[2] regarding the forces of Evil and death, even to views of the human realm as a "Middle Earth"* between higher Good forces (gods) and lower Evil forces ("subnature") that in the nine levels of *subnature* counterbalance by nine *supranature* higher hierarchies (Angel, Archangel, Archai, Exusiai, Dynamis, Kyriotetes, Thrones, Cherubim, Seraphim).*

*J. R. R. Tolkien used this term for Earth in his *Lord of the Rings* series.

While the two worldviews agree on nine chthonic levels of Evil in the inner Earth, Steiner's concept includes the change that occurred during the Mystery of Golgotha of Christ's first advent to "Middle Earth": now, the human being actively pursues a conscious path of *redemption of evil* in himself / herself for the sake of redeeming the forces / Beings ensnared in the lower chthonic realms of planet Earth.

We are no longer dealing with a simple depiction of Evil presenting as a demon with horns riding on our left shoulder. Evil in our era should be conceptualized as a complex of various subnature levels subject to technology. "Faith alone" without knowledge will not save us from modern technological Evil. We need our thinking caps, too—not the old thinking caps of the intellectual soul but the multilevel, imaginative thinking that Rudolf Steiner attributes to the consciousness soul born in the fifteenth century at the beginning of the fifth post-Atlantean epoch. The consciousness soul is individual, noninstitutional, metalogical, taking its lead from present-day, real-life experience as well as concepts carefully crafted in books. The consciousness soul attends to life's input ("destiny"), particularly events that answer to patterns (*synchronicity*[†]) often misinterpreted as "coincidence." Consciousness soul thinking utilizes analysis but goes beyond it in order to put concepts and perceptions together in a new perspective whose greater depth takes into account Great Chain forces, life experiences, and perceptions. Consciousness soul thinking is of a higher, more comprehensive order than intellectual soul thinking and encompasses the "spiritual discernment" described in the New Testament.[‡]

---

*See the chart in chapter 6, "The Weaponization of Magnetism."

†According to the *Psychology Today* website's "Basics" glossary, synchronicity is a phenomenon in which people interpret two separate—and seemingly unrelated—experiences as being meaningfully intertwined, even though there is no evidence that one led to the other or that the two events are linked in any other causal way.

‡Spiritual discernment: a kind of wisdom that comes from insight as much as from learned experience and knowledge; a gift from the Holy Spirit [an invisible Divine force, not a figment of the imagination]; the ability to determine the true nature of a situation, person, or thing; the quality of being able to grasp and comprehend what is obscure.

> Thus, three members must be distinguished in the soul as in the body, namely *sentient soul, intellectual soul* and *consciousness soul.* As the body works from below upwards with a limiting effect on the soul, so the spiritual works from above downwards into it, expanding it. The more the soul fills itself with the true and the good, the wider and the more comprehensive becomes the eternal in it. To him who is able to see the soul, the splendor radiating forth from a [human being] in whom the eternal is expanding is just as much a reality as the light that streams out from a flame is real to the physical eye. —Rudolf Steiner, *Theosophy,* chapter 1, "Body, Soul and Spirit," (GA 9)[3]

To grasp the nature or presence of Evil regarding technology, we must overhaul the assumption that graduate school thinking (analysis, weight of data, statistics, graphs, etc.) is "objective"; it is just more intellectual soul conditioning to prevent the masses from developing the consciousness soul, the past intellectual soul thinking being decidedly more easily controlled.

The presence of living, conscious forces helps us to consciously participate through the consciousness soul in the equation of a conscious universe. Scientism masquerading as science wants to keep us going from A to B and waiting for experts to tell us what to do. For example, we are told that Atlantis was a myth, but what if it actually was a world-famous flood ~11,700 years ago caused by an abuse of technology similar to what we have today? What if the sinking of the *Titanic* in 1912 and the 2,224 passengers aboard was really not an "accident" but to assassinate forty-eight-year-old John Jacob Astor IV to end his financial backing of Nikola Tesla?* Modern cover stories spun by the CIA and military intelligence abound, hiding modern "national security" experiments with Earth forces that cost

*The history of Astor and "the octopus" industrialist John Pierpont Morgan and their involvement with technology geniuses John Worrell Keely and Nikola Tesla has been tampered with, but the proof that Morgan had Astor ritually killed like a Norse lord with the April 12, 1912, sinking of the *Titanic* lies with the usual Freemason number game: Morgan's death on March 31, 1913, 13 days before the first anniversary of the *Titanic* sinking and 18 days before his seventy-fifth birthday. With Astor's death, Morgan took full control over Tesla and Keely. See Theo Paijmans' book *Free Energy Pioneer: John Worrell Keely* (Kempton, IL: Adventures Unlimited Press, 2004).

billions of dollars—the Philadelphia Experiment (magnetism), the Montauk Project (magnetism and mind control), the Roswell, New Mexico, crash (exotic propulsion), etc. Our era will not be remembered for its arts, spiritual insight, or societal progress but for its willingness to sacrifice humanity for the sake of global power. Can we think our way out of the silicon / carbon-based transhuman race now underway?

Resource-rich movers and shakers conspiring for planetary domination are hardly new. The macro- and micro-machinery by which to achieve *planetary*—not merely national or racial or religious—domination is new, or at least has not been seen since antediluvian times. Today, we are seeing attempts by the public to regain clean air in bills like the 2021 Rhode Island Clean Air Preservation Act (bill S0572), originally written by Susan Clarke and known as the "chemtrails / anti-geoengineering bill." It begins with:

23–23.8–2. Legislative intent and findings.

(a) It is the intent of the general assembly, by enactment of this chapter to preserve the safe, healthful, resilient and peaceful uses of Rhode Island's atmosphere for people, the environment, and agriculture, by regulating geoengineering, weather modification and other atmospheric activities and prohibiting those that are harmful.

(b) "Geoengineering" is defined as the intentional manipulation of the environment, involving nuclear, biological, chemical, electromagnetic and/or other physical-agent activities that effect changes to Earth's atmosphere and/or surface.

(c) The general assembly finds that geoengineering encompasses many technologies and methods involving hazardous activities that can harm human health and safety, the environment, agriculture, property, aviation, state security, and the economy.

(d) According to a 2020 report by the Environment Rhode Island Research & Policy Center, *Trouble in the Air*, "Air pollution is linked to health problems including respiratory illness, heart attack, stroke, cancer and mental health problems."

The bill continues in great detail and is available to read on the State of Rhode Island General Assembly website. Other states—Tennessee, Illinois, Kentucky, Minnesota, New Hampshire, Pennsylvania, and South Dakota—are also attempting to initiate similar bills, but what is the likelihood that federal weather manipulation and all the rest will bow to the public will?

If we are truly committed to developing the requisite skill by which to resist the planetary domination now intruding on the biological, we must first penetrate the deceit and secrecy of the energy potentates running world-class technologies like HAARP "sky heaters," satellites, nanotechnology, and synthetic biology that are overrunning what is natural with the synthetic and plugging human brains into AI systems via brain-computer interfaces (BCIs).

By the time I was writing *Geoengineered Transhumanism* in 2020, my consciousness soul thinking had realized that the Lyme and Morgellons I had studied in the early 2000s were precursors of the synthetic biology and nanotechnology now being delivered in Big Pharma Covid-19 serums being marketed to a fearful, ignorant world as "vaccines." Reading Arizona State University physicist Paul Davies's cleverly entitled book *The Demon in the Machine: How Hidden Webs of Information Are Solving the Mystery of Life* (Penguin Books, 2018) revealed to my consciousness soul that Davies's constant *repetition* of the term *demons* in relation to the body seemed to act like an *invocation* to forces slumbering in the reader's subconscious. His definition of the human soul on page 185 proved that at the very least Davies was not interested in consciousness soul development for the masses: "In popular Christian culture, [*res cogitans* (wispy mind-stuff)] has sometimes become conflated with the soul, an immaterial extra ingredient that believers think inhabits our bodies and drifts off somewhere when we die." Davies went so far as to state the "watergate"* by which Human 1.0 could be transmogrified into Human 2.0 "at the nano-scale":

> It turns out to be only at the nano-scale that the thermal energy in the temperature range [of life as we know it via water] is comparable to the chemical and mechanical energy of the biological machinery, and thus able to drive a wide range of transformations . . . Significantly, living organisms seem to be actual self-reproducing machines. We thus gain insight into the logical architecture of life by deliberating on the concepts of a universal computer (Turing machine) and a universal constructor (von Neumann machine).[4]

*A sluice or floodgate.

Biological machinery . . . self-reproducing machines . . .

The consciousness soul perceives the difference between an authentic human being—rich or poor, educated or not—and emperors like Davies. Let us not follow the intellectual soul path of the lords of the bedchamber!

So now the Emperor walked under his high canopy in the midst of the procession, through the streets of his capital; and all the people standing by, and those at the windows, cried out, "Oh! How beautiful are our Emperor's new clothes! What a magnificent train there is to the mantle; and how gracefully the scarf hangs!" In short, no one would allow that he could not see these much-admired clothes because, in doing so, he would have declared himself either a simpleton or unfit for his office. Certainly, none of the Emperor's various suits had ever made so great an impression as these invisible ones.

"But the Emperor has nothing at all on!" said a little child.

"Listen to the voice of innocence!" exclaimed his father; and what the child had said was whispered from one to another.

"But he has nothing at all on!" at last cried out all the people.

The Emperor was vexed, for he knew that the people were right, but he thought the procession must go on now! And the lords of the bedchamber took greater pains than ever to appear holding up a train, although, in reality, there was no train to hold.*

Once upon a time, our human brains were in sync with the Earth as reflected in the Schumann resonance of 7.83 Hz. Now, we are drowning in a sea of jagged electromagnetic frequencies and ubiquitous symptoms from ionized and non- ionized radiation being called "long Covid." Pulsed signals streak; power lines glow; toaster ovens and electric lights and hair dryers and refrigerators and washing machines talk to each other; billions of cell phones, laptops, and wireless devices pulse low-frequency emissions billions of times per second, destroying our natural resonance and invisibly assaulting us. Cancers and chronic autoimmune disorders multiply: lupus, fibromyalgia, chronic fatigue, Crohn's disease,

*From the Hans Christian Andersen tale, "The Emperor's New Clothes."

rheumatoid arthritis, celiac disease, and on and on. Human 1.0 struggles like a pinned butterfly under the onslaught of weapons of war called "conveniences." Not only does an endless plasma cloud cover blot out life-giving solar rays, but our eyeballs can no longer take in much light as neurological conditions increase.

If we continue down the path being carved out by the electromagnetics of those bent upon being the Masters of the Universe, our inner soul life will further weaken until we have no more original thoughts, no ideals, no high-cast dreams, no capability for insight or sustained intent—in short, until we are no longer human but transhuman. Haven't we watched passively from the sidelines long enough, oohing and ahhing as the titans pillage our planet and children? Are we done yet with complaining and trembling and waiting for artificial intelligence to take over our thoughts entirely, until, like our world, we are artificial, soulless, bereft of the mighty future that the Divine originally carved out for the grand human experiment in free will evolution on planet Earth? To accept the dictates of AI and its robotic bureaucrats is to sell the birthright of the unique human soul for baubles, our Divine potential leached out by an electromagnetic assault bent on controlling it all, now that nanobots infest one rung of life above another.

Human consciousness is not a numbers game per hundredth monkey or quantum entanglement. It is about a certain *quality* held by those who have the strength and discipline to hold to ideals and practices. Consciousness evolution is now dependent upon specific individual souls, not collective tribes or nations or "enlightened" corporations. A consciousness soul standard of a few exemplar human beings in our midst will suffice to direct the rest of us in the right direction.

When I was nine years old, I read Plato's "Allegory of the Cave" and glimpsed what it might mean to be truly "illumined," wherein the clear light of one's own wakeful perception—not the shadows dancing on the wall like one sees today on television and the iPhone—kickstarts real individual awareness. Quantum physicist Werner Heisenberg was convinced that the Uncertainty Principle points to the existence of multiple dimensions, one of which is the frequency where thought itself reveals itself in scalar holograms. Even atoms, whether as particles or waves, defy exact measurement in the presence of human consciousness because it is the nature of consciousness to precede all physical manifestations. Knowing and practicing such things in a gnosis praxis way leads to real spiritual freedom.

Perhaps this is why operating out of the consciousness soul calls for a modern review of old concepts like *æther*, *angels*, *demons*, *death*, etc.—forces that we can neither see nor weigh nor measure, but which can help us remain unsubsumed by materialism. By erasing the big black line drawn between science and spirit and the lie that the end justifies the means, we can remain open to possibilities far beyond what the powers-that-shouldn't-be have in mind for geoengineered transhumans. By practicing consciousness soul thinking, we can undermine the illusions that the Emperor seeks to bind us with and grow stronger in standing by our God-given free will so that satellite triangulation is unable to leach our own thinking from us. We say, "No, I will not comply" if it entails acts against one's humanity or the planet, whatever the cost, and the energies that are forces capable of serving the Good of the Earth support us.

Imagine a Spartacus uprising of human beings in sync with the Good defeating the synthetic juggernaut that wants to turn all of life into electromagnetic "nodes" in the AI-run Space Fence. First, we build a clear picture with consciousness soul thinking of how 5G/6G/7G electromagnetics manipulates all the nano-infested networks, then probe the role of synthetic biology in building transhumans. As I do in the synthetic biology followup, we review the Covid-19 years— all but 2.2 billion of the world's 7.951 billion population were inoculated in the three years from 2020 to 2022—and build a clear mental picture of an alternate approach to health based on living in accord with the four ethers and a geographic / environmental / terrain medicine approach able to relieve your area of chemtrails even as the four ethers and our human etheric bodies are revivified. We forge the beginnings of a Glossary for a biology forced to go digital, and a *Remedies* section to guide our liberation from the interlocked Big Medical / Big Pharma cartel. Esoteric insights of Rudolf Steiner continue to expand our consciousness soul thinking, guided by the rallying cry for freedom from medical tyranny.

In the earliest days of the Covid-19 assault, Drs. Klinghardt and Cowan quoted Rudolf Steiner's words from 100 years ago:[5]

> In the future, we will eliminate the soul with medicine. Under the pretext of a "healthy point of view," there will be a vaccine by which the

> human body will be treated as soon as possible directly at birth so that the human being cannot develop the thought of the existence of soul and Spirit. To materialistic doctors will be entrusted the task of removing the soul of humanity. As today people are vaccinated against this disease or that disease, so in the future children will be vaccinated with a substance that can be produced precisely in such a way that people, thanks to this vaccination, will be immune to being subjected to the "madness" of spiritual life. He would be extremely smart, but he would not develop a conscience, and that is the true goal of some materialistic circles. With such a vaccine you can easily make the etheric body loose in the physical body. Once the etheric body is detached, the relationship between the universe and the etheric body would become extremely unstable, and the human being would become an automaton, for the physical body must be polished on this Earth by spiritual will. So the vaccine becomes a kind of ahrimanic force so that the human being can no longer get rid of a given materialistic feeling. He or she becomes materialistic of constitution and can no longer rise to the spiritual.[6]

Let us think on these things—how we will commit our lives to a new approach to the challenge that lies before us.

APPENDIX 1

Invisible Mindsets

> *The case against science is straightforward: much of the scientific literature, perhaps half, may simply be untrue.*
>
> Richard Horton,
> editor in chief of *The Lancet*

Mindsets constitute a subtext to daily life that we often do not recognize as influencing our perceptions and decisions about what is what and who is who, and why. Each lifetime is a three-ring circus, a multitiered staged event filled to overflowing with hidden intentions and meanings there for the unraveling by those who desire to go to the trouble of wading through illusions, deceptions, and "false positives," to get to the truth behind one event or another. Make no mistake: it is working for the truth that develops the soul, and the lack thereof that wastes opportunities for spiritual development.

Be prepared for resistance when you begin to lift the veil over what you've been told is nonsense, conspiracy theories, or ancient history. In fact, the higher your level your education, the greater will be your resistance. The elite—depending on their advancement in the various secret societies—count on your intellectual resistance to seeing the truth behind the shadows.

The following are eight mindsets to be aware (beware) of.

SATANISM AND THE U.S. MILITARY

. . . at one point I defined evil as "the exercise of political power that is the imposition of one's will upon others by overt or covert coercion in order to avoid . . . spiritual growth."

M. Scott Peck, PhD, author of *People of the Lie: The Hope for Healing Human Evil* (1998)

The U.S. military (all branches) went through a transformative shift in the 1990s known as the Revolution in Military Affairs, abbreviated as RMA by acronym-loving bureaucrats. It wasn't just about shifting from ballistic to kinetic weapons (including nonlethals) and "asymmetric warfare"; it was a psychological shift born out of the CIA's Phoenix Program, trialed from 1968 to 1972 during the Vietnam "conflict." That program introduced the practice of employing paramilitary teams and their CIA "advisers" to routinely carry out torture, murder, and assassination as official government policy in which the ends justify the means.[1]

In 1980, Colonel Paul E. Vallely and Major Michael A. Aquino wrote a concept paper titled "From PSYOP to MindWar: The Psychology of Victory."[2] Aquino had been a highly decorated Green Beret psyops officer with top security clearance during the Phoenix Program, 1968–1972, "designed to identify and destroy the Viet Cong (VC) via infiltration, torture, capture, counter-terrorism, interrogation, and assassination" (Wikipedia).* The Phoenix counterinsurgency doctrine that began in Vietnam didn't end there, though; it was continued in Central America, and eventually constituted the core of the 1990s Revolution in Military Affairs (RMA) doctrine. The CIA Cold Warriors who conceived of Phoenix always intended it for domestic use.

What is strangely overlooked by military historians is that the decorated psyop war hero Aquino was not just a practicing Satanist—he was the high priest of the Temple of Set, a Satanic church on San Francisco's Russian Hill not that far from the Presidio U.S. Army base, where he was stationed after Vietnam. From *PSYOP to MindWar*, page 9:

*Phoenix ran concurrent with the CIA's Paperclip Nazi–run MK-Ultra and its 149 subprojects in the name of psyops and "behavioral control."

> The MindWar operative must know that he speaks the truth, and he must be PERSONALLY COMMITTED to it . . . For the mind to believe its own decisions, it must feel that it made those decisions without coercion. Coercive measures used by the MindWar operative, consequently, must not be detectable by ordinary means.

PSYOP to MindWar extends its application to human political and social interactions generally, identifying and refining what were previously only vague or unknown mental processes into "thought architecture." The paper even mentions ionizing the atmosphere as a method of controlling emotions: "An abundance of negative condensation nuclei ("air ions") in ingested air enhances alertness and exhilaration, while an excess of positive ions enhances drowsiness and depression. Calculation of a target audience's atmospheric environment will be correspondingly useful."[3]

Julianne McKinney, a former U.S. Army intelligence officer and the author of a December 1992 study titled "Microwave Harassment and Mind Control Experimentation," understood well the psyops "special access programs" (SAPs)* that would be used to create a societal "ethical and political revolution." Subtle, insidious Satanism was first on her list, and the "induced crime wave" produced by Satanic cults across the nation under Aquino's guidance was the last on her list: "Satanic cults, UFO cults, directed energy technologies, neurocybernetics / psychotechnologies, biotechnologies / experimental drugs, multinational government contractors and subsidiaries, investment portfolios and other financial inducements, imported foreign national scientists, a controlled and compliant media, decentralized U.S. government control, an induced crime wave . . ."†

Aquino was involved in the Zodiac killer serial murders in the late 1960s and early 1970s that occurred in the vicinity of the Presidio. Then in 1986 he was involved in the Presidio day care sex scandal in which sixty children were raped, sodomized, and traumatized. Children identified

*SAPs require signing a seventy-year nondisclosure agreement covering all things related to highly classified "national technical means" programs like UFOs, satellites, spy gear, antigravity exotic propulsion systems, etc.

†Correspondence dated January 8, 1995, from Julianne McKinney, director of the Electronic Surveillance Project, Association of National Security Alumni, to Steven Metz, PhD, and LTC James Kievit, cc: Col. John W. Mountcastle, director, Strategic Studies Institute, U.S. Army War College.[4]

"Mikey" as one of the men who had abused them, but the army protected Aquino by alleging the media were simply in a "satanic panic." The crime scene building was demolished, including the tunnels in which the children had been raped, and the army built a brand-new $2.3 million new day care center in its place. After twenty-two years as the army's resident Satanist specialist, Aquino was quietly discharged, having finally exposed too much about the new Phoenix military.*

In the spring of 1998, the U.S. War College's *Parameters* quarterly published "The Mind Has No Firewall."[5] Up until then, little had been allowed to get out to the public concerning such questionable tactics as "soft kill," "slow kill," "silent kill," and other so-called nonlethals,† as well as "conflict short of war" and behavior modification as legitimate tools of modern warfare. Thomas stressed that "to make RMA in conflict short of war would require fundamental changes in the United States—an ethical and political revolution may be necessary to make a military revolution."[6]

Michael Aquino wasn't the military's first Satanic favorite son. Marvel Whiteside "Jack" Parsons (1914–1952) was a self-taught genius jet propulsion rocket engineer from Pasadena who was involved in the founding of the Aerojet Rocketdyne Corporation and the Jet Propulsion Laboratory, which were funded by the Guggenheim family and administered by the California Institute of Technology (Caltech). British MI6 agent and Satanist Aleister Crowley chose Jack to head up the Pasadena chapter of the occult Ordo Templis Orientis. A crater on the dark side of the Moon has been named for Parsons, but otherwise he has been wiped from space program history.‡

---

*Cheri Seymour's book *The Last Circle: Danny Casolaro's Investigation into the Octopus and the PROMIS Software Scandal* (Trine Day, 2010) digs into the military's role in sustaining generational Satanic cults, due in part to Michael Aquino convincing military brass that they had lost the Vietnam War "not because we were outfought, but because we were outPSYOPed. PSYOPs—MindWar—had to be strengthened."

†Infrasound/VLF, neural inhibitors, hallucinogens, calmatives, neuroblockers, stun guns, pulsed high-power microwaves, nonpenetrating projectiles, laser rifles, flash-bang grenades, etc., are among the "nonlethals." In 1994, the Department of Defense and the Department of Justice formalized their cooperation in using such nonlethals.

‡For the astonishing story of the U.S. rocket program, Paperclip Nazi scientists, and Satanism, read John Carter's *Sex and Rockets: The Occult World of Jack Parsons* (Feral House, 1999). Yes, even L. Ron Hubbard of scientology fame was involved in Jack's black magick rituals.

Both Parsons and Aquino claimed to be the Antichrist.

> Both psyops and satanism involve using many levels of deception and coercion to manipulate the perceptions of others for ulterior motives, so these are skills that Aquino is very familiar with and very comfortable in using. He is also an expert in propaganda and skilled in techniques for disseminating misinformation/disinformation, and using misdirection, confusion tactics, isolation techniques, and revisionism. Therefore, he will certainly use these skills to defend himself against the truth of his actions.[7]

Such "skills" have now overtaken the entire nation: Washington, DC, politics, law enforcement, corporate culture, UN "peacekeepers," child "protection" agencies, and Hollywood. The "nonlethal" targeting of citizens in their homes and workplaces by their government is now considered legitimate. Aleister Crowley's admonition to "do what thou wilt shall be the whole of the law" has replaced "Do unto others as you would have them do unto you." Now it's *if* you can get away with it, go for it.

Movie and TV programming favor Satanism and pre-Christian belief systems that practice pedophilia, mind control, and blood rituals—from "Do what thou wilt" high rollers and church leaders to petty criminals and corporate flunkies, all are required to attend a blood ritual now and then in order to be kept traumatized and in line. Nietzsche's concept of the human will to power is the Satanist defense against weakening brotherly love. As for death and answering for one's deeds on Earth, those espousing Satanism are ever seeking earthly immortality, whether through adrenochrome from tortured children's blood or the brain-computer interface of Transhumanism.

Post 9/11, the Phoenix approach has justified all kinds of covert atrocities. Domestic NORTHCOM is the military component of the Department of Homeland Security, the post-9/11 spawn of Harvard's JFK School of Government, the Department of Defense, and the Department of Justice. Hitler's Germany had its *Geheime Staatspolezei* (secret state police, i.e., Gestapo); the U.S. has Homeland Security. The Phoenix philosophy of social control, along with the "public-private partnership" philosophy so in vogue now, has permeated American universities,[8] and I don't mean ongoing military recruitment on campuses. There is also the value of high-tech university labs. For example, the first contract ($42.5 million) between the U.S. Marine Corps and the Institute for Non-Lethal Defense Technologies under Penn State's Applied Research Laboratories, one of the U.S. Navy's top civilian research facilities, was to study the Schedule II drug fentanyl, a synthetic sedative eighty times stronger than morphine.[9]

But hey, it's the great game! Technology, finance, manufacturing, and military all in bed with intelligence and private security corporations, the door between government and transnational corporations spinning like a Russian roulette gun, "trusted partners" now scratching each other's back this week and stabbing each other in the back the next week. "Conflict of interest," like "monopoly," is passé, old school, having died along with honor, character, virtue, and being true to one's word. Arsenals of lies, state-sponsored terrorism, and remote electromagnetic torture ("nonlethals") are now big business.

Internet and social media corporations are in tight with the NSA's Special Source Operations and the FBI. Corporate executives and tech employees working abroad double as intelligence agents ("committing officers") with the guarantee of immunity for civil actions of data theft and "wetworks." People make the assumption that this is just how people are, but it's not. It's about mindsets allowed to run rogue for the sake of full-spectrum dominance, regardless of the cost.

MILITARY INFORMATION SUPPORT AND OPERATIONS (MISO) AND SMITH-MUNDT MODERNIZATION ACT OF 2012

Since 1961, the National Defense Authorization Act (NDAA) has enacted a series of federal laws that fund the Department of Defense. The most well known of these include allowing the government to wiretap American citizens without obtaining a warrant and imprisoning citizens indefinitely without charge in the name of "national security." Its other mission is to produce and disseminate propaganda to the world via the corporate media.

> The CIA, through Operation Mockingbird, started recruiting mainstream journalists and media outlets as far back as the 1960s in order to covertly influence the American public by disguising propaganda as a purposeful spin on news. The CIA even worked with top journalism

> schools to change their curricula in order to produce a new generation of journalists who would better suit the U.S. government's interests. The CIA objective of manipulating mainstream media and influencing American public opinion through propaganda came directly from the father of public relations Edward Bernays who put it succinctly: "The conscious and intelligent manipulation of the organized habits and opinions of the masses is an important element in democratic society. Those who manipulate this unseen mechanism of society constitute an invisible government which is the true ruling power of our country."[10]

In 2010, the Psychological Operations, or PSYOPS, of the U.S. Army Special Operations Command was rebranded as Military Information Support and Operations (MISO).

> MISO, previously known as Psychological Operations (PSYOP), uses themes and messages to reach target audiences in order to influence their emotions, motives, reasoning, and ultimately the behavior of foreign [and domestic] governments, organizations, groups, and individuals . . . If a MISO operator does not understand the depth of an individual's condition, concerns, fears, ambitions, and vulnerabilities, then MISO will not be effective.[11]

To protect "a key friendly center of gravity—to wit, the U.S. national will"[12]—the Pentagon normalized institutional lying by legalizing "the use of psychological operations through propaganda on U.S. civilian populations." This happened with the 2013 omnibus National Defense Authorization Act,[13] which revised the 1948 Smith-Mundt Act that banned domestic propaganda and psyops for sixty-four years.* President Obama lifted the ban in 2013 with the all-new Smith-Mundt Modernization Act of 2012. MISO and the Joint Force Military Information Operations (IO) have now officially railroaded American journalism with "the integrated employment of electronic warfare (EW), computer network operations (CNO), psychological operations (PSYOP), military deception (MILDEC)

*The NDAA is an annual Trojan horse of trouble. For propaganda and "narrative-building," the Global Engagement Center was established under Obama's EO13721 (April 2016), then made into law and buried in the 2017 NDAA.

and operations security (OPSEC), in concert with specified supporting and related capabilities to influence, disrupt, corrupt or usurp adversarial human and automated decision-making while protecting our own."[14]

The Broadcasting Board of Governors, an extension of the U.S. State Department, which itself appears to be an extension of the CIA, approves government narratives for Mockingbird mainstream media dedicated to silencing dissenting perspectives. Spin corporations like the Strategic Communication Laboratories (SCL)* are proud of their "influence operations" and "public diplomacy," and run simulations "from natural disasters to political coups" at exhibits like Defense Systems and Equipment International, the UK's largest showcase for military tech. For a price, SCL will run false flag events, help overthrow democratically elected governments in developing countries, and override national radio and TV broadcasts.[15]

The advent of the Joint Force Military Information Operations (IO) in public-private partnerships with shadow corporations like SCL announces loud and clear that human minds and brains are now the battlespace. No firewall means the mind is easy prey for IO psychotronic weapons, and we would be naive at this point to assume that "adversarial" refers only to foreign enemies, or that "protecting our own" refers to protecting the American people; for the post-RMA Joint Force military, the "U.S. national will" is the military will, not the people's.

SOFTWARE, HARDWARE, FIRMWARE, AND WETWARE PATENT LANGUAGE

The 2005 U.S. Patent 6965816-B2 mentioned in chapter 10, "PFN/TRAC system FAA upgrades for accountable remote and robotics control to stop the unauthorized use of aircraft and to improve equipment management and public safety in transportation," was filed shortly after 9/11. At first glance it seems to be about remote control of aircraft. But once the patent language is penetrated, it becomes evident that the PFN/TRAC (Point Focal Node/Trusted Remote Access Control) system is about *control over all moving things*, not just airplanes, by means of a large

*Strategic Communication Laboratories, founded in 1993, is "the first private-sector provider of psychological operations."

"machine-messaging matrix," either by sharing or replacing local and standard human-machine interfacing with remote control robotics. For this reason, this patent is regarded as the Internet of Things (IoT) patent, the key being in the language: "robust and accountable remote control for personal applications, stationary equipment and stand-alone functions, and coordinates them and interfaces them within the communication matrix." The same holds for the patent's reference to the "wetware"* status of human beings: "The technology incorporates existing technology as it exists in a present distributed architecture and coordinates and manages the essential function to stop and control an unwanted event and improve public safety. This requires hardware, software and wetware (people/the procedures and protocols)."

A month and a half before 9/11, another patent, Canadian Patent #2,397,911, "Protected accountable primary focal node interface" (July 26, 2001) references the term "firmware (software-embedded hardware)." Again, once the language is penetrated, the patent is actually referencing Human 2.0 cyborgs† with artificial organs, limbs, pacemakers, implants, and so forth—in other words, a human materials industry that includes nanobots (tiny machines) in the bloodstream and "wetware" brains that the Internet of Things/Internet of Nano Things/Internet of Medical Things tunes in to, thanks to three decades of GMO foods, chemical aerosols, and aggressive vaccination programs. With these clues we can now translate the following patent description in the true *synbio* "firmware" context that the patent addresses, while hiding the technology in plain sight.

The invention was always designed to remotely control machines, equipment, and vehicles through various levels of monitoring and remote control systems and networks. Much of the technology has been designed to marry preexisting devices and systems to, wherever possible, develop cost-effective enhancements to legacy systems and equipment. The PFN/TRAC system is further developed to integrate and consolidate components and functions through more efficient universal configurations

---

**Wetwork* is the CIA term for assassination.

†The term *cyborg* (cybernetic organism) was introduced in 1960 at the Psychophysiological Aspects of Space Flight Symposium in the context of preparing human beings for survival in "extraterrestrial environments," calling up a reference to the secret space program and even the Paperclip Nazi aviation medical doctor Hubertus Strughold (1898–1986), the so-called Father of Space Medicine.

of hardware, software, and firmware (software-embedded hardware, i.e., transhumans) to provide integrated accountable remote control and management for man and machine interfacing and to include full robotics by employing the latest developments, such as Systems on a Chip (SOC) technology. The systems and modalities of hardware, software, and firmware detailed in this application and related applications, including traceable communications and commands with individual and machine identity and integrity checks, are all part of PFN/TRAC system technology. As the patent language states, "The PFN/TRAC system is made up of individual nodes or units that communicate as part of an accountable machine-messaging network employing and managing various forms of communication computers and machine controls to aid humanity in the safe use of equipment while protecting the environmental and the Earth's resources." This describes the Internet of Things (IoT) and all its progeny to a *T.*

COMPUTERIZED PERSONALITY SIMULATION

The truth is that supercomputers and quantum computers cannot think; only a human with a mind—much more than a physical brain—can really think. But what AI can do is mind control. Would you recognize it if it were happening to your mind?

During the MK-Ultra years, Freudian psychiatrist Kenneth Colby (1920–2001) worked in Stanford Research Institute's computer science department on artificial intelligence vis-à-vis the human mind. He was deeply involved in developing chatterbots like ELISA, described by Robert Duncan, author of *The Matrix Deciphered,* this way: "ELISA was an artificial intelligence (AI) computer program demonstrated by the Stanford Research Institute back in the 70's. It acted like a psychologist and asked stupid questions and responded to answers, parsing and understanding sentences. This field is called *natural language processing (NLP)*—not to be confused with neurolinguistic programming."[16] ELISA fooled about half the people into believing it was a real human being on the other end.* Now, we are confronted with OpenAI's ChatGPT, a chatbot and virtual

*The Turing Test (telling the difference between a human being and an AI) is pretty much obsolete now.[17]

assistant based on large language models (LLMs), which means large amounts of data.

Of course, chatbots are dual use like everything else electromagnetic, which means they're not always your best friend. Fast-forward thirty-five years and instead of parsing natural language phrases from keystrokes, ELISA recognizes phonetic brainwaves and parses them into words and sentences for artificial intelligence and natural language processing for tortures and menticides scaling to entire populations: "The brainwave cognitive ELISA adds realism by inducing an empathetic emotion with the words the target feels, thus adding a new dimension of convincing the target that the synthetic mind virus is a real person."[18]

Colby came up with the following nuances of human thinking that chatterbots would have to imbibe:

- The credibility of a belief is based on the credibility of its source.
- Human personalities are based on belief systems concerning significant persons, including the self.
- Every psychological concept has specific significance (e.g., father, love, etc.)
- Input from others is evaluated and "colored" by mental [and emotional] patterns.
- A human mind changes with inner conflict, transforming beliefs to fit into an overall pattern.

From his observations, Colby deduced that:

- Capture a person's belief structures and they can be controlled.
- Unenlightened human minds are combinations of infantile beliefs and emotional patterns.
- Unenlightened human minds can be simulated by a computerized system.
- Through such systems, unenlightened people can be programmed and controlled.

Components of such "unenlightened human minds" can be captured and developed into accurate simulations that artificial intelligence technicians and political operatives can then use to manipulate us.

The Sentient World Simulation (SWS),* "a strategic geopolitical perspective that captures the interplay between military operations and the social, political, and economic landscapes,"[19] was presented as a military concept paper in 2006. It basically advances the idea that every person on the planet could be digitally represented as a software node and given a digital avatar, and the entire infrastructure erected as a blockchain, each person's consciousness cloned onto the avatar inside the Cloud/supercomputer/quantum computer. Direct links would exist between individuals and their avatars via brain-computer interface so that *everything done in the real world occurs in the computer simulation.* Even Earth herself would have a digital twin.

One read-through of this concept paper will show how the language has been contrived to be obscure to the uninitiated while being exact to the initiated—much like patent language. For example, terms like *granularities of access*, *coarse-grained* and *fine-grained*, *temporal and spatial granularities*, and *Joint Semi-Automated Forces (JSAF)* seem to point to the sensor nanotechnology essential to building virtual simulacra of the natural world and its residents. Read between the lines and "unbiased to specific outcomes" seems to refer to the amoral approach touted as "objective" in "agent-based" SEAS (Synthetic Environment for Analysis and Simulation) environments. All this ends up as the torture of millions of targeted individuals for the purpose of collecting endless biometric and behavioral data for "specific outcomes."

The targeting industry (defense contractors, private security corporations, etc.) provides guinea pigs for ongoing chatterbot and Sentient World Simulation experimentation. People's real lives are being exploited so as to:

> build a synthetic mirror of the real world with automated continuous calibration with respect to current real-world information . . . The ability of a synthetic model of the real world to sense, adapt, and react to real events distinguishes SWS from the traditional approach of constructing a simulation to illustrate phenomena . . . Basing the synthetic world in theory in a manner that is *unbiased to specific outcomes*

*SWS is definitely connected to metaverse twinning maneuvers as well as Rudolf Steiner's Eighth Sphere.

> [emphasis added] offers a unique environment in which to develop, test, and prove new perspectives.[20]

The end goal is to reverse our relationship with our avatar in the Sentient World Simulation: to make the feed now going from nanosensors and nanobots in our biological brains to AI collection points like fusion centers reverse so that a biological human being is being controlled by the virtual avatar inside the simulation. HAARP and Space Fence carrier waves synchronize these AI communications with SIM real world simulation.*

The remedy for this dystopian nightmare begins with turning off the TV and iPhone, reading and practicing consciousness soul thinking.

INORGANIC LIFE

With the latest psyop assault on planetary life beginning in 2020, we have begun to see in real time the dissolution of the big black line separating machines and everything biological—human beings, the rest of Nature, and planet Earth herself. Language as a reflection of how we perceive reality is changing, as well, with terms like *inorganic life* and *living machines* being added since the release of nanotechnology and gain-of-function altered biology. We need to understand how *synbio* has infiltrated life so that we can understand what is happening around and inside our bodies and brains. As a 2009 article published in *Systems and Synthetic Biology* points out,

> The difference between a non-living machine such as a vacuum cleaner and a living organism [such] as a lion seems to be obvious. The two types of entities differ in their material consistence, their origin, their development and their purpose. This apparently clear-cut borderline has previously been challenged by fictitious ideas of "artificial organism" and "living machines" as well as by progress in technology and breeding. The emergence of novel technologies such as artificial life, nanobiotechnology and synthetic biology are definitely blurring

*Initially, "SIM" referred to a video game that simulates an activity such as flying an aircraft or playing a sport, as in "a flight sim." Now, it refers to anything simulated.

> the boundary between our understanding of living and non-living matter.[21]

Lee Cronin of the University of Glasgow specializes in "digital chemistry." In 2011, he announced the first steps toward "creating 'life' from inorganic chemicals":

> All life on earth is based on organic biology (i.e., carbon in the form of amino acids, nucleotides, and sugars, etc.), but the inorganic world is considered to be inanimate. What we are trying to do is create self-replicating, evolving inorganic cells that would essentially be alive. You could call it inorganic biology . . . If successful, this would give us some incredible insights into evolution and show that it's not just a biological process. It would also mean that we would have proven that non-carbon-based life could exist and totally redefine our ideas of design.[22]

A decade later, here we are with synthetic mRNA "vaccines" delivering nanotech to our cells. At DARPA, the Engineered Living Materials program seeks to apply *synbio* not only to military logistics and warfare, but also to construction:

> In 2006, the Defense Advanced Research Projects Agency (DARPA) defined programmable matter as "an intelligent, or programmable, material that contains the actuation and sensing mechanisms to 'morph' into desirable/useful shapes under software control, or in reaction to external stimuli." DARPA was speculating about "InfoChemistry" [digital chemistry] or building information directly into materials, but now the concept has advanced from merely creating *shapes* to creating *effects*. DARPA's Engineered Living Materials (ELM) program, for example, seeks to create living materials that will construct and maintain buildings.[23]

But not to worry, we are constantly told. The advent of *synbio* may evoke images of science fiction cyborgs such as those populating *The Terminator* or *Blade Runner*, but at least for now "most inventions within the next decade will not resemble these humanoid cyborgs. Programmable living matter (PLM) instead resembles small- (or even micro- or nano-) scale components of the larger entities."[24]

SCIENTISM, NOT SCIENCE

In an undergrad biology course I took decades ago, I had to remind my professor that Darwinian evolution was a theory and not a law. Then in 1999, symbiogeneticist Lynn Margulis* was awarded the U.S. National Medal of Science in part because she pronounced neo-Darwinism and its theory of "survival of the fittest" dead. Margulis spoke out about how capitalists had exploited Charles Darwin's idea of "descent with modification" in order to justify their political and social policies of domination. Darwinism dispensed with the idea of God, and his theory of evolution was co-opted as a political-ideological expedient, a spike driven into the resistant heart of orthodox religion. Karl Marx viewed Darwin's evolutionary theory as the basis of natural history, and Hitler perceived that the theory of survival of the fittest supported the elimination of the weak. A higher race must prevail, it was thought: Nature decreed it.

A Skull and Bones biology class might have a dozen hand-picked students, with each weekly seminar taught by a JASON† scientist recruited from top programs and labs, all approved by the British Royal Society. None of the assigned readings would be mainstream, following John Comenius's 1668 manifesto, *The Way of Light*, which says that for science to "secure the empire of human mind over matter," a clearinghouse is needed to accredit and disseminate only the knowledge the public should know and prevent the dissemination of anything the public should not know. Thus was born the British Royal Society in the late seventeenth century, composed entirely of Freemasons and elitists whose duty was to set the mold for future regulators of knowledge. *Scientia est potentia*, "Knowledge is power," was their manifesto. In dictating to what "scientific" would and

*New organelles, bodies, organs, and species arise from symbiogenesis. Margulis wrote *Acquiring Genomes: A Theory of the Origin of Species* (Basic Books, 2003).

†JASON is an independent scientific advisory group under contract with the Department of Defense to provide consulting services on matters of defense science and technology. Similar to the story that HAARP has shut down, the story is that JASON has been disbanded after fifty-nine years of service. This is most certainly not true, though in "going black," it has no doubt had a "face change."[25]

wouldn't mean, God—the incorporeal quality of being behind magnitude, figure, motion, cohesion, and firmness—was the first to go. Next would be æther. Meanwhile, the physical laws of *Mater Natura*, Mother Nature, would suffice for the masses, at least until matter itself—namely, the study of atoms—would be revealed to be the supreme mystery.

Our Skull and Bones class might begin with a quick overview:

"In the beginning was lifeless matter, followed by spontaneous generation. *Natura*, being a sentient sovereign being—self-creating, self-sustaining, self-regenerating—created itself, going from matter to life, the first tenet of anthropomorphic mysticism. Thus, the living and nonliving are inseparable. Spontaneous generation and the driving doctrine of natural selection set Western science free from the image of God thanks to the ancient mystery religion of Freemasonry. For centuries, *Natura* was the guide, the force responsible for our thoughts and actions. Religion is no longer needed. Now the superman is within our reach thanks to the royal art of science and technology."

The deeper into the physical one probes, the closer spirit is. The deeper one goes into quantum physics, the closer the alchemical point of view is to those of the adepts of old who understood how to converge the occult (esotericism) and science, the Mysteries and matter. From the Hermetica discourses:

> Willing then that humankind should be at once creatures of nature and capable of immortality, God compacted the human of these two elements, the one divine, the other mortal; and being as it is thus compacted, it accords with God's purpose that the human is not only better than all mortal beings, but also better than the gods, who are made wholly of immortal elements.[26]

The Royal Society began in the eighteenth century as the Lunar Society, which was founded by Charles Darwin's grandfather Erasmus Darwin. Members were scientists of renown at the time, "merchants of light" (from Francis Bacon's *New Atlantis*)—Wilkinson, Watts, Boulton, Priestly, Wedgewood, Benjamin Franklin—and met only on the full moon. All were Freemasons, as was the philosopher John Locke (1632–1704) Freemason and Fellow of the Royal Society T. H. Huxley, grandfather of Aldous Huxley, spoke for the reclusive Charles Darwin when he insisted that the occult evolutionary idea of *becoming* was the new science.* The Lunar Society was intimately bound up with the machinations behind the French Revolution.† Freemason Benjamin Franklin was a member of both the Nine Muses (French) and Lunar Society (English), playing the part of international envoy between French and English Freemasons.

Nazism—National Socialism—was based on Marx: subordinate the individual to the collective, which itself is subordinate to a small, elite central control. Then consolidate all capital in a monopoly, either government (communism) or corporate (fascism), communism and fascism being two sides of the same coin, two oligarchies founded on the same Darwinian worldview. Unfortunately, none of this is ancient history: it's all still going on in secret societies at the pinnacle of power in society.

Should the Hippocratic Oath of "First, do no harm" be added to PhD science requirements?

OCCULT MEANS HIDDEN

When you flip a switch on the wall, somewhere inside the wall, metal components make energy move and light appears. For the most part, people simply accept that light appears and appliances work, after which they think no more about the effects of electricity.

The presence of invisible, occult (hidden, concealed, covered-over, secret) realities and events made to happen for political, societal, or personal agendas is a constant in modern times, but is disguised, lied about, and denied for equally obscure reasons (at least to the majority), and therefore goes unnoticed. Because the Earth forces that science seeks to

---

*G. A. Bondarev, "in basic outline, the system, the organism of cosmic life extends from the Life-Spirit of Christ to the elementary beings of the solid mineral realm of the Earth and on to the organic life of the human being. He is woven entirely of beings and their interactions, for life cannot be a finished fact that 'has become.' When it becomes a fact, it dies. It is wholly process, an ongoing *realization* of itself. The real human being is therefore only possible in *becoming*."[27]

†Lord Acton (1834–1902), in *Essays on the French Revolution*: "The appalling thing in the French Revolution is not the tumult but the design. Through all the fire and smoke, we perceive the evidence of calculating organization. The managers remain studiously concealed and masked, but there is no doubt about their presence from the first."[28]

manipulate are all but invisible (electricity, gravity, magnetism, etc.), it should be no surprise to learn that the scientists who have occult knowledge of scientific esoterica have an advantage over conventional scientists who are not "insiders" and therefore know nothing about such esoterica, and even think it is complete nonsense. While I could write a whole book on this slippery topic, a few examples of the mindset of movers and shakers versed in the occult will have to suffice.

In chapter 6, "The Weaponization of Magnetism," I reference military occultists being aware of how the fact that the geometry of hexagons can "keep an electric charge moving in an interrupted circuit" has to do with "doors of perception* between dimensions." Then in chapter 10, "5G/6G, the Internet of Things, and the Wireless Body Area Network (WBAN)," I discuss "Europe's Hexa-X, the 'flagship for 6G vision and intelligent fabric of technology enablers connecting human, physical, and digital worlds'" that alludes to the metaverse and Rudolf Steiner's Eighth Sphere, *and* the fact that the double-meaning language describing Hexa-X seems occult, as well.

In 1865, German chemist Friedrich August Kekulé dreamed of six dancing atoms forming a snake eating its own tail (the ouroboros). It was the hexagonal ring structure of benzene hand-delivered to Kekulé from another dimension. The benzene ring in close proximity to a nitrogen molecule with two or three carbon molecules† between produces potent hallucinogenic substances, as if the hexagon were a "door of perception" to other dimensions. While we are conditioned to think of psychedelic substances as inducing hallucinations in the physical brain, the mind is not physical and is likely what accesses molecular geometry, straddling dimensions as Kekulé did.[29] Not every mind experience is a hallucination.

The geometrical hexagon shape, whether microcosmic (molecular structure) or macrocosmic (insect structures, stone formations, telescope mirrors,‡

*An allusion to "insider" Aldous Huxley's autobiographical book about his May 1953 mescaline experience, *The Doors of Perception and Heaven and Hell* (Harper and Brothers, 1954).

†Carbon has six electrons, six protons, and six neutrons, thus, the maligned number 666.

‡The optical Hobby-Eberly Telescope in the Davis Mountains of Texas has a honeycomb of ninety-one hexagonal mirrors in a fixed position for its studies of spectroscopy, the interaction between matter and electromagnetic radiation.

Saturn's north pole), is some sort of interlocking agent. What is geometry, anyway? What kind of power does it hold? Why did ancient Greeks like Plato and Pythagoras study geometric forms so carefully?

In 1995, a CRAY C90 supercomputer discovered a 1,500-mile diameter hexagonal close-packed crystalline structure aligned with Earth's spin axis that is growing in the temperature-pressure extremes at the core of Earth.[30] For techno-magicians, the Earth is a crystalline electrical Being with a North Pole (anode or positive charge) and a South Pole (cathode or negative charge), enveloped in a giant protective ether cocoon bathed in transmissions by seven planetary organs—the Sun, Moon, Mercury, Mars, Venus, Jupiter, and Saturn—each infusing the electromagnetic chakra system of Earth that runs from pole to pole along the north-south "spine" of mountains throughout the Western Hemisphere as rich in minerals and energy as the human spine, the extraordinary transceiver antenna that receives, transmits, and resonates to frequencies and pulses, the crown center serving as a relay station for individual consciousness as well as the impersonal will-to-power via the geomagnetic grid.*

The medical definition of *occult*—"not accompanied by readily discernible signs or symptoms"—certainly applies to Covid-19 and makes me ask if the entire global psyop was a forced three-year initiation rite for humanity, given the endless repetition of certain memes via mainstream media. Makia Freeman writes about this extraordinary three-year phenomenon as an "occult corona-initiation ritual" forced by global elites and what Rudolf Steiner calls "shadow Brotherhoods" whose families ("bloodlines") have for thousands of years practiced such rites as drugs, ritualized quarantine, lockdown, social distancing (isolation), hand-washing, mask-wearing, and terror to separate people from their normal lives so as to break them down, engender submission, reprogram them before returning them to a "new normal."[32] In anthropology, *liminality* (L., *limen*, threshold) is the quality of ambiguity or disorientation that occurs in the middle stage of a rite of passage, when participants no longer hold their pre-ritual status but have not yet begun the transition to the status they will hold when the rite is complete.

At the core of the science that studies the essence of the ground of

---

*The peak voltage of the spinal column is 100 MHz (the FM radio band is 87.5–108 MHz) impacting the central nervous system and weakening the blood brain barrier.[31]

being (such as the nature of matter) are Mysteries* that blur the distinction between physics and metaphysics. Nanoparticles and the instruments that can perceive and measure them are now leading the pack in the Mystery of matter. The greatest fallacy perpetrated by Western secret societies like the Freemasons against the profane (*pro-fane*, meaning "outside the Temple," i.e., uninitiated) has been the hard division between the physical and the spiritual. Those who have dis-spelled (and it is a spell) this false division are primarily the truly science-minded rather than the scientism elite.

THE EGREGORES OF CORPORATIONS, FOUNDATIONS, AND BROTHERHOODS

While in London two decades ago, I encountered an Italian who had worked at the Vatican in a high secular position during the reign of Pope John Paul II (1978–2005). He shared with me a term used broadly in esoteric circles: *egregore*. Let's take a look at what the Freemasons, very well-acquainted with egregores, have to say:

> An egregore is a kind of group mind which is created when people consciously come together for a common purpose. Whenever people gather together to do something an egregore is formed, but will dissipate rather quickly unless maintained deliberately by those who know the techniques of how to do so, after which the egregore will continue to grow in strength for centuries, its effectiveness greater than the mere sum of its individual members, given that it continuously interacts with its members, influencing them and being influenced by them. The interaction works positively by stimulating and assisting its members, but only as long as they behave and act in line with its original aim. It will stimulate both individually and collectively all those faculties in the group which will permit the realization of the objectives of its original program. If this process is continued for a long time, the egregore

*"From the Greek verb *muien*, 'to speak through pressed lips, to communicate in secret.' Extensive teachings about the creation of the world and the interaction between the Gods and humanity, preserved and disseminated through cult-centers of spiritual training associated with specific planets and celestial beings, and differing in character according to the race and geographic location of the people they served. . . . The Mysteries were the universities of spiritual training of former times."[33]

> will take on a kind of life of its own and can become so strong that even if all its members die, it will continue to exist on inner dimensions where it can be contacted even centuries later by a group of people prepared to live the lives of the original founders, particularly if they are willing to provide the initial input of energy to get it going again.
>
> If the egregore is concerned with spiritual or esoteric activities, its influence will be even greater. People who discover the keys can tap into a powerful egregore representing, for example, a spiritual or esoteric tradition, [and] will, if they follow the line described above by activating and maintaining such an egregore, obtain access to the abilities, knowledge, and drive of all that has been accumulated in that egregore since its beginnings. A group or order which manages to do this can, with a clear conscience, claim to be an authentic order of the tradition represented by that egregore.[34]

For those conditioned to believe that thoughts and feelings are ephemeral and fleeting, it may be difficult to grasp how such a psychological and mental *entity* could be formed by "mere" individuals' thoughts and feelings and impulses—and yet what have we recently witnessed? The creation of a worldwide egregor based on fear of a virus that doesn't even exist, all due to the World Economic Forum, controlled corporate media, and 5G/6G systems! Entities like Google, Facebook, the Centers for Disease Control, and the World Health Organization march to the rhythm of the egregore; the same goes for those who belong to useful secret societies, foundations, and think tanks. And thousands are still lining up for shots, despite all the damning evidence about what is really going on.

We are surrounded by egregores devouring whomever they can, people who could have become individuals of honor, virtue, honesty, courage, and free will, but instead have chosen to feed entities that can't be seen. *Our challenge is that our minds exist not just here on Earth but in spiritual dimensions.* Egregors are not physical, and yet the thoughtforms they produce have power, whether they be good and constructive or evil and destructive. This is what is meant in the ancient spiritual books that say human beings were created in the image of God: we are creators not just with our hands and industry, but with our minds.

Think of demons as negative thoughtforms with the power to destroy. As the apostle Paul puts it in the New Testament, *For we wrestle not against flesh and blood, but against principalities, against powers, against the rulers of the darkness of this world, against spiritual wickedness in high places* (KJV, Ephesians 6:12). Our struggle as human beings is with powers and principalities, including egregores whose central nervous systems are fed by information-carrying electrical impulses transmitted by carefully monitored electromagnetic fields; whose sensory organs are microwave scanners, satellites, microphones, cameras, and synthetic telepathy; whose nerves are radio waves from ELF to superhigh microwaves; whose digestive system devours real lives in both the psyche-soul sense and the orgone/etheric bodily sense.

APPENDIX 2

Quantum, Scalar, and Hyperspace

The Divine is not only Transcendent but also Cosmic and finally individual. Its power and force must be brought down into the physical world in order to effect any real transformation of the earth-consciousness.

Sri Aurobindo (1872–1950)

Unlike the nineteenth century, the twentieth century drew a big black line between occult and science, magic and technology, alchemist and scientist. Physicists under the spell of materialism prefer to think of forces like gravity and magnetism as lifeless mechanical forces set in motion by a Big Bang universe or absent Clockmaker God, whereas occultists steeped in ancient traditions know better.

Plasma, electrical vapor, luminous magnetism and luminous æther, the all-pervading fifth element, have fascinated scientists for millennia. The Nazis called ether *vril,* a term borrowed from *The Coming Race* by Edward Bulwer-Lytton (1803–73), Grand Patron of the Societas Rosicruciana in Anglia, mother lodge of Philadelphia. *Vril,* the free force filling all of space.* Wilhelm Reich, MD (1897–1957), renamed *vril, orgone.* Rockets like the old V-2, *lift* weight, but etheric force (vril, orgone, *ki,* etc.) *negates* weight, the Nazi objective having

*Science fiction was the favored preprogramming vehicle in the nineteenth century. Now, box office hits like the 2002 film *The Minority Report* from a Philip K. Dick short story provide the public's preprogramming while the "pre-crime" tool, the Future Attribute Screening Technology (FAST), tests for who has "hostile intent."[1]

been to harness ether, then command the Earth's electromagnetic grid.

Imagine the dark liquid space beyond our atmosphere as a labyrinth of holes and tunnels, bubbles and webs of energy rising and falling, winding and unwinding in ceaseless activity, a breathing sponge of ether—not the old uniform, contiguous model of planets and stars hanging suspended in a vacuum of emptiness. A living, conscious universe! Space is not empty,* and our world is a 3D slice out of an infinite dimensional superworld, dynamic and elastic, bending and stretching. An infinity of spacetimes with different topologies and geographies interconnecting as waves, as exciting as it is frightening, including the living world of plasma embodiments visible to Human 1.0 in the infrared.

Heisenberg's Uncertainty Principle pointed the way as early as 1927. Particles appeared where they had no right to be, measurements disturbed the system and left only probabilities, uncertainty in momentum p times uncertainty in positions larger than Planck's constant. The universe that quantum mechanics wants to measure won't stay still because it is in motion, ever transforming. By observing the process, observation being in motion, we change its terms. So when quantum theory says that an observed world behaves as particles and an unobserved world behaves as waves, *observation* becomes the pivotal question. *(The sound of one hand clapping? The silent tree falling in the middle of the forest?)*

Spacetime moves about and changes due to the presence of matter and energy, a truly quantum landscape of infinite spacetimes with different shapes and topologies fitting together not as things in themselves or particles or even linear events, but as waves of intersecting themes interfering with each other, synchronicity seemingly being a wave strength of different strands of spacetime now and then *communicating* with our perceptions.

Appearance, Reality, and Theory. ART, Immanuel Kant's (1724–1804) paradigm. Appearances as perception, both inner and outer; Reality lying behind Appearance as cause; Theory as the stories we tell about the other two.

Thousands of years ago, the wise Upanishads said Time is the mysterious condition that orders Space, Saturn-Chronos being the portal by which Being enters Space in order to know itself—first contracting, then densifying to become audible and at last visible. From unity into form

*A vacuum is a plenum—pure potential—not an emptiness or nothing.

and multiplicity, like Chronos dismembering his Father or physicist David Bohm's Implicate order becoming Explicate order. Bohm's Zero Point is Brahmin, fusion of Time and Consciousness into Timelessness, when Time-energy vibrates so intensely that Time seems to stand still and consciousness experiences transcendence or *immanence.* Graph this Moment in Time and the amplitude and vertical distance between wave crests becomes narrower and narrower, until the points of rest overlap and the oscillating wave becomes a straight line. This is the Zero Point of stillness, arrival, *satori, nirvana*, the born-again second birth of the conscious Self.

Bohm discovered wavelengths of 10^{-34}. One cubic centimeter of Zero Point energy contains more energy than all the matter in the known universe. In other words, Emptiness or Void does *not* represent the highest spiritual realization. Fullness to overflowing, yes! Emptiness, no. As Sri Aurobindo said, nothing can arise only from nothing, just as something can arise only from something. Vibrating the primordial *Om* (the Schumann resonance) once opened Ali Baba's door, but entry to the Luminiferous Ether's Zero Point is no longer entirely mechanical.

Enter scalar physics.

After the elimination of the concept (and therefore reality) of the ether and so many other scientific truths having to do with our planet Earth's relationship with *etheric hyperspace*—gravitational propulsion, psychic energy, thought, feelings, neuro-physiological effects, etc.—the extraordinary Tesla weapon system started up in 1908 with Wardenclyffe Tower on Long Island, New York, and the Tunguska, Russia, explosion, then graduated to the microwave Soviet Woodpecker, and finally to the HAARP phased array network that more correctly should be referenced as a *scalar weapon network.*

The lead-up to the removal of ether from Western scientism had directly to do with the same shadow Brotherhood removal of hyperspace references from James Clerk Maxwell's (1831–1879) Quaternion equations.[2] Maxwell had discovered fields and wave forms existing *invisibly* inside the electromagnetic spectrum at a specific number of right-angle *(orthogonal)* rotations separate from normal electromagnetic field components subject to time and space. That he died young is instructive, as is the fact that the ability to manipulate hyperspatial components *not* subject to time and space allows one to gain power over human biology and consciousness.

In 1903, as the concept of the ether was being forced to "go black," British physicist E. T. Whittaker (1873–1956) mathematically proved[3] that the *stress* in spacetime—basically made up of vector waves—could be split ("decomposed") into two scalar waves. Thus, every *transverse wave* (electromagnetic) is actually two coupled scalar waves until separated, at which point they become *longitudinal (scalar) waves.*

From that point on, the scalar foundations of *Nature's* electromagnetics (as opposed to the *subnature* electromagnetics) as well as the complete Unified Field Theory dependent upon Maxwell's equations became inaccessible to scientists consigned to teaching the masses. Truths like the interdependence of electromagnetism and gravitation "went black" while electromagnetics was reduced to four dimensions instead of five by removing gravity. Electromagnetics remained subject to the laws of conservation for the "profane" (scientists who didn't belong to certain secret societies) while JASON-level scientists continued to work with the unadulterated Maxwell equations toward what Oxford scholar Joseph P. Farrell categorizes as the elite "breakaway civilization."[4]

The final nail in Maxwell's Quaternion coffin may have occurred in the 1960s when hertz (Hz) replaced cycles per second (cps) for the entire electromagnetic spectrum, ignoring the fact that only infrared (IR), ultraviolet (UV), X-ray, and gamma are hertzian (vector) waves, while the following frequencies above infrared are scalar, not hertzian, *as are all of the waves in Nature and biosystems*:

Scalar wave (non-hertzian) frequencies

ELF: extremely low frequency
ULF: ultralow frequency
LF: low frequency
MF: middle frequency
HF: high frequency
VHF: very high frequency
UHF: ultrahigh frequency
EHF: extremely high frequency
IR: infrared (hertzian / vector)
UV: ultraviolet (hertzian / vector) . . .

The hidden scalar component may be the core reason for keeping the entire atmosphere ionized and wireless. It certainly bears upon what retired Lt. Col. Thomas E. Bearden, PhD (1930–2022), called:

> . . . the awful secret that Tesla partially discovered by 1900 . . . Thus, by transmitting a scalar standing wave into the earth, he could easily tap the fiery scalar fields produced in the molten core of the planet itself, turning them into ordinary electromagnetic energy. In such a case, a single generator would enable anyone to put up a simple antenna and extract all the free energy desired.[5]

Non-hertzian scalar / Tesla waves, therefore, are longitudinal waves (sometimes called *gravitational* or *standing waves*) in an ether matrix of dipole electrostatic potential, scalar electrostatic potential being "a measure of the intensity of the virtual state flux through a 4-dimensional spacetime point."[6] Though it may initially be difficult to imagine two kinds of electromagnetism—one a hertzian vector wave born of *subnature*, the other non-hertzian and non-vector born of Nature—the fact that electrostatic potential exists as an "internal stress" in all of matter* within "the fabric of spacetime" leads to a somber realization of the weapon potential of natural scalar waves if the technology for such power is built.

Cometh physicist Bernard Eastlund's (1938–2007) patents for the High-frequency Active Auroral Research Project/Program (HAARP).†

*Trapped EM energy = not only mass but gravitational potential.

†Note how Eastlund is discredited on the CIA's Wikipedia website in order to obscure the scalar biowarfare weapon that HAARP is: "Eastlund authored 53 peer reviewed scientific papers and 23 US patents for applications such as well-drilling, sterilization of medical devices, high intensity lighting, and atmospheric plasma heating. One of Eastlund's patents (US4686605 A) described an adaptation of concepts first proposed by Nikola Tesla. Eastlund's 'Method and apparatus for altering a region in the earth's atmosphere, ionosphere, and/or magnetosphere,' described as 'grandiose,' proposed a 40-mile square radio transmitter that used Alaskan natural gas to generate current to create electromagnetic radiation that would excite a section of the ionosphere. Eastlund's patent speculated on 'possible ramifications and potential future developments' including magnetotelluric surveys, local weather modification, and missile defense. Eastlund later claimed that HAARP was built using his patents, prompting conspiracy theorists such as Nick Begich to claim that HAARP is capable of secretly controlling the weather. According to HAARP program manager John L. Heckscher, 'HAARP certainly does not have anything to do with Eastlund's thing, that is just crazy. What we have here is a premier scientific research facility with military applications.'"

(Wardenclyffe come of age?) Bearden calls earthquake fault zones scalar interferometer due to the piled up piezoelectric stresses and charges waiting to be triggered and released. We may get an inkling of how two interfering beams from ionospheric heaters like HAARP can be made to transmit straight through the Earth to meet at a target area prepared in advance with an atmospheric chemical witch's brew delivered by so-called chemtrails.

But there is more. Bearden issued a clear warning to future terrain or geographical medicine practitioners, which I will explore in the sequel regarding how Transhumanism's synthetic biology is dependent upon electromagnetics *and* scalar weaponization:

> The scalar EM potential is an organized spacetime lattice of perfectly ordered EM energy, passing through it in a Whittaker wave structure. Living systems alter this internal EM Whittaker channel (this spacetime lattice structure) and communicate through it for mind, thought, personality, long-term memory, and deep control of cellular and body functions. All the deepest biological control systems are in the Whittaker channel. To understand environmental-EM effects, you've got to understand how the local environmental EM (both internal and external) affects and alters the Whittaker channel, and hence the deep biocontrol systems utilizing it.[7]

Among our natural biological systems—integumentary (skin and its appendages), musculoskeletal system, respiratory, circulatory, digestive, excretory, nervous, endocrine, and reproductive—only the nervous system is not entirely of Nature, given that it is dependent upon the electromagnetics of *subnature*, as described in the epilogue.* Bearden wrote about an advanced scalar weapon called a *quantum potential weapon*[8] that can mimic the signature / frequency of diseases *and re-create them on scalar carriers*:

> . . . any disease can be imprinted onto our cellular system using frequencies ranging from ultraviolet to infrared. Whole populations can have new diseases and death induced as well as latent diseases

*This dependence will be clarified in the synthetic biology book.

> being activated with quantum potential diseases in targeted areas. Manufactured symptoms of radiation poisoning, chemical poisoning, bacterial infection and even the effects of many kinds of drugs, including hallucinogenic ones, can be induced with these very subtle scalar waves which flow in hyperspace or the sea of ether. They become embedded right into the immune system or etheric counterpart of the physical body.[9]

This is exactly what must be investigated in the sequel on synthetic biology, given what the real story of 2020–2023 and post-2023 entails as to the role of 5G/6G electromagnetics, hardware, software, and wetware folded in with the scalar secret, subject to neither time nor space.

APPENDIX 3

Trinity

In the Arabian Tales I read how genii transported people into a land of dreams to live through delightful adventures. My case was just the reverse. The genii had carried me from a world of dreams into one of realities. What I had left was beautiful, artistic and fascinating in every way; what I saw here was machined, rough and unattractive . . . 'Is this America?' I asked myself in painful surprise.

Nikola Tesla

Long before Marie and Pierre Curie began their quest to discover the radioactivity in thorium, polonium, and radium, alchemists had been manipulating matter in crucibles to provoke the radiation force field. The Curies' uranium salts glowed with an unearthly light. Uranium emitted heat 250,000 times that of coal, and radium, millions of times more active than uranium, rapidly ejected electrons and alpha particles as it transmuted into lead.

Nuclear fission was discovered twenty-seven years after New Zealand atomic physicist Ernest Rutherford (1871–1937) created the present model of the atom. In 1932, British physicist James Chadwick at the Cavendish Laboratory in Cambridge, England, discovered the neutron. But putting nuclear fission to the test required a war with an antagonist evil enough to justify developing and using such an infernal technology, so Hitler was appointed chancellor of Germany in 1933, the very year the Curies allegedly discovered artificial radiation. The following year, Madame Curie died at the age of sixty-six, and physicist Enrico Fermi bombarded the nucleus with the neutron and induced radioactivity in twenty-two different elements using fluorine.

In his book about the Manhattan Project, *Brighter Than Ten Thousand Suns*, nuclear historian and antinuclear activist Robert Jungk wrote, "Among the young atomic scientists, some looked upon their work as a kind of intellectual exercise of no particular significance and involving no obligations, but for others, their researches seemed like a religious experience."[1]

The age of the four horses of the apocalypse* commenced with the ritual detonation of three atomic bombs *after* World War II had officially ended on May 8, 1945.† During the war, Japan's attempts to surrender were ignored so that the Manhattan Project could continue to its intended conclusion: nuclear experimentation over Japan on a genocidal level. The men involved in the Manhattan Project were for the most part well-placed members of secret societies dedicated to either working against Nature and human evolution, or else misled pawns. On July 16, 1945, at Trinity Site 60 miles north of White Sands National Monument, Manhattan Project director Robert Oppenheimer uttered the famous words, "Now I am become Death, the destroyer of worlds." Thereafter, scientism eliminated the life force (æther) and respect for life from scientific inquiry.

President Harry Truman, a 33° Freemason, had signed off on White Sands, Hiroshima, and Nagasaki, and called Oppenheimer a "crybaby" when the physicist—too late—awakened to what he had been party to. In his diary of the Potsdam, Germany, conference, President "True Man" wrote of wandering through Berlin, bombed to utter ruin, thinking about empires that had fallen—Atlantis, Carthage, Babylon, Rome—and rulers like Alexander, Darius, and Genghis Khan.

The American monument to the advent of the nuclear age is Trinity Site, located on four-thousand-square-mile White Sands Missile Range in New Mexico. After the deadly mushroom cloud rose into the atmosphere

*Revelation 6:8: "And I looked, and behold a pale horse: and his name that sat on him was Death, and Hell followed with him. And power was given unto them over the fourth part of the earth, to kill with sword, and with hunger, and with death, and with the beasts of the earth."

†Caveat lector: Google has changed the date of World War II's end to September 2, 1945, so as to justify the bombing of Japan.

on July 16, 1945, the wind turned the soil into a green sea of trinitite* as the charred remains of greasewood plants leaned as far away from the blast as they could. Trinitite was hauled away in barrels for more experiments.

The United States tested a hydrogen bomb in 1952; the Soviet Union did the same in 1953. Tests in the Bikini Atoll in 1954 caused radiation sickness among crew members on the Japanese tuna boat *Lucky Dragon*. The Atomic Energy Act of 1954 gave the newly created Atomic Energy Commission (AEC) carte blanche over the production and ownership of fissionable materials, along with the power to issue licenses to corporations to build and operate nuclear power plants while promoting and regulating them. The AEC was eventually split into the Department of Energy and the Nuclear Regulatory Commission.

Cows ate fallout grass, and children raised on fallout milk had alpha, beta, gamma, and X-ray "halos" around their heads. Ionized subatomic particles caused atoms to lose their orbital electrons or break their nuclei. Radioactive I-131 (half-life eight days) settled into thyroid glands; strontium-90, cesium-137, zirconium, and other radioactive isotopes seeped into the tissues of all natural biological life. Two generations later, the average American IQ had sunk as Wilhelm Reich's deadly orgone energy clouds† devoured the orgone/æther life force on the American continents.

In the 1954 film *Godzilla*, the monster had radioactive breath—no doubt a reference to the air we would henceforth be breathing.‡ The neutron bomb was first tested in the United States in 1962. In 1964, a satellite fell to Earth and dispersed 2.1 pounds of plutonium-238, fatal at one particle of less than a millionth of a gram. In September 1972, the first scientific examination of Trinity Site since 1955 revealed that plutonium was still migrating deeper and deeper into the ground. The Trinity explosion was hotter than either Hiroshima or Nagasaki because it had

*Also called atomsite and Alamagordo glass (due to its proximity to the town of Alamagordo), trinitite is the name given to the soil that fused into a glass-like consistency by the heat from the Trinity test.

†In 1951, psychoanalyst Wilhelm Reich, MD, conducted the ORANUR Experiment (ORgonomic Anti-Nuclear Radiation) to determine the effect of nuclear radiation on life forms. He acquired two 1 mg units of radium, each in a separate half-inch lead container. His helpers at Organon, Reich's research center in Maine, became ill.

‡Other nuclear films: *Hiroshima Mon Amour* (1954); *Record of a Living Being* (1955); *Dr. Strangelove* (1964); *Black Rain* (1988); *Rhapsody in August* (1991).

not been exploded two thousand feet in the air, but on the ground. In 1975, White Sands became a National Historic Site. Theoretical physicist Werner Heisenberg (1901–1976) observed that with the advent of nuclear, "The space in which man's spiritual being develops is in a different dimension from that in which it was moving in previous centuries."

The Bomb had changed "the space in which man's spiritual being develops."

Death upon death followed the nuclear ritual. The first U.S. Secretary of War (now ironically called Defense), James Forrestal (1892–1949), died by defenestration from the sixteenth floor of the National Naval Medical Center in Bethesda, Maryland, on May 22, 1949. Julius and Ethel Rosenberg, accused of nuclear spying for the Soviet Union, were electrocuted on June 19, 1953. CIA scientist Frank Olson died by defenestration from the tenth floor of the Statler Hotel in Manhattan on November 28, 1953; and Wilhelm Reich was eliminated with a heart attack in Lewisburg Federal Penitentiary, Pennsylvania, on November 3, 1957—one week before his release.

Like all secret societies, the Satanic religion known as Nazism enjoyed two tiers of power: the outer tier of the *exoteric* National Socialist German Workers Party, and the inner tier of the *esoteric* Schutzstaffel (the Nazi SS), whose mind-controlled "knights" had been programmed as an inversion of the ancient Knights Templar, a military order of the Catholic Church. To this day few realize that Hitler was MK-Ultra-style controlled with drugs, hypnosis, and rituals conducted by the Nazi inner esoteric circle.

The Los Alamos birthplace of the Manhattan Project bomb was over a sacred Native American kiva just west of the Sangre de Cristo ("Blood of Christ") Mountains. The core of the bomb, as big as an orange, was transported in the backseat of a sedan from Los Alamos through Santa Fe ("City of Holy Faith") to the Trinity Site via the Oscura ("dark," "hidden") Mountains of the Badlands (*El Malpais*), the Navajo legend being that Twin War Gods had slain the monster Yé'iitsoh on Mount Taylor and the blood had run down the mountain, then coagulated and solidified to form the black lava flows of *El Malpais*.

Historical accounts, military projects and operations, and mainstream media accounts regarding staged events that centuries-old shadow brotherhoods deem essential for controlling the public's consciousness are thick with numbers, names, symbols, and cyphers for psychological operations

(PSYOPS), which are basically black magick.* The blast at Trinity Site, located at latitude 33°, on Monday, July 16, 1945, was just such an operation. In her 1993 book *A Chorus of Stones: The Private Life of War*, Susan Griffin stresses that the explosion is still happening inside us on a cellular, immunological level, transforming the chemicals in our air and water, its radioactive half-life continuing to rend the fabric of life and the matrix of meaning.

The "Little Boy" bomb exploded on August 6, 1945 (the eve of the Christian feast of the Transfiguration), over Hiroshima, located at latitude 34°, and the "Fat Man" bomb detonated on August 9, 1945, over Nagasaki, located at latitude 32.7°. All in all, White Sands, Hiroshima, and Nagasaki were the three prongs of a triune Plutonic (Pluto rules the subconscious) initiation. The "sacrifice" involved 318,000 Japanese vaporized and untold numbers wounded or made sick with cancer. Dorothy Day, a leading figure in the Catholic Worker Movement, countered jubilant newspaper accounts of the annihilation with a description much like that which Susan Griffin would write a half century later:

> Our Japanese brothers scattered, men, women and babies, to the four winds, over the seven seas. Perhaps we will breathe their dust into our nostrils, feel them in the fog of New York on our faces [shades of 9/11?], feel them in the rain on the hills of Easton . . . A cavern below Columbia [the occult name for America] was the bomb's cradle, born not that men might live but that men might be killed. Brought into being in a cavern and then tried in a desert place in the midst of tempest and lightning, tried out, and then again on the eve of the Feast of the Transfiguration of our Lord Jesus Christ, on a far-off island in the eastern hemisphere, tried out again, this new weapon which conceivably might wipe out mankind and perhaps the planet itself.[2]

Thousands of atmospheric, aboveground and underground nuclear "tests" have been carried out since, in part to discover the bomb's exact geometric trigonometry in sync with spacetime and planetary positions, in part to infuse the Earth with the force known as radiation.

---

*See Michael Hoffman's books *Secret Societies and Psychological Warfare* (2001) and *Twilight Language* (2021). For example, *onomancy* is the ancient science of divination by names.

APPENDIX 4

The HAARP Patents of Bernard Eastlund, PhD

And the concrete reality is that the intellectual thoughts evolved inwardly by men today will in time to come creep over the Earth like a spider's web, wherein human beings will be enmeshed, if they will not reach out to a world lying beyond and above their shadowy thoughts and concepts.

RUDOLF STEINER,
"A PICTURE OF EARTH-EVOLUTION IN THE FUTURE"
(GA 204), MAY 15, 1921

Assigned to Advanced Power Technologies Inc., Los Angeles, California, and Washington, DC:

- **U.S. Patent 4686605:** Method and apparatus for altering a region in the Earth's atmosphere, ionosphere, and/or magnetosphere. Issued Aug. 11, 1987, filed Jan. 10, 1985
- **U.S. Patent 5038664:** Method for producing a shell of relativistic particles at an altitude above the Earth's surface. Issued Aug. 13, 1991, filed Jan. 10, 1985
- **U.S. Patent 4712155:** Method and apparatus for creating an artificial electron cyclotron heating region of plasma. Issued Dec. 8, 1987, filed Jan. 28, 1985

- **U.S. Patent 5068669:** Power beaming system. Issued Nov. 26, 1991, filed Sept. 1, 1988
- **U.S. Patent 5218374:** Power beaming system with printer circuit radiating elements having resonating cavities. Issued June 8, 1993, filed Oct. 10, 1989
- **U.S. Patent 5293176:** Folded cross grid dipole antenna element. Issued March 8, 1994, filed Nov. 18, 1991
- **U.S. Patent 5202689:** Lightweight focusing reflector for space. Issued April 13, 1993, filed Aug. 23, 1991
- **U.S. Patent 5041834:** Artificial ionospheric mirror composed of a plasma layer which can be tilted. Issued Aug. 20, 1991, filed: May. 17, 1990
- **U.S. Patent 4999637:** Creation of artificial ionization clouds above the Earth. Issued March 12, 1991, filed May 14, 1987
- **U.S. Patent 4954709:** High resolution directional gamma ray detector. Issued Sept. 4, 1990, filed Aug. 16, 1989
- **U.S. Patent 4817495:** Defense system for discriminating between objects in space. Issued April 4, 1989, filed July 7, 1986
- **U.S. Patent 4873928:** Nuclear-sized explosions without radiation. Issued Oct. 17, 1989, filed June 15, 1987

APPENDIX 5

Substances Used in Chemical Spraying Operations*

Arsenic
Bacilli and molds
Barium
Cadmium
Chromium
Desiccated human red blood cells
Ethylene dibromide
Enterobacter cloacae
Enterobacteriaceae
Human white blood cells-A
Restrictor enzyme used in research labs to snip and combine DNA
Lead
Lithium
Mercury
Methyl aluminum
Mold spores
Mycoplasma

---

*I apologize for the lack of a dependable internet source, but the so-called chemtrail program categorized as geoengineering is highly classified (and therefore lied about constantly), and because the government program is a public-private partnership with Big Pharma, the chemicals vary over varied regions, depending upon the military-industrial-intelligence-Big Pharma agenda.

Nano-aluminum-coated fiberglass
Nitrogen trifluoride (aka "chaff")
Nickel
Polymer fibers
Polyvinyl alcohol (PVA)*
Pseudomonas aeruginosa
Pseudomonas florescens
Radioactive cesium
Radioactive thorium
Selenium
Serratia marcescens
Sharp titanium shards
Silver
Streptomyces
Strontium
Submicron nanoparticles containing live biological matter
Unidentified bacteria
Uranium
Yellow fungal mycotoxins

*Scientist Clifford Carnicom hypothesizes that polyvinyl alcohol points to synthetic organ development and replication of DNA.

APPENDIX 6

A Meeting

A meeting[1] has taken place recently between an investigative researcher and a well-placed military source. The identity of both parties is to be protected. The source has intimate knowledge of at least one aspect of the aerosol operations and asserts the following:

1. The operation is a joint project between the Pentagon and the pharmaceutical industry.
2. The Pentagon wishes to test biological diseases for war purposes on unsuspecting populations. It was stated that SARS is a failure as the expected rate of mortality was intended to be 80 percent.
3. The pharmaceutical industry is making trillions on medications designed to treat both fatal and nonfatal diseases given to populations.
4. The bacteria and viruses are freeze-dried and then placed on fine filaments for release.
5. The metals released along with the diseases heat up from the sun, creating a perfect environment for the bacteria and viruses to thrive in the air supply.
6. Most countries being sprayed are unaware of the activities and they have not consented to the activities. He states that commercial aircraft are one of the delivery systems.
7. Most of the "players" are old friends and business partners of the senior Bush.
8. The ultimate goal is the control of all populations through directed and accurate spraying of drugs, diseases, etc.

9. People who have tried to reveal the truth have been imprisoned and killed.
10. This is the most dangerous and dark time that I have experienced in all of my years of serving this country.

This information is relayed without qualification, as I am knowledgeable in the level of integrity of the researcher that has made this information available to the public. There is both risk and restraint that has been exercised in the preparation of this statement.

APPENDIX 7

Visitors to www.Carnicom.com, Aug. 26, 1999

Let it be noted that some of the visitors to this website[1] included*:

1. Desert Research Institute in Nevada (weather modification research institution) (repeat visits)
2. Fort Lewis Army Military Base in the State of Washington (home of Special Forces air squadron)
3. Lockheed Martin (aviation and space defense contractor) (repeat visits) (repeat repeat visits)
4. Los Alamos National Laboratory (repeat visit)
5. Allergan Pharmaceutical Corporation (Allergy Pharmaceutical Research Company)
6. Alliant Techsystems (Space and Strategic Defense Systems contractor)
7. Raytheon Defense Systems (defense contractor) (repeat visit) (repeat repeat visit) (repeat repeat repeat visit)
8. BOEING AIRCRAFT COMPANY (100 visits minimum)
9. United States Defense Logistics Agency (supplies and support to combat troops)
10. Davis-Monthan Air Force Base, Tucson, AZ (home of 355th Wing) (repeat visits) (repeat repeat visits) (repeat repeat repeat visit)
11. Dept of Defense Naval Computer and Telecommunications Area Master Station

*Let it also be noted that United States government computer systems are to be used for official purposes only.

12. U.S. Naval Sea Systems Command
13. Western Pacific Region of the Federal Aviation Administration, Lawndale, CA (repeat visit) (repeat visit) (repeat visit)
14. National Aeronautics and Space Administration Langley Research Center (10 visits minimum)
15. United States Environmental Protection Agency (20 visits minimum)
16. St. Vincent Hospital, Santa Fe, New Mexico
17. HEADQUARTERS UNITED STATES AIR FORCE, THE PENTAGON
18. United States Department of the Treasury (repeat visit) (repeat visit)
19. United States Department of Defense Educational Activity
20. ANDREWS AIR FORCE BASE, PROUD HOME OF AIR FORCE ONE
21. United States Federal Aviation Administration
22. United States Naval Research Center, Washington, DC
23. Rockwell-Collins (U.S. defense contractor)
24. Honeywell (U.S. defense contractor) (repeat visit)
25. Wright-Patterson Air Force Base, Dayton, OH (repeat visit) (repeat repeat visit)
26. Kadena Air Force Base, Okinawa, Japan
27. Camp Pendleton, United States Marine Corps (mandatory U.S. Defense anthrax vaccination program described at www.cpp .usmc.mil) (repeat visit) (repeat visit)
28. Ames Research Center, NASA (one of their primary missions is to research ASTROBIOLOGY, i.e., the study of life in outer space) (repeat visit)
29. Space Dynamics Laboratory, Utah State University, North Logan, Utah
30. Merck (Pharmaceutical Products and Health Research) (repeat visit)
31. McClellan Air Force Base, Sacramento, CA (The Sacramento Air Logistics Center at McClellan Air Force Base, California performs depot maintenance on the KC-135 Stratotanker aircraft and is heavily involved in space and communications—electronics.) (repeat visit)
32. TRW (U.S. defense contractor) (repeat visit)
33. Teledyne Brown Engineering (U.S. defense contractor)
34. United States Navy Medical Department

35. Air National Guard, Salt Lake City, Utah
36. Monsanto Company (chemical, pesticide, and pharmaceutical products) (repeat visit) (repeat repeat visits)
37. U.S. Department of Veterans Affairs
38. Arco Chemical Corporation
39. Sundstrand Aerospace (U.S. defense contractor)
40. National Oceanic and Atmospheric Administration Aeronomy Laboratory (conducts fundamental research on the chemical and physical processes of the Earth's atmosphere)
41. Allied Signal Corporation (chemical, aerospace, energy) (repeat visit) (repeat repeat visit) (repeat repeat repeat visit) (repeat repeat repeat repeat visit)
42. Aviation Weather Center, National Oceanic and Atmospheric Administration
43. United States Army Medical Department (repeat visit)
44. NASA Goddard Space Flight Center
45. Applied Physics Laboratory, a research division of John Hopkins University, which supports the U.S. Defense Department
46. United States Naval Health Research Center, San Diego, CA
47. HEADQUARTERS, UNITED STATES ARMY, THE PENTAGON
48. United States General Accounting Office (The General Accounting Office is the investigative arm of Congress. GAO performs audits and evaluations of government programs and activities.)
49. Bristol-Myers Squibb Company (pharmaceutical research and development)
50. United States Naval Criminal Investigative Service (A worldwide organization responsible for conducting criminal investigations and counterintelligence for the Department of the Navy and for managing naval security programs)
51. National Computer Security Center (NCSE) (Involved in advanced warfare simulation)
52. The Mayo Clinic (repeat visit) (repeat repeat visit) (repeat repeat repeat visit)
53. The Federal Judiciary (home of the United States Supreme Court)
54. United States Federal Emergency Management Agency (Controls a comprehensive, risk-based, emergency management program of mitigation, preparedness, response, and recovery) (repeat visit)

55. United States Naval Surface Warfare Center, Crane, IN (repeat visit) (repeat repeat visit)
56. United States National Guard Public Affairs Web Access (no public access to this site)
57. UNITED STATES SENATE (repeat visit) (repeat repeat visit) (repeat repeat repeat visit)(repeat repeat repeat repeat visit)
58. Headquarters, United States Air Force Reserve Command
59. Kaiser Permanente health organization
60. United States Naval Warfare Assessment Station
61. Air University, United States Air Force
62. United States Naval Research Laboratory (repeat visit)
63. Enterprise Products Partners L.P. (MTBE production)
64. United States Navy Naval Air Weapons Stations, China Lake, CA
65. California Pacific Medical Center
66. United States Defense Information Systems Agency (mission: "To plan, engineer, develop, test, manage programs, acquire, implement, operate, and maintain information systems for C4I and mission support under all conditions of peace and war.")
67. San Francisco Department of Public Health
68. BJC Health System, St. Louis, Missouri
69. United States Open Source Information Systems (OSIS) (an unclassified confederation of systems serving the intelligence community with open source intelligence) OSIS sites include: (AIA) Air Intelligence Agency, Kelly AFB, San Antonio, TX IC-ROSE (CIA) Central Intelligence Agency, Reston, VA (DIA) Defense Intelligence Agency, Washington, DC (NSA) National Security Agency, Ft. Meade, Laurel, MD (NIMA) National Imagery & Mapping Agency, Fairfax, VA (NAIC) National Air Intelligence Center, Wright-Patterson AFB, Dayton, OH (NGIC) National Ground Intelligence Center, Charlottesville, VA (MCIC) Marine Corps Intelligence Center, Quantico, VA (NMIC) National Maritime Intelligence Center, Office of Naval Intelligence, Suitland, MD (ISMC) Intelink Service Management Center, Ft. Meade, Laurel, MD (repeat visit)
70. New Mexico Department of Health
71. United States Space and Naval Warfare Systems Command (SPAWAR)
72. United States McMurdo Research Station, Antarctica

73. Orlando Regional Healthcare System, Florida
74. United States Andersen Air Force Base, Guam
75. United States Misawa Air Base, Japan
76. United States Hickam Air Force Base, Hawaii
77. United States Osan Air Force Base, Korea
78. Royal Air Force, Lakenheath, Suffolk
79. United States Scott Air Force Base
80. United States F.E. Warren Air Force Base
81. United States Air Force News Agency
82. United States Langley Air Force Base (repeat visit)
83. United States Tinker Air Force Base
84. United States McConnell Air Force Base
85. United States Charleston Air Force Base
86. United States Randolph Air Force Base
87. United States Air Force Reserve Command
88. United States Seymour Johnson Air Force Base
89. United States Bolling Air Force Base, Washington, DC
90. Keesler Air Force Base, MS
91. United States Hill Air Force Base
92. United States Vandenberg Air Force Base, California
93. United States Minot Air Force Base, North Dakota
94. United States Eielson Air Force Base, Alaska
95. ANDREWS AIR FORCE BASE, PROUD HOME OF AIR FORCE ONE (repeat visit)
96. HEADQUARTERS UNITED STATES AIR FORCE, THE PENTAGON (repeat visit) (Visitors 75–96 arrived within a 24-hour period (09/23/99))
97. United States Cannon Air Force Base, New Mexico
98. United States McGuire Air Force Base
99. United States Beale Air Force Base (home of the U-2 fleet of reconnaissance aircraft)
100. United States Department of Justice—Federal Bureau of Prisons
101. Metnet—United States Navy (associated with weather reporting system and SPAWAR)
102. TRADOC—United States Army Training and Doctrine Command, Fort Monroe, VA
103. *Newsweek* magazine

104. United States Defense Advanced Research Projects Agency
105. Massachusetts Medical Society, Owner—Publisher: *New England Journal of Medicine*
106. OFFICE OF THE SECRETARY OF DEFENSE: THE OFFICE OF WILLIAM S. COHEN, SECRETARY OF DEFENSE (repeat visit)
107. HEADQUARTERS UNITED STATES AIR FORCE, THE PENTAGON (repeat repeat visit)
108. UNITED STATES JOINT FORCES COMMAND (reports to U.S. Secretary of Defense) (repeat visit)
109. Naval Warfare Assessment Station, Corona, CA
110. Los Angeles County Emergency Operations Center
111. Commander in Chief, United States Pacific Fleet, United States Navy
112. HEADQUARTERS UNITED STATES AIR FORCE, THE PENTAGON
113. Defense Logistics Agency, Administrative Support Center in Europe
114. United Stated Department of Defense Network Information Center, Vienna, VA (repeat visits)
115. Office of the Assistant Secretary of the Army
116. Headquarters, United States Air Force, The Pentagon (repeat visit)
117. *U.S. News and World Report*
118. Naval Air Warfare Center—Aircraft Division (repeat visits)
119. New Zealand Parliament
120. HEADQUARTERS UNITED STATES AIR FORCE, THE PENTAGON (multiple repeat visits)
121. NIPR—Department of Defense Network Operations (NIPRNet); The Defense Information Systems Agency (DISA) has established a number of NIPRNet gateways to the Internet, which will be protected and controlled by firewalls and other technologies. (repeat visits)
122. Peterson Air Force Base, Colorado Springs, CO (home of NORAD and SPACECOM)
123. Raytheon (visits immediately after introduction of HAARP implications)
124. United States Army War College
125. Lawrence Berkeley National Laboratory
126. Fermi National Accelerator Laboratory

APPENDIX 8

Movers and Shakers

The technotronic era involves the gradual appearance of a more controlled society. Such a society would be dominated by an elite unrestrained by traditional values. Soon it will be possible to assert almost continuous surveillance over every citizen and maintain up-to-date complete files containing even the most personal information about the citizen. These files will be subject to instantaneous retrieval by the authorities.

ZBIGNIEW BRZEZINSKI (1928–2017),
INVETERATE COLD WARRIOR AND NATIONAL
SECURITY ADVISER TO FOUR DEMOCRATIC PRESIDENTS

On July 1, 1960, President Eisenhower (1953–1961) signed an executive order that moved U.S. space operations* from the U.S. Army to NASA, a civilian agency not answerable to Congress. Equipment then valued at $100 million† and 4,700 civilian employees, plus Operation Paperclip scientist Wernher von Braun and his German rocket team were transferred from the U.S. Army's Redstone Arsenal in Huntsville, Alabama, and Cape Canaveral, Florida, to NASA's Marshall Space Flight Center.

Since then, the revolving door of the military-industrial-intelligence complex—now referred to as "public-private partnerships"—has become more like a Gorgon's head sprouting devouring snakes. Going up against

*Referring to the secret space program, "secret" because it's been buried for "national security" reasons.

†One dollar in 1960 was worth eight dollars today.

such a Gorgon means running the risk of being turned into stone, which is to say professionally ostracized, impoverished, or killed under mysterious circumstances. To answer the perennial question as to who is behind the geoengineered transhuman juggernaut, "they" include public and private organizations like the following.

ADVANCED RESEARCH PROJECTS AGENCY FOR HEALTH (ARPA-H)

The Advanced Research Projects Agency for Health, or ARPA-H, merges military and medical technology under the jurisdiction of the National Institutes of Health (NIH). As its clever acronym makes obvious, it is modeled after the Defense Advanced Research Projects Agency, DARPA. The NIH site suggests that ARPA-H is more of the same research into weapons innovations as are going on at DARPA, but more specifically having to do with a synthetic biology aimed at Transhumanism, or in their obfuscatory language, "leveraging research advances for real world impact . . . to drive biomedical breakthroughs ranging from molecular to societal."

ASILOMAR CONFERENCE

It could be said that the Asilomar Conference, a privately organized confab "on signals, systems, and computers," held annually on the Monterey Peninsula in California, are planning sessions for elite technocrats intent on establishing the pivotal systems that assure global control over the environment, all organic life forms, and human consciousness. In particular,

- The February 27, 1975, Asilomar Conference on Recombinant DNA lifted the moratorium on genetic engineering and allowed the alteration of the genetics of all life on Earth.
- The March 22–26, 2010, Asilomar Conference on Climate Intervention Technologies decided, based on the biotechnology decisions made at the 1975 conference, to focus on changing the world environment by means of geoengineering.
- The January 5–8, 2017, Asilomar Conference on Beneficial AI confirmed that the AI system is in place, from satellites to nanoparticles.

BAIN & COMPANY

Bain & Company was founded in 1973 with Mormon wealth in the person of Mitt Romney. This management consulting firm has been called "the KGB of consulting."[1] In this case, management consulting translates to taking over companies, gutting them, then placing its own people in power. Private equity firms like Bain, the Carlyle Group, and BlackRock are basically CIA fronts for the Fortune 500 corporations the CIA serves and behave as cuckoo birds that lay their eggs in other birds' nests so that when the host bird's fledglings hatch out, they are immediately devoured.

Bain & Company handles* clients such as Halliburton, Raytheon, Monsanto, Microsoft, Starbucks, and the Bill & Melinda Gates Foundation. It serves as the "fixer" in potentially troublesome situations, such as making sure celebrities like Kanye West stay in line via mind control, that social engineering accompanies the commodities it "handles" (like Starbucks' fake meat burgers), and that backdoor access computers are omnipresent.

BLACKROCK

The world's biggest shadow bank asset manager is BlackRock. BlackRock monitors more than $18 trillion in assets for two hundred financial firms, including the U.S. Federal Reserve and European central banks, and interfaces directly with UN Agenda 2030 to bring about the World Economic Forum's "Great Reset." BlackRock runs the Biden crime family's regime from the U.S. to Ukraine. Brian Deese, director of the National Economic Council, was once BlackRock's global head of sustainable investing; before that he'd replaced John Podesta as senior adviser to President Obama and negotiated the "global warming" deception at the 2015 UN Conference on Climate Change that resulted in the Paris Agreement. The deputy secretary of the treasury, Nigerian-born Adewale Adeyemo, hails from BlackRock, as does Vice President Kamala Harris's senior economic adviser, Michael Pyle. F. William Engdahl, a geopolitical analyst, strategic

---

**Handle/handler* has two basic meanings: to oversee or deal with certain commodities, and to train or have charge of an animal (or person). In the case of Bain *&* Company (the CIA is known as "the Company"), the term is intended to cover both senses.

risk consultant, author, professor, and lecturer, describes BlackRock as "a law unto itself."[2]

COUNCIL OF GOVERNORS

The acronym for the Council of Governors, COG, is the same as the acronym for the "continuity of government" created by President John F. Kennedy on February 12, 1962, so as to shield the essential infrastructure of the United States government from destruction, permitting its continued operation and authority in a time of crisis. The mandate of the Council of Governors sounds similar—"to strengthen further the partnership between the Federal Government and State Governments to protect our Nation against all types of hazards"[3]—but has, since the 2020 Covid-19 global "emergency" and subsequent curtailment of human bodily sovereignty and personal freedoms, proven to be quite different, with the state governors seeming to get their marching orders from the Deep State.

DEPARTMENT OF HOMELAND SECURITY

Before September 11, 2001, there was no Department of Homeland Security (DHS), nor had the United States ever been referenced as a "homeland," a borrowed term from Nazi Germany. The irony is that Hitler appropriated the term "homeland" from the 1920s and 1930s Zionist movement's goal to create a Jewish "homeland" in the Middle East. In 1934, at the Nazi party's big coming-out event at one of its most famous Nuremberg rallies,* the term "homeland" was introduced. Prior to that Germans had always referred to Germany as "the Fatherland" or "the Motherland" or "our nation." But Hitler's handlers wanted the German people to think of themselves with the kind of semi-tribal passion that the Zionists had for Israel.[4]

The DHS is actually a civilian agency answering to the Office of the Director of National Intelligence in its task of overseeing fusion centers, black projects, and COG (continuity of government) operations, including federal detention camps, which are often located on "closed" military bases

*As seen in the 1935 Leni Riefenstahl film *Triumph of the Will.*

or Bureau of Land Management lands. The DHS touts itself as a domestic and international terrorism watchdog and is the third-largest federal department, overseeing a gigantic bureaucracy of twenty-two different federal agencies.

A particularly telling example of the DHS role in its mandate of overseeing international terrorism is the Friday, April 13, 2007, "Agreement between the Government of the United States of America and the Government of the Kingdom of Sweden on Cooperation in Science and Technology for Homeland Security Matters."[5] The agreement gave Sweden access to billions of National Security Agency (NSA) dollars for Swedish biotech corporations, institutions, universities, and laboratories—in other words, nuclear, biological/chemical R&D, underwater breathing techniques, border control, nanosensors and microprocessors, search and surveillance, and targeting. Four years later, at the end of 2011, then-DHS secretary Janet Napolitano met with Swedish ministers and signed the "Preventing and Combating Crime Agreement" for the fluid exchange of biometric and biographic data on citizens "to bolster counterterrorism and law enforcement efforts."[6]

FEDERAL ACCOUNTING STANDARDS ADVISORY BOARD (FASAB)

You probably haven't heard of this government body, nor of its October 4, 2018, announcement "Statement of Federal Financial Accounting Standards 56 (FASAB 56)"[7] that basically legalizes lying about federal budget line items, or as investigative reporter Matt Taibbi put it in 2019, "a system of classified money-moving."[8] The U.S. Securities and Exchange Commission (SEC), mandated to keep watch on government transparency, basically looked the other way as FASAB 56 became policy and thus overrode the U.S. Constitution to, in the words of Catherine Austin Fitts, take "a large portion of the U.S. securities marker dark."[9]

And yet at the same time companies are being forced to reveal their ESG (environmental, social, and governance) investments under the UN's "Principles for Responsible Investment" in the wake of the "Climate Risk Disclosure Act" sponsored by Senator Elizabeth Warren. The SEC now requires publicly traded companies to disclose "climate change–related risks," including companies' direct and indirect greenhouse gas emissions, the total amount of fossil fuel–related assets they own or manage, and their management strategies related to physical risks posed by "climate change."

GOOGLE

Behind the friendly face of Google, the internet has been weaponized. Google can do things the CIA cannot do, namely, operate as "corporate intervention in foreign affairs at a level that is normally reserved for states."[10]

Google Earth and its search functionality marked the beginning of Google's military-intelligence-security expansion into Big Data. Now Google is a committed military-intelligence for-profit corporate contractor in service to virtually all the three-letter agencies—the NSA, CIA, DIA, FBI, NGA—the whole alphabet soup. Its parent company is in fact named Alphabet Inc.[11] Google has *always* been about intelligence. Back in the 1990s, Google cofounders Larry Page and Sergey Brin were at Stanford University when the CIA's own venture capital investment firm In-Q-Tel bailed Stanford out so it could purchase the Key Hole spy satellite, while the National Geospatial-Intelligence Agency (NGA, another Department of Defense agency) could pay for Key Hole to be "tailored" to meet intelligence gathering needs via CIA-assisted technology like Google Earth. Google started at Stanford and has been in bed with the NGA ever since.

Google is the exclusive provider of geospatial intelligence services to America's military and intelligence agencies.[12]

Google went on to become the first "cloud-based" provider for the U.S. government with federal security classification for nonclassified data that led to contracts with the U.S. Naval Academy and U.S. Coast Guard Academy, the U.S. Army, the Department of Defense, the Department of Interior, as well as nuclear labs, state and municipal governments,

FACEBOOK

Facebook, as everyone who has seen the movie *The Social Network* knows, was supposedly started by a scruffy Harvard undergrad named Mark Zuckerberg, a CIA controlled person if there ever was one. Some say

Facebook is DARPA's data-mining Total Information Awareness program on steroids.* Facebook's emergence with advanced tech, along with its Harvard connection, scream CIA, particularly the CIA's In-Q-Tel, which helped to fund the Facebook startup, along with PayPal and venture capital firms Palantir Technologies, Founders Fund, and Accel Partners.

Facebook not only collects intel on everyone on its network; its algorithms try to predict who will do what (e.g., will you wear the mask out of fear?) and uses that data to help its intelligence partners create messaging that formulates desired mindsets. Consider the small print in Facebook's terms of use and privacy policy:

> By posting member content to any part of the website, you automatically grant, and you represent and warrant, that you have the right to grant to Facebook an irrevocable, perpetual, non-exclusive, transferable, fully paid, worldwide license to use, copy, perform, display, reformat, translate, excerpt and distribute such information and content, and to prepare derivative works of, or incorporate into other works, such information and content, and to grant and authorize sublicenses of the foregoing . . . Facebook may also collect information about you from other sources, such as newspapers, blogs, instant messaging services, and other users of the Facebook service through the operation of the service (e.g. photo tags) in order to provide you with more useful information and a more personalized experience. By using Facebook, you are consenting to have your personal data transferred to and processed in the United States.

THE HIGHLANDS FORUM

The Highlands Forum, "has operated as a bridge between the Pentagon and powerful American elites outside the military since the mid-1990s. . . . [It] has for 20 years provided an off the record space for some of the most prominent members of the shadow intelligence community to convene

*Remember TIA's all-seeing eye logo? Both Sergey Brin (cofounder of Google) and Mark Zuckerberg (cofounder of Facebook) were groomed in the "exceptional children" program of the Center for Talented Youth run by Johns Hopkins University—mind control programming at its best.

with senior US government officials, alongside other leaders in relevant industries."[13]

Not at all coincidentally, the Highlands Forum, HAARP, and the Project Cloverleaf aerosol delivery program all went active around the same time, 1995. With the shift from analog to digital and the resurrection of the Strategic Defense Initiative Star Wars program thanks to HAARP, I have no doubts that the Highlands Group, the founding organization behind the Highlands Forum and similar forums, played a pivotal role in Lockheed Martin's Space Fence, now classified under information warfare "perception management." Certainly the Highlands Forum, funded by the Department of Defense, DARPA, the National Science Foundation, all branches of the U.S. military, Harvard University, and, of course, the CIA, has overseen the global reconfiguration of the socio-political landscape since 9/11. The Highlands Group spins the Washington, DC, roulette wheel between government and elites in business, industry, finance, and mainstream media—thus, a "shadow network of private contractors," intelligence, and Pentagon influence over the private sector metastasized.[14] In short, the Highlands Forum pushes the doctrine of full-spectrum dominance at every level of postmodern life.

IN-Q-TEL

An entire book could be written on the CIA's role in geoengineering and its manifold operations moving us all on the conveyor belt toward a transhumanist future.[15] Here, I draw attention to its "venture capital arm," In-Q-Tel, founded in 1999 to cover research and development for the seventeen agencies of the intelligence community, along with the National Geospatial-Intelligence Agency (NGA), the Defense Intelligence Agency (DIA), and Homeland Security's Security Science and Technology Directorate, plus Google, Oracle, IBM, Lockheed Martin—well, the list just goes on and on.[16]

So what is the point of these interwoven (one might say incestuous) relationships? The underlying element that permeates these tech companies is that the federal government is systematically involved in their funding, purpose, objectives, acquisitions, and equity participation. Is this the new normal for the economy? Looks like it is, but such an organizational structure does not conform to the standards of free enterprise.

This impetus for invention does not have the same importance as the Manhattan Project did in its day, but it may very well create the kind of technology that will have the same or an even greater risk of creating our extinction.

Is Google a real stand-alone company? Is Facebook the social network department of the NSA? Is the wealth of Bill Gates, the equity positions of Larry Page and Sergey Brin, or the tax dodge shares of Mark Zuckerberg a true reflection of their actual ownership stakes in their companies, or are they mere fronts for the shadow government? Just how many Eric Schmidt types are embedded in high-tech pulling the strings for the intelligence community?

The In-Q-Tels of this world act as the JP Morgans of the twenty-first century. Nikola Tesla's free electricity wireless distribution was killed by the robber baron. Today the role of inventive genius is managed and contained by technocrats following the directions of spooks who do the bidding of the supra elite . . .[17]

Today, A-teams like Delta Force and SEAL Team 6 conduct quiet kill-or-capture missions everywhere while the CIA's private corporate arm, the venture capital entity In-Q-Tel, diversifies throughout the transnational corporate world. In this way, the CIA has melded to itself the FBI, military contractors, and police forces while *conducting*—not combatting—"domestic terrorism" (by targeting U.S. citizens with electromagnetic weapons) in partnership with Homeland Security, military contractors, private security firms, and telecom corporations.

INTERNATIONAL CULTIC STUDIES ASSOCIATION

Shadow organizations and agencies play a shell game with their techno ergot, acronyms, and renaming of hidden-in-plain-sight projects. In 2005, the American Family Foundation (AFF) merged with the International Cultic Studies Association,[18] purportedly an anti-cult organization in the mold of the False Memory Syndrome Foundation (FMSF) founded in 1992 as a cover for pedophiles and Satanic ritual abuse perpetrators that casts doubt on children's testimony.

The AFF board was composed of ex-CIA advisers and former MK-Ultra psychiatric personnel. Funded by top Wall Street players such as Morgan Stanley Investment Advisors, Richard Mellon Scaife, and the Bodman and Achellis Foundations, the AFF was administered by the New York City law offices of Morris and McVeigh (intelligence and banking families), with overlapping directorates. Wikipedia reports that in 2000, Bodman gave the AFF $50,000 specifically for "the development and marketing of Citizenship and Character, instructional material to supplement American government and history classes in U.S. high schools."*

LIFEBOAT FOUNDATION

Lifeboat is a nonprofit nongovernmental organization (NGO) that proclaims it is "dedicated to encouraging scientific advancements while helping humanity survive existential risks and possible misuse of increasingly powerful technologies, including genetic engineering, nanotechnology, and robotics/AI, as we move towards a technological singularity." Besides being a futurist proponent of the Singularity University at NASA's Ames Research Center, which is sponsored by both NASA and Google (Ray Kurzweil, transhumanist mouthpiece, is on Lifeboat's scientific advisory board), Lifeboat builds underground bunkers, sponsors the development of the NanoShield that will "protect Earth against attacks by nanoweapons,"[19] and otherwise serves as a nexus for Fortune 500 corporations investing in the secret space program's mandate to prepare Earth for full-spectrum dominance.

Lifeboat's SecurityPreserver program looks for ways to "provide early warning of attacks before such attacks can be fully designed, planned, developed, deployed, let alone launched."[20] The program says that surveillance "is the best way to handle the threat of existential threats that will soon be in the hands of small groups of people."[21]

Pimp to the elite Jeffrey Epstein sat on Lifeboat's financial board, and whistleblower Edward Snowden was given the Lifeboat Foundation Guardian Award. Like other good globalists, Lifeboat plays both sides of the fence.

*According to the 1961 Annual Report of the [CIA] Human Ecology Foundation and 1941–1960 correspondence of John Clare Whitehorn, director of the Department of Psychiatry, Johns Hopkins University, the Scottish Rite Foundation was also a funder, but secret society references in general have been expunged from the internet.

LOCKHEED MARTIN

Lockheed Martin—once Lockheed Missiles and Space Company in Sunnyvale, California—is a foremost player in the secret space program. As early as 1987–1992, Lockheed launched thirty satellites into geosynchronous orbit out of the brand-new Schriever Air Force Base just ten miles east of Peterson Air Force Base in Colorado Springs. Lockheed now runs and owns the patents for the entire calibrated planetary Space Fence.

Defense contractors like Lockheed Martin, the world's largest defense contractor and the prime cyber-security and information technology supplier to the federal government, are not subject to congressional oversight or citizen demands—which is why big defense corporations run private security hitmen teams with direct access to sophisticated technology in order to keep Information Operations (IO) matters out of public view, given that many operations are pure, unadulterated mind control. Lockheed Martin "hitmen" have tentacles into every security and law enforcement agency in the nation, including state and local police and seventy-two regional "fusion centers" administered by the U.S. Department of Homeland Security. According to company literature, Lockheed Martin has operations in forty-six of the fifty states.

Lockheed Martin also has operational command and control over a U.S. government microwave radio frequency weapon system deployed on cell tower masts throughout the United States—quite handy for silently torturing, impairing, and electronically incarcerating certain targeted individuals. The nexus of this American "torture matrix"* appears to be Lockheed Martin's Mission and Combat Support Solutions control command center in Norristown, Pennsylvania, which employs several thousand workers. The defense contractor's global headquarters is in Bethesda, Maryland, just outside the nation's capital.

Much as Raytheon owned the patents and oversaw HAARP for the CIA, Lockheed Martin oversees the Space Fence lockdown infrastructure and the development of the covert Information Operations (IO) microwave assault on the public. Each day, a nationwide scalar electromagnetic radiation "multifunctional" radio frequency directed-energy weapon

*The CIA Torture Paradigm Matrix is set up to (1) cause self-inflicted harm; (2) cause sensory disorientation; (3) attack individual fears; and (4) attack cultural identity.[22]

attack system employing phased-array cell tower antenna transmitter/receivers and GPS satellites, under the administration of military contractor Lockheed Martin, is used to silently and invisibly torture, impair, subjugate, and degrade the physical and neurological health of untold thousands of American citizens who have been extrajudicially "targeted" by a hate- and ideology-driven domestic "disposition matrix" as "dissidents" or "undesirables."[23]

MITRE CORPORATION

MITRE Corporation, a U.S. Navy nonprofit founded in 1958 that supports various U.S. government agencies in the aviation, defense, healthcare, homeland security, and cybersecurity fields, produced the first draft of the Environmental Impact Statement in February 1993, then the "Electromagnetic interference impact of the proposed emitters for the High Frequency Active Auroral Research Project (HAARP)" on May 14, 1993. MITRE's environmental impact statement bypassed Congress and went straight to the Environmental Protection Agency (EPA) before being announced in the Federal Register on July 23, 1993.

MITRE works in concert with the intriguing JASON Group, a secretive scientific organization with top security clearance established in 1960. Note that JASON's Wikipedia entry reads, "For administrative purposes, JASON's activities are run through the MITRE Corporation, a nonprofit corporation in McLean, Virginia, which contracts with the Defense Department."

SCIENCE APPLICATIONS INTERNATIONAL CORPORATION (SAIC)

Did you notice that SAIC is CIAS spelled backward?

Men who organize and perpetrate violence on others for profit are a particular kind of secret society—like the Fraternal Order of Police, but with top security clearance and diplomatic channels. Such *omertà* Brotherhoods constitute a criminal class of disaster capitalism in service to black programs protected by military contractor corporations like Science Applications International Corporation (SAIC).

With 44,000 employees, SAIC is larger than the departments of Labor,

Energy, and HUD combined. Not counting its "black" contracts, SAIC holds thousands of active no-bid federal contracts. As a "lead systems integrator" for the army's Future Combat Systems Program, SAIC designed the Iraqi Reconstruction and Development Council two months before the 2003 assault on Iraq, packing the council with Iraqi journalists, police, and military. Through the Bureau for International Narcotics and Law Enforcement Affairs, SAIC set up international law enforcement academies around the world—10,000 officers in fifty countries—to "ensure a safe environment abroad for market economies"—"market economies" now referring to disaster capitalism.

SAIC is a for-profit company with a multibillion-dollar annual revenue, most of which comes from classified contracts with the U.S. military. Because it is employee-owned, there are no outside stockholders and therefore no outside controls on it. If you leave the company, you have to sell your shares in it. SAIC's board of directors reads like a who's who of the military-industrial-security complex (former secretaries of defense, spy agency heads, and other Deep State functionaries). SAIC's website features "Homeland Security."[24]

SAIC bought out Network Solutions Inc. (NSI) in 1995 and basically took control of internet domain names; in 2009, the internet domain name registry was supposedly transferred from the NSA and British Telecommunications (BT) to Nominet UK and the nonprofit ICANN, including a quiet agreement to open up internet development and oversight to transnational corporations and governments. NSI is still going strong, and since former President Obama didn't renew control over ICANN on September 30, 2016,[25] SAIC probably has carte blanche over that, too, in great part *because ICANN issues IP addresses to Internet of Things (IoT) devices.* (IPV6 (Internet Protocol version 6) means digital slavery.)[26]

Besides internet control over the IoT, SAIC is busy setting up global dual-use communications for the military, FBI, and IRS, handling electronic voting machines for rigged elections, training foreign militaries, targeting citizens, mind control—a real devil's kitchen.

SENIOR EXECUTIVE SERVICE

Begun in 1978 under President Jimmy Carter, who was a member in good standing of the Council on Foreign Relations, the Senior Executive

Service (SES) is said by Wikipedia to be "a position classification in the United States federal civil service equivalent to general officer or flag officer rank in the U.S. Armed Forces," but actually serves as the shadow government with the power of the purse (including salaries) over seventy-five government agencies that we call the Deep State. In fact, the Office of Personnel Management (*opm.gov*)—the agency that makes sure the eight thousand SES slots are filled with career executives (salaries starting at $250,000 per annum), the majority of whom are with the departments of Defense and Justice—views the SES through its insignia or emblem as *the keystone*, "the center stone that holds all the stones of an arch in place."[27] Seventy-five percent of the Security Exchange Commission are SES members, as are 70 percent of the Nuclear Regulatory Commission. SES members cannot be fired and are accountable to no one given that its members have the highest security clearance and are not subjected to background checks. SES has a black budget.

On December 18, 2015, President Obama signed Executive Order 13714, "Strengthening the Senior Executive Service,"[28] which, when examined, seems to be tightening up operations and encouraging the search for yet more "talent" for the advent of an imminent challenge such as the 2016 presidential "changing of the guard."[29]

SERCO INC.

Few know that all our most crucial administrative functions have been outsourced overseas, to British conglomerate Serco. It has a staff of 40,000, a budget of $4 trillion per annum, and they are in thirty-eight countries. Serco has its fingers in everything—trains, satellites, hospitals, schools, missile defense systems, American bases, border screening, intelligence, the U.S. departments of State and Homeland Security—the list goes on and on.

> Serco provides and controls communications and transportation systems worldwide, and otherwise is in positions of key importance that impact the security of the United States and other nations, as well. It is an instrument of "the global control network," the vast interlocking trust directorate overseen in the U.S. by the highly paid, largely faceless, un-elected Senior Executive Service (SES) discussed above.[30]

The SES feeds endless lucrative contracts to Serco's seventy subsidiaries around the world, many in service to British Aerospace and Lockheed Martin. Serco manages U.S. FEMA Region 9 (Arizona, California, Hawaii, Nevada, and the Pacific Islands); oversees the Laboratory Corporation of America (LabCorp), whose annual revenue is $6 billion; and perhaps most surprisingly is the gatekeeper over the U.S. Patent and Trademark Office through Serco's subsidiary located in Herndon, Virginia (6,000 employees, annual revenue of $1 billion). Control over U.S. patents and visas (the National Visa Center) in the Tesla era we are now in is a powerful position—as powerful as, say, the good old boy peer review system that decides which scientific papers will get published and which won't, and the gain-of-function patents that may impact everyone who has undergone the mRNA gene transfer as transhumans that no longer qualify for human rights. Trade secrets, copyrights, trademarks, and patents must pass through the Patent Office in their quest for legitimization, security, and funding. *Copyrights and patents are the only two property rights specifically protected by the U.S. Constitution.*

Needless to say, the patent system (like the peer review system) has been hijacked.

Serco also seems to quarterback for exchange-traded funds racketeers like BlackRock, Vanguard,[31] and State Street, a globalist triumvirate with $18 trillion in assets and the inside track on SIGINT (signals intelligence is practically a synonym for mind control from space), along with Five Eyes racketeers like Maximus ("to protect health, support families, strengthen workforces, and streamline government services"), Pfizer, and Microsoft.

Glossary

Acoustic gravity: Infrasound low frequency (20–100 Hz) modulated by ultra-low infrasonic waves (0.1–15 Hz). Many of the groans and hums heard around the world are acoustic gravity waves.

Antimatter: The natural oscillation of energy in the cell is due to the constant interaction of the matter-antimatter cycle (pulsing at 144,000 cps). The antimatter energy signature is in all life forms; we need antimatter in order to live in the physical realm. Discovered in 1955, antimatter is unstable and powerful; in fact, drawing energy from antimatter realms leads to violence and chaos. It must be contained. Given that it attracts entities from quantum dimensions, it also has to do with the paranormal, our thoughts and feelings.

Artificial atoms: A new branch of chemistry in a second periodic table for flat atoms (2D, like graphene) that can have thousands of electron states.

Atmosphere: Earth's atmosphere is composed of the troposphere (4–12 miles above the surface), the stratosphere (12–31 miles), the mesosphere (31–50 miles), and the thermosphere (50–429 miles).

B/CI: Brain/cloud interface

BCI: Brain-computer interface

Carbon nanotubes (CNTs): Hollow tubes of pure carbon as wide as a strand of DNA; stronger and lighter than steel; high conductivity. Delivered by chemical trails, CNTs have been found in Parisian children's lungs (Mike Williams-Rice, "Nanotubes found in lungs of French kids," *Futurity.org*, October 19, 2015)

Cation: An ion or group of ions having a positive charge

Chelation: A type of bonding of ions and molecules to metal ions

Climate: Long-term weather; the balance between incoming short waves from the Sun (light) and loss of outgoing long-wave radiation (heat)

Cloud condensation nuclei (CCN): Tiny particles that promote formation of water droplets

Clouds: Before Project Cloverleaf, clouds were generally visible aggregates of tiny water droplets or ice crystals or a mixture of both suspended in the air; now, clouds are plasma (gases and particles heated by electromagnetics) that absorb radiation from the Sun and Earth and radiate it back into space.

Cloud-seeding: Unlike geoengineering, cloud-seeding is done locally on a small scale by transforming super-cooled water droplets into ice crystal clouds with silver iodide, dry ice (frozen CO_2), and other chemical agents.

CME: Coronal mass ejection; large expulsions of plasma and magnetic field from the Sun's corona

CMOS: Complementary metal oxide semiconductor

CPU / QPU: Central processing unit; quantum processing unit

Curie: A unit of radiation; amount of any nuclide that undergoes 3.7×10^{10} radioactive disintegration per second. The so-called safe standard of radioactivity is 100 billion curies, a "safe dose" being 100 microcuries per liter.

Cyborg: a portmanteau of *cybernetic* and *organism*

dB: Decibel: a measurement used in acoustics and electronics, such as gains of amplifiers, attenuation of signals, and signal-to-noise ratios

Dielectric: Insulating as opposed to conductive

Doppler: A specialized radar that bounces a microwave signal off a target to analyze how the target's motion has altered the frequency of the return signal. The highly accurate Doppler effect is useful for aviation, satellites, weather systems, radar guns, and targeting. NEXRAD (next generation weather radar) is a Doppler radar system, and so is 3D video detection and ranging (ViDAR).

Dusty plasma: Nanometer- or micrometer-sized dust particles suspended in plasma. Dust particles may be charged and therefore behave as a plasma that can create liquid and crystalline states and plasma crystals.

Effective radiative power (ERP): ERP measures the combination of the power emitted by a transmitter and the ability of the antenna to direct that power in a given direction. The ERP of HAARP is extremely powerful.

EHF: Extremely high frequency whose transmissions travel in a straight line (a beam)

Electrochemical: The relation of electricity to chemical changes and interconversion of chemical and electrical energy

Electrojet: A charging electric current 56–93 miles in the ionosphere; the equatorial electrojet (magnetic equator) and the auroral electrojets of the Arctic and Antarctic are ionospheric fluctuations

Electro-optics: A branch of physics that deals with the effects of an electric field on light traversing it

ELF: Extremely low frequency, 3 to 300 Hz

ENPs: Engineered nanoparticles as opposed to natural nanoparticles

Fullerenes: Carbon spherical nano-molecules; hollow cages of atoms that use silica, carbon, graphene, or other polymer or metallic materials to construct, repair, assemble and mutate at the cell level; utilizes cell DNA replication mutations. Thorium, strontium, barium, aluminum, lithium, nano-silver, styrene, polymers, liposomes, and hydrogels act as transport mechanisms. Buckminsterfullerene = buckyballs.

Gauss: Measurement of magnetic flux density or magnetic induction

G-force: Gravitational force exerted on pilots who jink at close to 500 mph

GPR: Ground-penetrating radar

Graphene: An allotrope of carbon consisting of a single layer of atoms arranged in a hexagonal lattice nanostructure, also known as "black goo"

Gyrotron: High-power linear-beam vacuum tubes that generate millimeter-wave electromagnetic waves by stimulating the cyclotron resonance of electrons in a strong magnetic field

Harmonics: The mathematics that connect wavelengths of matter, gravity, and light

Hertz (Hz): A measure of frequency. 1 Hz wave = 186,000 miles long (the speed of light = 186,000 miles per second). Hertz replaced cycles per second (cps) in the 1960s.

Heterodyne: An engineering term meaning to mix signals

HPM: High-power microwave

Hydroscope: An optical device for viewing objects far below the surface of water

Hygroscopy: A substance's ability to attract and hold water molecules from the surrounding environment. A hygroscope is an instrument showing changes in humidity.

Ice-forming nuclei (INI): Tiny particles that promote the formation of ice crystals; much less abundant than CCN, active only at temperatures well below freezing

IMINT: Imagery intelligence (NGA)

Infrared: A wavelength from 800 nanometers to 1 millimeter, between visible light and microwaves. Far infrared (fIR) radiation has a wavelength of 15 micrometers to 1 millimeter, corresponding to a range of about 20 THz to 300 GHz.

Infrasound: Less than < 20 Hz

Ion: An atom, group of atoms, or molecule that has acquired or is regarded as having acquired a net electric charge by gaining or losing electrons

Ionization: The process by which ions are formed, (1) passing radiation through matter, and (2) heating matter to high temperatures to give the atoms energy and force the electrons to leave the atom to become free electrons (negative charge). Whatever is left of the atom becomes a positive ion.

Ionizing radiation: Radiation of energetic charged particles such as alpha and beta rays, as well as nonparticulate radiation such as X-rays and neutrons

Ionosphere: From the top of the mesosphere (50 miles) to the top of the thermosphere (429 miles)

Interferometry: An interference of waves. HAARP is a scalar interferometer weapon. See appendix 2.

Isotope: Atoms that have the same number of protons and a different number of neutrons

Kilometer: 1 kilometer = 0.621371 miles

Laser: An acronym for *light amplification by stimulated emission of radiation.* A laser is an artificially created very narrow beam of light that can travel a long way and concentrate a very powerful energy on a small area.

LEO: Low earth orbit

LiDAR: An acronym for *light detection and ranging* or *laser imaging detection and ranging.* LiDAR is a remote sensing method that uses light in the form of a pulsed laser to precisely measure ranges (variable distances) and surfaces in three-dimensions.

Magnetofection: A transfection utilizing magnetic fields to concentrate nanoparticles containing vectors to target cells in the body

Magnetosphere: Overlaps the ionosphere and extends into space to 37,280 miles toward the Sun

Magnetron: A barrier microwave technology that operates nonionizing radiation at the extremely high frequency (EHF) end of the spectrum by generating electrons with a cathode-anode cylinder to combine a magnetic field with an electrical field. A large magnetron can generate a consistent EHF beam of microwave pulses equaling 10 million watts per pulse. Microwave ovens have magnetrons.

MASER: Acronym for *microwave amplification by stimulated emission of radiation.* A "microwave laser" device that uses the natural vibrations of atoms or molecules to generate electromagnetic radiation at a single microwave frequency. *Troposcatter* uses microwave transmitters to do what the gyrotron resonance maser does.

MASINT: Measurement and signature intelligence (DIA= Defense Intelligence Agency): surveillance and mind control via satellite

Materials science or physics: A syncretic nanomaterials construction discipline hybridizing metallurgy, ceramics, solid-state physics, and chemistry

MEMS: Acronym for *micro electro mechanical systems.* Engineered sensors gather information and communicate with other nodes in the network. Includes GEMS (global environmental MEMS sensors) and NEMS (nanoelectromechanical systems); BioMEMS are engaged in replacing Human 1.0 cells.

Metamaterial: From the Greek, *meta,* beyond; engineered to have a property that is not found in Nature.

Micron: 1 micrometer = one millionth of a meter (0.000039 inch) or 1,000 nanometers, or 1/70 the thickness of a human hair

Microrobot 3D printing: Nanobots ("functional nanoparticles") that self-propel, chemically power, and magnetically steer are created with 3D printing.

Millimeter (mm): 1 millimeter is one-thousandth of a meter (0.039 inch), the International System Units (SI) base unit

Molecular nanotechnology: Engineered nanoscale machines operating on the molecular level. Example: the *molecular assembler,* a machine that produces structures atom by atom via mechanosynthesis. Molecular machine systems are not the same as the manufacture of nanomaterials like CNTs. Nanosystems installed in human bodies are hybrids of silicon technology and biological molecular machines.

Nanoantennas: Nanoscale engineered structures for sending and transmitting—for example, chlorophyll molecules arranged in antenna complexes or bio-inspired short strands of DNA

Nanohybrid (NH): Multicomponent assemblies in which two or more pre-synthesized nanomaterials are conjugated

Nanometer (nm): 1 nanometer = 1 billionth of a meter (0.000000001 m)

Nanophotonics: Behavior of light on the nanoscale

Nanoplasmonics: Study of optical phenomena in nanoscale metal surfaces

Nanoscience/nanotechnology: Study and application of extremely small things used in other science fields such as chemistry, biology, physics, materials science, and engineering; the ability to work (see, measure, manipulate) at the atomic, molecular, and supramolecular levels in the 1–100 nm range.

NanoTopes: Nanotechnology and radio isotopes

Nanotubes: Rolled up sheets of interlocking carbon atoms, tubes so strong, they could be used to tether satellites to a fixed position above the Earth

Nanowire: 450 atoms wide, silicon nanowire grows from gold catalyst particles after disilane (a silicon-rich gas) liquefies it; a solid silicon crystal forms and grows into a nanowire.

NBIC: Nanotech, Biotech, Info tech, Cognitive tech

NGA: National Geospatial-Intelligence Agency

NEO: Near-earth orbit

Neutrino: A type of scalar wave having to do with internal communication; an electrically neutral subatomic particle with a mass close to zero

Nonionizing: Nonthermal radiation; includes electric and magnetic fields, radio waves, microwaves, and infrared, ultraviolet, and visible radiation

Nuclei: Tiny solid/liquid particles of matter on which condensation or deposition of water vapor takes place; products of forest fires, volcanic eruptions, soil erosions, saltwater spray, industrial discharge, and engineered nanoparticles

Nuclide: Any atomic nucleus specified by atomic number, atomic mass, and energy state

Optogenetics: The use of light-sensitive proteins to stimulate specific neurons in response to light delivered to the brain—for example, via fiber optics

Origami DNA: Manipulated strands of DNA bound into shapes outside the traditional double helix; DNA nanotechnology entails origami computers that fold and unfold strands of DNA (Arun Richard Chandrasekaran and David A. Rusling, "Triplex-forming oligonucleotides: a third strand for DNA nanotechnology," *Nucleic Acids Research*, February 16, 2018).

OTH: Over-the-horizon radar

PM: Particulate matter; particles that are 10 microns or less

ppm: Parts per million

Permittivity: Ability of a substance to store electrical energy in an electric field

Photon: A quantum of electromagnetic energy generally regarded as a particle with no mass or electric charge

Physics: The study of matter and energy, including electricity, magnetism, and electromagnetism

Piezoelectric: Electricity resulting from pressure and latent heat, such as in crystals

Plasma: Ionized (electrically conductive) gas in which atoms have lost electrons and exist in a mixture of free electrons and positive ions; positively charged electrons and ions governed by electric and magnetic forces possessing collective behavior

Polar vortex: A manmade deep freeze: Designer chemicals in dry powder aerosols that produce a *differential buoyancy aerosol spraying (DBAS)* to melt the Arctic, two tanker jets with separate chemicals attack the moisture system rolling in from the South Pacific, one heating with aluminum sulfate, the other cooling with aluminum fluoride.

Positron: Antimatter counterpart of the electron

Programmable matter: Has the ability to change its physical properties (shape, density, moduli, conductivity, optics, etc.) in a programmable fashion, based upon user input or autonomous sensing. (Wikipedia)

Quantum dots (QDs): Colloidal semiconductor nanocrystals with unique optical properties, including fluorescence and the entire color spectrum; discovered in 1980, nicknamed "artificial atoms." QTs tune optical and electronic properties by changing the crystallite size or internal structure; they behave more like organic molecules than metal nanoparticles.

Quark: Any of three hypothetical subatomic particles having electric charges of 1/3 or 2/3 the magnitude of an electron

Qubit: A quantum bit (particle in superposition)

Radar: Acronym for *radio detection and ranging*; a synchronized radio transmitter and receiver that emits radio waves and processes their reflections for display and is used especially for detecting and locating objects

SAR: Acronym for *specific absorption rate of radiation* (usually ionizing), 1.6 W/kg (FCC)

Scalar waves: Natural longitudinal waves as opposed to subnature electromagnetic waves, which are transverse; having only magnitude, not (vector) direction

SDI: Strategic Defense Initiative "Star Wars" program

Sentient World Simulation (SWS): The name given to the SEAS (Synthetic Environment for Analysis and Simulations) "continuously running, continually updated mirror model of the real world that can be used to predict and evaluate future events and courses of action." metaverse, twinning, Eighth Sphere

SIGINT: Signals intelligence

SMART: Secret militarized armaments in residential technology

Spectrometry: Measurement of the interactions between light and matter, and the reactions and measurements of radiation intensity and wavelength

Spectroscopy: Study of the interaction between matter and electromagnetic radiation

Spectrum analyzer: Device for analyzing oscillations, especially sound

SRM: Solar radiation management

Stochastic sensors: Nanosensors that detect DNA, RNA by their frequency signatures

Synergy: Increased intensity caused by combining two or more substances

Telemetry: Recording and transmitting readings of an instrument

Transfection: Method of inserting foreign nucleic acid into eukaryotic cells

Transgenic: Describing an organism containing genetic material into which DNA from an unrelated organism has been artificially introduced

Transmitter: Electronic device that uses radio waves to transmit data by converting energy from the power source into radio waves then sent to receivers

TTWS: Through-the-wall surveillance or sensor

UAP: Unidentified aerial phenomena, aka UFOs

ULF: Ultralow frequency

Ultrasound: > 20,000 Hz

Utility fog (foglets): Micro-mechanical shapeshifters, computer-controlled swarms working together to simulate macro-scale machines. By combining virtual telepresence technologies with nanotech utility fog, consciousness can be remotely projected and we can interact in remote environments through distant, artificial bodies made of utility fog. "Ghost in the machine."

Vacuum: A vacuum is a plenum—pure potential—not an emptiness or nothing.

V2V: Vehicle to vehicle

VLF: Very low frequency (20–35 kHz)

VTRPE: The Variable Terrain Radio Parabolic Equation (VTRPE) is a computer radio frequency propagation program that enables the U.S. Navy's Radio Frequency Mission Planner (RFMP) system to visually view the target battlefield in 3D on a monitor as well as determine vector from scalar radiation wave radar propagation.

Watts: Measure of electrical power. For direct current (DC), watts = volts × amps; for alternating current (AC), calculating watts is more complicated but if using the root mean squared values for voltage and current, watts = volts x amps. Normal U.S. household electricity is 120 volts, so if your appliance draws 2 amps, watts = 120 volts × 2 amps = 240 watts.

Weather: Short-term climate, including temperature, humidity, precipitation, cloudiness, visibility, and wind

X-ray: 0.01 to 10 nm (shorter than UV waves, longer than gamma waves)

Yottabyte: According to NSA expert James Bamford, the yottabyte is equivalent to *septillion bytes*, a number so large that no one has yet coined a term for the next higher magnitude.

Notes

Many of the sources listed below have been subject to censorship. Articles that have been retracted and websites that have become inoperative at the time of publishing have been noted.

INTRODUCTION

1. Kurzweil, Ray, *How to Create a Mind: The Secret of Human Thought Revealed* (New York: Viking Press, 2013), discussing the cross-section of human and artificial brains. Kurzweil also authored *The Age of Spiritual Machines* (New York: Viking Press, 1999).
2. Murkowski, Lisa, "Murkowski Committee Q&A on HAARP Future," Senator Lisa Murkowski website, Press Videos, YouTube, May 14, 2014.
3. Skidmore, Mark, "DOD and HUD $21 Trillion Missing Money," interview by Catherine Austin Fitts, September 28, 2017, Solari Report website; see also, Farrell, Joseph P., *Covert Wars and Breakaway Civilizations: The Secret Space Program, Celestial Psyops and Hidden Conflicts* (Kempton, IL: Adventures Unlimited Press, 2012) positing that a hidden system of finance for a "breakaway civilization" has been in place since World War II so as to solve the Deep State's "triple strategic problem": (1) Nazi survival, (2) Communist bloc, (3) UFOs—all of which lead to the secret space program that geoengineering serves.
4. Fitts, Catherine Austin, "The Black Budget: What Does It Mean to U.S. Federal Budget, the Economy and You?," Secret Space Program Conference, 2014 San Mateo, Solari Report website.

1. ATMOSPHERIC NANOSCIENCE

1. House, T. J., J. B. Near Jr., W. B. Shields, R. J. Celentano, and D. M. Husband, "Weather as a Force Multiplier: Owning the Weather in 2025," U.S. Department of Defense, Defense Technical Information Center website, August 1, 1996; David, Leonard, "Military Wants the Weather on Its Side," NBC News website, October 31, 2005.
2. See the works of J. Marvin Herndon on the Nuclear Planet website.
3. Herndon, J. Marvin, "Evidence of Coal-Fly-Ash Toxic Chemical Geoengineering in the Troposphere: Consequences for Public Health," *International Journal of Environmental Research and Public Health* 12, no. 8 (2015): 9375–90. This article has since been retracted; see Herndon's response to the retraction, "Concerted Efforts to Unwarrantedly Cause Retractions of Peer-Reviewed and Published Public Health Scientific Papers" on the Nuclear Planet website.
4. Watson, Steve, "CIA Is Funding Government-Led Chemtrail Project: Spy Agency to Help Study 'Security Impacts' of Geo-engineering," Global Research website, July 20, 2013.
5. "Project Cloverleaf Timeline, 1994 to 2001," Biblioteca Pleyades website, formerly on IndyMedia website (no longer active), March 2004.
6. Manobianco, John, "Global Environmental MEMS Sensors (GEMS): A Revolutionary Observing System for the 21st Century Phase I Final Report," prepared for NASA Institute for Advanced Concepts (Cocoa Beach, FL: ENSCO, 2002).
7. Sutter, John D., "'Smart Dust' Aims to Monitor Everything," CNN website, May 3, 2010.
8. Kautz-Vella, Harald, and Kristin Hauksdottir, "The Chemistry in Contrails: Assessing the Impact of Aerosols from Jet Fuel Impurities, Additives and Classified Military Operations on Nature," (lecture at the Open Mind Conference, Oslo, October 27, 2012), Academia.edu website.
9. Becker, Robert O., *The Body Electric: Electromagnetism and the Foundation of Life* (New York: William Morrow, 1998).
10. Searle, Mike, "Australian Student Confirms That Giant Plasma Tubes Are Floating above Earth," Before it's too Late website, June 7, 2015, originally published as Chris Pash, "There Are Giant Plasma Tubes Floating Above Earth," *Business Insider*, March 29, 2018; see also, CAASTRO "Cosmic Cinema: Astronomers Make Real-Time 3D Movies of Plasma Tubes Drifting Overhead," YouTube website, May 31, 2015.

11. Zinkova, Mila, "Rare and Unusual (The Novaya Zemlya Effect) Sunset Mirage with Green Flashes and Whales Spouts," YouTube website, September 24, 2016.
12. Crew, Bec, "NASA Space Probes Have Detected a Human-made Barrier Surrounding Earth," ScienceAlert website, May 18, 2017.
13. Tanenbaum, Michael, "U.S. Air Force Wants Drexel to Help Plasma Bomb the Skies," Philly Voice website, August 9, 2016.
14. Hodgkins, Kelly, "The U.S. Air Force Wants to Detonate Plasma Bombs in the Sky," Digital Trends website, August 12, 2016.
15. "Plasma Energy Space Station Continues," (video) *BPEArthWatch* website, August 11, 2017, detailing what may be a plasma generator on the International Space Station (ISS) creating what appear to be plasma orbs seen guiding exotic propulsion craft and cruising around directed energy weapon (DEW) targeted individuals.
16. University of California–San Diego, "UCSD Researchers Fabricate Tiny 'Smart Dust' Particles," UCSD press release, EurekAlert website, September 2, 2002.
17. Stanley, Jay, "The Last Mile to Civilization 2.0: Technologies from Our Not too Distant Future," TechSpot website, December 13, 2017.
18. Rowinski, Dan, "Connected Air: Smart Dust Is the Future of the Quantified World," ReadWrite website, November 14, 2013; and Mukbopadbyay, Rajendrani, "Magnetic Dust Mobilizes Droplets," *Analytical Chemistry* 77, no. 3 (2005): 55A.
19. Spry, Jeff, "Amazing Magnetic Spray Turns Tiny Inanimate Objects into Insect-Scale Robots," SyFy website, November 21, 2020.
20. Lee, Jamie (Aplanetruth3), "Paradise Lost #51: Fire Embers Are Programmable Swarming Nanobots," YouTube website, July 28, 2019.
21. Lawrence, Cate, "Is Smart Dust the IoT Vector of the Future?," ReadWrite website, August 20, 2016.
22. Thomas, Liji, "Mesh Electronics Could Make Brain Stimulation the New Therapeutic Norm," News–Medical Life Sciences website, September 6, 2019; Patel, Shaun R., and Charles M. Lieber, "Precision Electronic Medicine in the Brain," *Nature Biology* 37, no. 9 (September 2019): 1007–12.
23. Freeman, Makia, "Nanochips and Smart Dust: The Dangerous New Face of the Human Microchipping Agenda," Activist Post website, September 23, 2023; Farrell, Joseph P., "The Latest in Mind Manipulation Technology: Neural Smart Dust . . .?" The Giza Death Star website, August 7, 2016.

24. Watson, Robin D. P., "Nano Blenders Thesis, Parts 1 & 2," Academia website, November 17, 2015. For how vibration/spinning/rotation works, see de Andrade, V. C., and J. G. Pereira, "Torsion and the Electromagnetic Field," *International Journal of Quantum Physics* 8, no. 2 (1999): 141–51. The spin of nanoparticles (particularly the paramagnetic) can create electrical charges discharged as micro-lightning or micro-arcing that damages DNA, mitochondria, etc.
25. Watson, Rob, "Illuminati Plan to Fry Humanity?," HenryMakow website, August 25, 2016.
26. Watson, Robin D. P., "Nano Blenders Thesis Part 2: Nanoparticle Excoriation Promotes Thrombogenesis—A Possible Pathophysiologic Mechanism," Academia website, November 17, 2015.
27. Hodson, Hal, "20 Billion Nanoparticles Talk to the Brain Using Electricity," New Scientist website, June 8, 2015; Guduru, R., P. Liang, J. Hong, A. Rodzinski, A. Hadjikhani, J. Horstmyer, E. Levister, and S. Khizroev, "Magnetoelectric 'Spin' on Stimulating the Brain," *Nanomedicine* 10, no. 13 (2015): 2051–61.
28. University of Alaska Fairbanks, "About HAARP," University of Alaska Fairbanks HAARP website, accessed July 10, 2024.
29. U.S. Department of Defense, "Electromagnetic Spectrum Superiority Strategy," October 2020.
30. Begich, Nick, and Jeane Manning, *Angels Don't Play This HAARP: Advances in Tesla Technology* (Anchorage, AK: Earthpulse Press, 1995, 2002). I go more into detail in my earlier books, as well.
31. Zhang, N., et al., "A Detection Performance Analysis of Sanya Incoherent Scatter Radar Tristatic System," *Radio Science* 56, no. 5 (2021).
32. Vey, Gary, "The Shape of the Future," View Zone website, April 4, 2012; Vey, Gary, "The Great Grid of China: A Technological Wonder of the World," April 12, 2012, describes hundreds of square miles of buried fiber optic cable laid out in fractal patterns in north China; Dockrill, Peter, "China and Russia Have Run Controversial Experiments That Modified Earth's Atmosphere," ScienceAlert website, December 19, 2018.
33. Cooper, Charles, "Collapse in Earth's Upper Atmosphere Stumps Researchers," CBS News website, July 16, 2010.
34. Keith, David, "Photophoretic Levitation of Engineered Aerosols for Geoengineering," *PNAS* 107, no. 38 (2010): 16428–31. In the article he references the "magnetic materials" that we are breathing.

35. Eastlund, Bernard J., U.S. Patent 4686605, issued August 11, 1987.
36. Fontenot, Christopher, "Phased Array Antenna (HAARP) Breaks the Rules," A Microwaved Planet website, May 18, 2016 (no longer available).
37. Carnicom Institute, "Clifford E. Carnicom," Carnicom Institute website.
38. Carnicom, Clifford, "Atmospheric Conductivity II," Carnicom Institute website, May 7, 2003; Carnicom, Clifford, "Atmospheric Conductivity," Carnicom Institute website, July 9, 2001.
39. See Watson, Steve, "CIA Is Funding Government-Led Chemtrail Project: Spy Agency to Help 'Security Impacts' of Geo-engineering," Global Research website, July 20, 2013.
40. Carnicom, Clifford, "Air Force Spokesman Is 'Master Intelligence Officer,'" Carnicom Institute website, June 29, 2005.
41. Freeland, Elana, *Under an Ionized Sky* (Port Townsend, WA: Feral House, 2018) goes into great detail about the construction of the Space Fence, thanks to Billy "The HAARP Man" Hayes, who worked as a tower erecter at the 260 sites around the world dedicated to Space Fence infrastructure.
42. Crew, Bec, "There's a Detectable Human-Made Barrier Surrounding Earth," ScienceAlert website, April 5, 2018.
43. For background on Enron, read *Fortune* reporters Bethany McLean and Peter Elkind, *The Smartest Guys in the Room* (New York: Penguin 2004) or Palast, Greg, *The Best Democracy Money Can Buy* (New York: Penguin, 2003); or watch the 2005 documentary *Enron: The Smartest Guys in the Room.*
44. Freeland, Elana, *Chemtrails, HAARP, and the Full Spectrum Dominance of Planet Earth* (Port Townsend, WA: Feral House, 2014).
45. Lockheed Martin, "Space Fence: The World's Most Advanced Radar," Lockheed Martin website.
46. Lee, James Franklin, Jr., "The Russian Woodpecker, Chernobyl Meltdown, and Ionospheric Heating over the USA (1983–1986)," ClimateViewer News (blog), May 4, 2018.
47. Carnicom, Clifford, "Potassium Interference Is Expected," Carnicom Institute website, May 15, 2005.
48. See Pall, Martin, "5G: Great Risk for EU, U.S. and International Health! Compelling Evidence for Eight Distinct Types of Great Harm Caused by Electromagnetic Field (EMF) Exposures and the Mechanism That Causes Them," Peace in Space (blog), May 17, 2018.

49. See Kinjo, T. G., and P. P. M. Schnetkamp, "Ca2+ Chemistry, Storage and Transport in Biologic Systems: An Overview," Madame Curie Bioscience Database (online) (Austin, TX: Landes Bioscience, 2000–2113).
50. Buckley, Julia, "Transatlantic Airplanes Are Flying at the 'Speed of Sound' Right Now. Here's Why," CNN website, November 1, 2023. Mach 1 is 717 miles per hour.
51. Nordrum, Amy, "Arctic Fibre Project to Link Japan and U.K.," IEEE Spectrum website, December 29, 2014.
52. European Space Agency, "Scientists Discover Massive Jet Streams Flowing Inside the Sun," Press Release, European Space Agency website, August 29, 1997.
53. Kincaid, Paul W., "USAF Redeploys X-37B Climate Chaos, Earthquake & Tsunami Inducing Orbital HAARP Weapon," Liberty Beacon website, October 16, 2015.
54. Kotlikoff, Laurence, "Is Our Government Intentionally Hiding $21 Trillion in Spending?," *Forbes*, July 21, 2018; also see the Secret Space Program Conference 2015. For the "breakaway civilization," see Farrell, Joseph P., *Covert Wars and Breakaway Civilizations: The Secret Space Program, Celestial Psyops and Hidden Conflicts* (Kempton, IL: Adventures Unlimited Press, 2012). Farrell is an Oxford-educated historian who specializes in alternative history, World War II, and Secret Technologies.
55. Schulte, David Joseph, U.S. Patent 20130015260-A1, "Concept and model for utilizing high-frequency or radar or microwave producing or emitting devices to produce, effect, create or induce lightning or lightspeed or visible to naked eye electromagnetic pulse or pulses, acoustic or ultrasonic shockwaves or booms in the air, space, enclosed, or upon any object or mass, to be used solely or as part of a system, platform or device including weaponry and weather modification," application October 7, 2004, issued July 22, 2014.
56. Chow, Denise, "Swirling Mass of Plasma Raining Electrons Observed above Earth for First Time," NBC News, Yahoo website, March 4, 2021.
57. Grey, Eva, "Battlefield 2050: Direct Energy Weapons Meet the Forcefield," Army Technology website, May 10, 2016.
58. Science Mission Directorate, "The Earth's Radiation Budget," NASA Science website, 2010.
59. Fontenot, Christopher, A Microwaved Planet website (no longer active). Fontenot is a former U.S. Navy nuclear propulsion electrical engineer's mate who held a secret security clearance.

60. Whale, T. F., M. Rosillo-Lopez, B. J. Murray, and C. G. Salzmann, "Ice Nucleation Properties of Oxidized Carbon Nanomaterials," *Journal of Physical Chemistry Letters* 6 (2015): 3012–16; see The HAARP Report, "Polar Vortex Secret Disclosure, Best Science, DBAS Aerosol Spray," YouTube website, November 12, 2019, for one of the best videos on how geoengineered weather is made to work.
61. Quadrelli, Marco, "Orbiting Rainbows: Optical Manipulation of Aerosols and the Beginnings of Future Space Construction," NASA website, August 28, 2014. (Note "future space construction"—reference to the secret space program?)
62. Quadrelli, "Orbiting Rainbows."
63. Ebner, Daniel M., "The 'Primeval Code': The Ecological Alternative to the Controversial Genetically Engineered Seeds of the International Agro-multinationals" (lecture, World Mysteries Forum, Basel, Switzerland, 2008), Rex Research website.
64. Ebner, Daniel, "'Primeval Code' Reactivated: Living Archetypes of Plants and Animals Created at the Laboratory," Rex Research website.
65. Manley, Scott, "Rare Lightning Traveling from Ground to Clouds in Slow Motion," YouTube website, August 17, 2020.
66. HeartMath Institute, "Global Coherence Research: The Science of Interconnectivity," HeartMath Institute website.
67. Gargurevich, Ivan A, "Future Strategic Issues/Future Warfare [Circa 2025] from NASA Langley Research Center," Academia.edu website, 2001.
68. Mack, Eric, "Earth's Magnetic North Pole Is Shifting Dramatically from a Powerful Tug of War," *Forbes* website, May 7, 2020.
69. University College London, "Magnetic Rope Observed for the First Time between Saturn and the Sun," UCL Mathematical and Physical Sciences website, July 6, 2016.

2. THOSE WHITE LINES IN THE SKY

1. Ritchie, G., K. Still, J. Rossi 3rd, M. Bekkedal, A. Bobb, and D. Arfsten, "Biological and Health Effects of Exposure to Kerosene-Based Jet Fuels and Performance Additives," *Journal of Toxicology Environmental Health* 6, no. 4 (2003): 357–451; see also Lee, Jim, "Geoengineering and Weather Modification Exposed," Climate Viewer News website, January 3, 2013, updated January 31, 2020.

2. Schactman, Noah, "Legendary CIA Airline Now in Danger of Crashing," Wired magazine website, June 2, 2011.
3. Hale, John C., U.S. Patent US781962B2, "Enhanced aerial delivery system," issued October 26, 2010.
4. See Huff, Ethan, "Carbon Dioxide Levels Have Nothing to Do with Global Temperatures, Top Scientist Says," Natural News website, September 30, 2022; Richard, Kenneth, "Physicist: CO2 Retains Heat Only 0.0001 Seconds, Warming 'Not Possible,'" Climate Change Dispatch website, October 18, 2019.
5. Mawdsley, Stephen E., "Burden of Proof: The Debate Surrounding Aerotoxic Syndrome," *Journal of Contemporary History* 57, no. 4 (2022): 959–74.
6. Nogreenpass Palermo, "Geoingegneria: intervista ad Enrico Giannini," YouTube website, July 2, 2022.
7. Schweber, Bill, "Aerosol Delivers 5G as Mist, Solves Countless Technical Problems," Electronic Design website, March 31, 2020 ("easy-to-manufacture nanoparticles in the shape of the 5G designation, internally code-named 5G-hype nanoparticles . . .").
8. Hardell, L., and M. Carlberg, "Health Risks from Radiofrequency Radiation, Including 5G, Should Be Assessed by Experts with No Conflicts of Interest," *Oncology Letters* 20, no. 4 (2020): 15.
9. Thompson, Avery "Space Startup Apollo Fusion Wants to Use Mercury as a Fuel," *Popular Mechanics* website, November 20, 2018.
10. Awake Souls, "Fuel Hoax Airbus A380 Exposed Free Energy Truth," YouTube website, February 8, 2018.
11. Cook, Nick, *The Hunt for Zero Point: Inside the Classified World of Antigravity Technology* (New York: Broadway Books, 2001); also see Tugmakvu, Muang'Akili, "Who Was Viktor Schauberger?," Muangakili Medium website, September 16, 2021.
12. NASA Technology, "Aircraft Geared Architecture Reduces Fuel Cost and Noise," NASA *Spinoff* magazine, 2015, 40.
13. Forsythe, Michael, and Ronen Bergman, "To Evade Sanctions on Iran, Ships Vanish in Plain Sight," *New York Times* website, July 2, 2019.
14. See Gorag, Istvan, U.S. Patent 3630594, "Holographic scan converter," issued December 28, 1971; De, Lang H., U.S. Patent 3810687A, "Apparatus for reconstructing an image of an object which image has been recorded in holographic form," issued May 14, 1974; Velzel, Christiaan, U.S. Patent 3653736A, "Holographic multiple image formation with astigmatism correction," issued April 4, 1972.

15. "15,000 Planes, 1 Image: Stunning Satellite Map Shows Jet Signals Worldwide," RT news website, May 8, 2015.
16. "How They Plan to Control Everything in Your Life," Patriots for Truth website, January 12, 2018.
17. Galeon, Dom, "Mind Control: New System Allows Pilots to Fly Planes Using Their Brainwaves," Futurism website, November 22, 2016, updated November 18, 2016.
18. DeMott, Paul J., et al., "Sea Spray Aerosol as a Unique Source of Ice Nucleating Particles," *Proceedings of the National Academy of Sciences*, December 17, 2015.
19. Gopalakrishnan, V., et al., "Intermediate Ion Formation in the Ship's Exhaust," *Geophysical Research Letters* 32 (May 2005): L11806.
20. Gopalakrishnan, et al., "Intermediate Ion Formation in the Ship's Exhaust."
21. The science-minded might want to read Nagoya University, "Wave-Particle Interactions Allow Collision-Free Energy Transfer in Space Plasma," Phys.org website, September 19, 2018.
22. Tracy, S. M., J. M. Moch, S. D. Eastham, and J. J. Buonocore, "Stratospheric Aerosol Injection May Impact Global Systems and Human Health Outcomes," *Elementa* 10, no. 1 (2022): 00047.
23. Palit, Carolyn Williams, "What Chemtrails Really Are," November 9, 2007, Gang Stalking, Mind Control, Cults website, posted November 3, 2017.
24. Narayanan, S. N., R. Jetti, K. K. Kesari, R. S. Kumar, S. B. Nayak, and P. G. Bhat, "Radiofrequency Electromagnetic Radiation-Induced Behavioral Changes and Their Possible Basis," *Environmental Science and Pollution Research* 26 (2019): 30693–710.
25. See Pall, Martin, "5G: Great Risk for EU, U.S. and International Health! Compelling Evidence for Eight Distinct Types of Great Harm Caused by Electromagnetic Field (EMF) Exposures and the Mechanism That Causes Them," Peace in Space (blog), May 17, 2018.
26. Kıvrak, E. G., K. K. Yurt, A. A. Kaplan, I. Alkan, and G. Altun, "Effects of Electromagnetic Fields Exposure on the Antioxidant Defense System," *Journal of Microscopy and Ultrastructure* 5, no. 4 (2017): 167–76.
27. Gill, Ruth, "Biological Warfare with Psychotronic Weapons," Psychophysical-Torture website; also see Karlstrom, Eric, "'Microwave Harassment and Mind-Control Experimentation, 1994 (The McKinney Report),' by Julianne McKinney," 9/11 New World Order website, February 25, 2014.

28. Wang, Huanhua, et al. "Engineered Nanoparticles May Induce Genotoxicity," *Environmental Science & Technology* 51, no. 9 (2013): 4831–40.
29. Wald, Matthew L., "Uranium Leak at Tennessee Laboratory Brings Fears of an Accidental Chain Reaction, *New York Times* website, November 25, 1994.
30. Carnicom, Clifford, "The Expected Composition," Carnicom Institute website, March 28, 2002.
31. Carnicom, Clifford, "Atmospheric Magnesium Disclosed," Carnicom Institute website, June 10, 2001.
32. Witt, George, "Noctilucent Cloud Observations," *Tellus* 9, no. 3 (1957): 365–71.
33. "Water Fluoridation Data and Statistics," Centers for Disease Control website.
34. NASA, U.S. Patent 3813875A, "Rocket having barium release system to create ion clouds in the upper atmosphere," issued June 4, 1974.
35. Temple, James, "Geoengineering Researchers Have Halted Plans for a Balloon Launch in Sweden," MIT Technology Review website, March 31, 2021.
36. Azurdia, J. A., J. Marchal, P. Shea, H. Sun, X. Q. Pan, and R. M. Laine, "Liquid-Feed Flame Spray Pyrolysis as a Method of Producing Mixed-Metal Oxide Nanopowders of Potential Interest as Catalytic Materials," *Chemistry of Materials* 18, no. 3 (2006): 731–39; also see "Air Pharmacology II: Spray Pyrolysis and Chemi-ionization," in my book *Under an Ionized Sky*, 65–72.
37. Edwards, Lin, "New Method of Incorporating Fluoride into Drugs," Phys.org website, September 6, 2013.
38. Proudfoot, A. T., S. M. Bradberry, and J. A. Vale, "Sodium Fluoroacetate Poisoning," *Toxicology Review* 25, no. 4 (2006): 213–19.
39. Fluoride Action Network, "TSCA Fluoride Lawsuit," Fluoride Alert website.
40. Fluoride Action Network, "Sources of Fluoride: Pesticides," Fluoride Alert website.
41. Fluoride Action Network, "EPA Lawsuit," Fluoride Alert website.
42. Wang, Zhang Lin, et al., "On the Electron Transfer Mechanism in the Contact-Electrification Effect," *Advanced Materials* 30, no. 15 (2018): 1706790.
43. McKeown, L. A., "German COVID-19 Autopsy Data Show Thromboembolism, 'Heavy' Lungs," tctMD website, May 11, 2020.
44. Hirai, Toshiro, et al., "Amorphous Silica Nanoparticles Size-Dependently Aggravate Atopic Dermatitis-like Skin Lesions Following an Intradermal Injection," *Particle and Fibre Toxicology* 9, no. 3 (Feb. 2012).
45. American Institute of Physics, "For Future Chips, Smaller Must also Be Better," Science Daily website, Oct. 18, 2010.

46. Kautz-Vella, Harald, and Kristin Hauksdottir, "The Chemistry in Contrails: Assessing the Impact of Jet Fuel Impurities, Additives and Aerosols from Classified Military Operations in Nature" (lecture at the Open Mind Conference, Oslo, Norway, October 27, 2012), Academia.edu website.
47. Johnson, Dexter, "Spray-on Nanoparticles Mix Turns Trees into Antennas," IEEE Spectrum website, February 16, 2012 (noting, "The technology was originally intended for military applications").
48. "What Coltan Mining in the DRC Costs People and the Environment," The Conversation website, May 29, 2022; also see Vialls, Joe, "Operation Crimson Mist: Electronic Slaughter in Rwanda," May 29, 2003, Gang Stalking, Mind Control, posted July 16, 2020. Vialls died in 2005 in hospital at sixty-one of a "heart attack."
49. Miridzhanian, Anait, "Conflict Uproots Record 6.9 Million People in Congo–IOM," Reuters website, October 30, 2023.
50. Burks, Robin, "The Metals Used to Make Smartphones Could Run Out Soon," Tech Times website, March 26, 2015.
51. Korybko, Andrew, "Lithium, a Strategic Resource: Here's Why the U.S. Wants to Break Bolivia to Bits with Hybrid War," Global Research website, October 26, 2019; also see Roth, Sammy, "Drilling for 'White Gold' Is Happening Right Now at the Salton Sea," *Los Angeles Times* website, November 15, 2021.
52. Rao, Joe, "NASA Rocket to Spark Light Show over U.S. East Coast Tonight," Space website, January 29, 2013; Fillmore, Ann, "Aerosol Experiments Using Lithium and Psychoactive Drugs over Oregon," Positive Health Online website, February 2016.
53. Trafton, Anne, "New Clue to How Lithium Works in the Brain," MIT News website, July 7, 2016.
54. Machado-Vieira, R., H. K. Manji, and C. A. Zarate Jr., "The Role of Lithium in the Treatment of Bipolar Disorder: Convergent Evidence for Neurotrophic Effects as a Unifying Hypothesis," *Bipolar Disorder* 11 (Suppl 2) (2009): 92–109.
55. Herndon, J. Marvin, "Human and Environmental Dangers Posed by Ongoing Global Tropospheric Aerosolized Particulates for Weather Modification," *Frontiers in Public Health* 4, no. 139 (2016), article retracted by publisher; Herndon replies, "The author considers the retraction to be unwarranted and therefore does not agree." Also see my book, *Under an Ionized Sky* (Feral House, 2018), 60–63, for more on Herndon's work. All of Herndon's papers are listed at Nuclear Planet website.

56. Herndon, "Human and Environmental Dangers."
57. Hvistendahl, Mara, "Coal Fly Ash Is More Radioactive Than Nuclear Waste," *Scientific American* website, December 13, 2007.
58. Nelson, Loren, U.S. Patent 3659785-A, "Weather modification utilizing microencapsulated material," issued May 2, 1972.
59. Shaw, C. A., S. Seneff, S. D. Kette, L. Tomljenovic, J. W. Oller Jr., and R. M. Davidson, "Aluminum-Induced Entropy in Biological Systems: Implications for Neurological Disease," *Journal of Toxicology*, October 2, 2014: 491316 (epub).
60. Pakrashi, Sunandan, et al., "*Ceriodaphnia dubia* as a Potential Bio-Indicator for Assessing Acute Aluminum Oxide Nanoparticle Toxicity in Fresh Water Environment," *Plos One* 8, no. 9 (September 5, 2013): e74003.
61. Herndon, "Human and Environmental Dangers."
62. Galalae, Kevin, "Water, Salt, Milk Killing Our Unborn Children," Slide Share website, September 21, 2012. Galalae is a multibook author whose bio states: "Galalae is a Canadian human rights activist, author, historian, journalist and the world's foremost independent authority on covert geopolitical programs and policies with respect to globalization and depopulation."
63. John's Hopkins Medicine, "Early-Onset Alzheimer's Disease," Hopkins Medicine website.
64. Exley, C., and E. Clarkson, "Aluminum in Human Brain Tissue from Donors without Neurodegenerative Disease: A Comparison with Alzheimer's Disease, Multiple Sclerosis, and Autism," *Scientific Reports* 10, no. 1 (2020): 7770.
65. Exley and Clarkson, "Aluminum in Human Brain Tissue"; Mold, M., D. Umar, A. King, and C. Exley, "Aluminum in Brain Tissue in Autism," *Journal of Trace Elements in Medicine and Biology* 46 (2018): 76–82. This subject is covered extensively by Robert F. Kennedy Jr.'s organization, Children's Health Defense.
66. Exley and Clarkson, "Aluminum in Human Brain Tissue." Christopher Exley is an expert in the effect of aluminum on the human body.
67. Sifferlin, Alexandra, "Doctor Who Opposed Vaccines Found Dead in Apparent Suicide," *Time* magazine website, June 27, 2015.
68. Davis, Iain, "GcMAF and the Persecution of David Noakes, Lyn Thyer & Immuno Biotech, UKColumn website, May 28, 2019.
69. Smits, Jeanne, "Accomplished Pharma Prof Thrown in Psych Hospital after Questioning Official COVID Narrative," LifeSiteNews website, December 11, 2020.

70. "Growing List of Assassinations Of COVID-19 Researchers and the Mysterious Deaths of Two African Presidents Who Called Covid a Hoax," Humans Are Free website, Dec. 21, 2020. No longer accessible.
71. Jack Murphy, "Doctors Who Discovered Cancer Enzymes in Vaccines All Found Murdered." Neon Nettle website, February 10, 2016. No longer accessible.
72. Fight4Freedom, "What's Really Going on? Dietrich Klinghardt," YouTube website, January 26, 2020.
73. Freeman, Makia, "Aluminum, Fluoride, Glyphosate and EMF: The Deliberate Concoction to Shut You Down," The Freedom Articles website, May 19, 2020.
74. Department of Energy, Lawrence Berkeley National Laboratory, "Scientists Discover New Pathway to Forming Complex Carbon Molecules in Space," SciTechDaily website, September 8, 2019.
75. Department of Energy, Lawrence Berkeley National Laboratory, "Scientists Discover New Pathway to Forming Complex Carbon Molecules in Space"; Zhao, Long, et al., "Molecular Mass Growth through Ring Expansion in Polycyclic Aromatic Hydrocarbons via Radical-Radical Reactions," *Nature Communications* 10 (August 15, 2019): 3689.
76. Birnbaum, Michael, "The World's Biggest Plant to Capture CO2 from the Air Just Opened in Iceland," *Washington Post* website, September 8, 2021.
77. Castle, Mike, "Chemtrails—Bio-Active Crystalline Cationic Polymers," Rense website, July 14, 2003.
78. Freeland, Elana, *Geoengineered Transhumanism: How the Environment Has Been Weaponized by Chemicals, Electromagnetism & Nanotechnology for Synthetic Biology* (self-pub., 2021).
79. Gray, William M., "Feasibility of Beneficial Hurricane Modification by Carbon Black Seeding," Department of Atmospheric Science, Colorado State University, MountainScholar website, April 1973.
80. Heller, Michael J., and Richard H. Tullis, "Self-Organizing Molecular Photonic Structures Based on Functionalized Synthetic Nucleic Acid (DNA) Polymers," *Nanotechnology* 2 (1991): 165–71.
81. Carlsson, Jan-Otto, and Peter M. Martin, "Chemical Vapor Deposition," chap. 7 in *Handbook of Deposition Technologies for Films and Coatings*, 3rd ed, edited by Peter Martin (Amsterdam: Elsevier, 2010).
82. Bernstein, Michael, "Helping the Carbon Nanotube Industry Avoid Mega-mistakes of the Past," American Chemical Society press release, ACS website, August 20, 2007.

83. Lauridsen, M. J., and B. C. Ancell, "Nonlocal Inadvertent Weather Modification Associated with Wind Farms in the Central United States," *Advances in Meteorology*, August 6, 2018: 2469683 (epub); Reilly, John, "Too Much Wind and Solar Raises Power System Costs. Deep Decarbonization Requires Nuclear," Utility Dive website, December 2, 2019.
84. Cohen, Bonner, "Chinese-Owned Wind Farm in Devils River, Texas, Threatens Power Grid and More," CFACT website, August 18, 2020. Dr. Cohen is a policy adviser with the Heartland Institute, a senior policy analyst with the Committee for a Constructive Tomorrow, and an adjunct scholar at the Competitive Enterprise Institute.

3. THEY DO IT WITH SMOKE AND MIRRORS

1. Menzel, D. H., and E. H. Taves, *The UFO Enigma: The Definitive Explanation of the UFO Phenomenon* (Garden City, NY: Doubleday, 1977).
2. Johnstone, A. K., *UFO Defense Tactics: Weather Shield to Chemtrails* (Hancock House, 2002).
3. Johnson, Dexter, "Spray-on Nanoparticles Mix Turns Trees into Antennas," IEEE Spectrum website, February 16, 2012.
4. U.S. National Archives and Records Administration, "Military Records: Project BLUE BOOK—Unidentified Flying Objects," National Archives website.
5. Wilson, R. Mark, "Mapping Magnetite in the Human Brain," Physics Today website, August 30, 2018.
6. "The Emergence of Maitreya the World Teacher," Share International website.
7. Swehat, Yassin, "'Clients' in Syria and Egypt: How Economic Openness Keeps Authoritarian Elites in Power," *Al-Jumhuriya*, Assafirafabi website, November 18, 2022.
8. News of the Weird, "News of the Weird for December 29, 1998," UExpress website, December 29, 1998.
9. YogaEsoteric, "Project Blue Beam: The Technology for Mass Holographic Deception and Psychological Manipulation Has Existed for Decades," YogaEsoteric website, May 21, 2020.
10. Goodman, Jason, "Did Serco ConAir Patents Cause California Wildfires via CIA Nanowaves? Special Guest David Hawkins," Alt Censored website, November 28, 2018. Hawkins is a forensic economist with a Cambridge

University background in the Science of Waste and Chaos. For more on the jack-of-all-trades Serco, see appendix 8, "Movers and Shakers."

11. Firstenberg, Arthur, "One Million Satellites Planned," Cell Phone Task Force website, October 17, 2023.
12. Keller, John, "Air Force Asks Industry to Develop New Electro-optical and Electronic Materials for Electronic Warfare (EW)," Military Aerospace Electronics website, February 7, 2023.
13. Brooks, Michael, "Is Quantum Physics behind Your Brain's Ability to Think?" New Scientist website, December 2, 2015.
14. Giles, Martin, "Explainer: What Is a Quantum Computer?," MIT Technology Review website, January 29, 2019.
15. Lahey, Susan, "Quantum Computers Could Be a 'Superhighway' to Experiencing Our Other Selves in the Multiverse," *Popular Mechanics* website, May 12, 2023.
16. "JASON (advisory group)," Wikipedia website, accessed July 7, 2024.
17. Goldberg, Carey, "Heaven's Gate Fit in with New Mexico's Offbeat Style," *New York Times* website, March 31, 1997.
18. Jeff P., "DEW Weaponized Sun Simulator News 24 Caught in Action Paradise Fire Dutchsinse," YouTube website, January 4, 2018; Look for THE Life of Love, "Finally Undisputable Proof of Sun Simulator—UK Nov 2018—Jeff P channel," YouTube website, November 17, 2018; Jeff P., "Cell Phone Captures 2 Suns/Flickering Sun/Sun Simulator/Jeff P," YouTube website, October 20, 2021; Rayces, Juan Luis, U.S. patent 3247367A, "Solar simulator," issued April 19, 1966.
19. See Browning, Oliver, "Two Bright 'Suns' Seemingly Appear in Sky over China in Rare Phenomenon," *Independent* website, July 26, 2023.
20. "Xenon Arc Lamp," Wikipedia website, accessed July 7, 2024.
21. Chen, Stephen, "China's Sun Simulator Sheds New Light on the Ravages of Space Travel," SCMP website, July 13, 2016.
22. Jeffrey, Colin, "Ultrathin Metasurface Lenses Do Things Conventional Optics Can't," New Atlas website, September 8, 2015.
23. Rayces, U.S. patent 3247367A, "Solar simulator."
24. Allen, Laura, "New Device Can Harvest Clean Energy from Humid Air Anywhere," Science News Explores website, August 18, 2023.
25. "Carbon capture, utilization & storage [CCUS]," Aramco website, accessed July 7, 2024.

26. "Rectenna," Wikipedia website, accessed July 7, 2024.
27. Brown, Hannah, "A Massive Floating Solar Farm Orbiting in Space Could Soon Be Reality. Here's How It Would Work," Euro News website, October 13, 2022.
28. Wang, Ucilia, "Solaren to Close Funding for Space Solar Power," GreenTechMedia website, December 4, 2009.
29. FTI Consulting, "PG&E: Restoring Power, Restoring a Company," FTI Consulting website, January 14, 2022.
30. Egan, Matt, "Secretive Energy Startup Backed by Bill Gates Achieves Solar Breakthrough," CNN Business website, November 19, 2019.
31. Solaren, "Solaren Space Solar Power Overview," Solaren website, accessed July 7, 2024.
32. Jeff P., "DEW Weaponized Sun Simulator News 24 Caught in Action Paradise Fire Dutchsinse," YouTube website, January 4, 2018.
33. Universal News Media, "Enormous Crater in Red Planet on Alaskan FAA Weathercam. Planets Are Closing In," YouTube website, March 21, 2018.
34. Universal News Media, "Red Beam Projected from the Sun, Caught on FAA Weathercams. More Planets, Eclipses," YouTube website, March 13, 2018.
35. Universal News Media, "What Are These Round, Rimmed, Concave Disks Seen on Many FAA Weathercams. What Is It?," YouTube website, March 24, 2018.

4. GEOENGINEERED "WILDFIRES," DEWS, AND THE CLIMATE CRISIS DECEPTION

1. Peretz, R., E. Sterenzon, Y. Gerchman, V. Kumar Vadivel, T. Luxbacher, and H. Mamane, "Nanocellulose Production from Recycled Paper Mill Sludge Using Ozonation Pretreatment Followed by Recyclable Maleic Acid Hydrolysis," *Carbohydrate Polymers* 216 (2019): 343–51.
2. Lauri, Arto, "Arto Lauri 169: Latest Chemtrail," YouTube website, February 24, 2016.
3. Johnson, Dexter, "Spray-on Nanoparticles Mix Turns Trees into Antennas," IEEE Spectrum website, February 16, 2012; Irving, Michael, "Sprayable Antennas Turn Surfaces into Ultra-thin, Transparent Transmitters," New Atlas website, September 24, 2018.
4. California Department of Forestry and Fire Protection, "2017 Incident Archive," Fire.Ca.gov website.

5. California Department of Forestry and Fire Protection, "2018 Incident Archive," Fire.Ca.gov website.
6. Lee, James W., *Paradise Lost: The Great California Fire Chronicles* (self-pub., 2019).
7. Dutchsinse, "6/02/2023—ALL of S. Quebec Canada just Erupted into Fires—Canada under Attack? DEW or People?!," YouTube website, June 2, 2003 (noting the fires all erupted at exactly the same time, and "controlled burns" cannot be seen from space).
8. European Space Agency, "Counting Wildfires across the Globe," European Space Agency website, March 8, 2023.
9. Lee, *Paradise Lost: The Great California Fire Chronicles*; for the swarms of nanobots ("embers"), see Got Reality, "PARADISE LOST #51 52,000 Missing vs. 88 Reported Dead, a 'planetruth' video," YouTube website, January 13, 2019 (where you will also find all of Jamie Lee's "Paradise Lost" videos).
10. Takle, Eugene S., "Biomass burning contributes to stratospheric sulfate particles," 2003; from Notholt, J., et al., "Enhanced Upper Tropical Tropospheric COS: Impact on the Stratospheric Aerosol Layer," Science 300, no. 5617 (2003): 307–10.
11. Taylor, P. J., "Federal Government Thanks Wildfires for Clearing Path for Trans Mountain Pipeline," *The Beaverton* website, August 20, 2018.
12. Bishop, Todd, "Bill Gates Encouraged by U.S. Climate Initiatives as 'Mega-region' Grapples with Long-Term Challenges," Geek Wire website, September 13, 2022.
13. Gruss, Mike. "AMOS Conference JSpOC Upgrade on Track for 2016, although Parts of the Overhaul Face Delays," Spacenews website, September 11, 2014.
14. Hitachi Social Innovation, "The New Smart Grid in Hawaii: JUMPSmartMaui Project," Hitachi website, February 18, 2015.
15. Chandler, Craig C., Jay R. Bentley, "Forest Fire as a Military Weapon," May 1, 1970, Defense Technical Information Center website, released May 10, 1983; also see "Wildfire Public Information Map," ESRI website.
16. "Retired USDA Biologist Francis Mangels on Geoengineering," interview by John Whyte at Chemtrails Conference, YouTube website, September 1, 2012.
17. Mills, Denis, "Aluminum Dust from Geoengineering Fueling Super Wildfires According to Author," PRNewsWire website, September 6, 2018; South Canterbury Sky Watch, "Inferno in the Making," YouTube website, November 21, 2019.

18. Grundvig, James, "The White-Hot Fuel of the Maui Firestorm," Burning Post website, August 18, 2023.
19. Reese, Greg, "CCP Satellites Over Maui at Time of Fires," Reese Report website, September 14, 2023.
20. See "DEW & Blue Beam Energy Plasma Fires—Extensive Coverage—Jeff Snyder," Bitchute website, January 31, 2023, as an example of plasma fire simultaneously immolating and melting 73 cars while leaving the grass intact.
21. Blue Panther, "TR3B Caught Bombing Beirut Lebanon," Bitchute website, August 11, 2020; see Gautreaux, Sean, *What Is in Our Skies* (self-pub., CreateSpace, 2012), part 1, Introduction, to better understand Geoengineering Operation #7.
22. Dunn, Natasha, "Portugal's Longest Running Wildfire 'Dominated' for Second Time as Questions Swirl," Portugal Resident website, August 17, 2022.
23. Tsimitakis, Matthaios, "The Greek Wildfires: What Went Wrong and What Can Be Fixed?," Aljazeera website, August 19, 2021.
24. Wood, Judy, *Where Did the Towers Go? Evidence of Directed Free-Energy Technology on 9/11* (self-pub., New Investigation, 2005).
25. Wood, *Where Did the Towers Go?*
26. Wood, *Where Did the Towers Go?*
27. Livingston, Vic, "'Microburst' Winds Again Hit Home of Journo Exposing Electromagnetic Radio Weapon Crimes," VicLivingston.blogspot website, November 13, 2012; regarding raising winds to drive fires, see Duginski, Paul, "The Fire That Devastated a Sierra Town Created a Pyrocumulus Cloud. What Does That Mean?," *Los Angeles Times* website, November 19, 2020.
28. Livingston, "'Microburst' Winds Again Hit Home of Journo Exposing Electromagnetic Radio Weapon Crimes."
29. Wired Staff, "Say Hello to the Goodbye Weapon," Wired website, December 5, 2006.
30. "9/11: the Controversial Story of the Remains of the World Trade Center," The Conversation website, September 7, 2021.
31. Strawman Ofoz, "Shasta County Board of Supervisors Hearing on Chemtrails, July 15, 2014 (edited)," September 9, 2014.
32. Podkul, Cezary, "Inflated Home Appraisals Drain Billions from Government Insurance Fund," *Wall Street Journal* website, November 16, 2018.
33. Hoffman, Michael, *Twilight Language* (Coeur d'Alene, Idaho: Independent History and Research, 2021).

34. Matheson, Rob, "Wireless Communication Breaks through Water-Air Barrier," MIT News website, August 22, 2018.
35. Varma, A., A. S. Mukasyan, A. Rogachev, and K. V. Manukyan, "Solution Combustion Synthesis of Nanoscale Materials," *Chemical Reviews* 116, no. 23 (2016): 14493–586; Li, F., J. Ran, M. Jaroniec, and S. Zhang Qiao, "Solution Combustion Synthesis of Metal Oxide Nanomaterials for Energy Storage and Conversion," *Nanoscale* 42 (2015): 17590–610.
36. Williams, Suzanne, "Argonne's New Combustion Synthesis Research Facility Heats Up High-Throughput Manufacturing of Nanomaterials," Argonne National Laboratory press release, Argonne National Laboratory website, September 4, 2018; also, Freeland, "Project Cloverleaf," chapter 1, in *Under an Ionized Sky.*
37. Wood, *Where Did the Towers Go?*, chapter 17, "The Tesla-Hutchison Effect."
38. Gammon, Katharine, "'Fire-Breathing Dragon Clouds': A Wildfire-Fueled Phenomenon explained," The Guardian website, August 6, 2022.
39. Duginski, "The Fire That Devastated a Sierra Town Created a Pyrocumulus Cloud."
40. 60 Minutes, "The Fire Was Growing . . . at a Rate of One Football Field a Second," YouTube website, December 2, 2018.
41. MIT Professional Education, "5 Ways the New Space Economy Can Improve Human Life on Earth," MIT Professional Programs website, accessed July 7, 2024.
42. Zuverza-Mena, N., D. Martínez-Fernández, W. Du, J. A. Hernandez-Viezcas, N. Bonilla-Bird, M. L. López-Moreno, M. Komárek, J. R. Peralta-Videa, and J. L. Gardea-Torresdey, "Exposure of Engineered Nanomaterials to Plants: Insights into the Physiological and Biochemical Responses—A Review," *Plant Physiology and Biochemistry* 110 (2017): 236–64. "Carbon-based and metal-based engineered nanomaterials (ENMs) . . . have the potential to build up sediments and biosolid-amended agricultural soils."
43. McCarthy, Wil, "Ultimate Alchemy," Wired website, October 1, 2001; Ventura, Tim, "Smart Dust, Utility Fog & Virtual People: The Future of Programmable Matter," Medium website, December 7, 2019.
44. Ewens, Michelle, "Future Foglets of the Hive Mind," Scribd website, March 24, 2011; Mouteva, G.O., et al., "Black Carbon Aerosol Dynamics and Isotopic Composition in Alaska Linked with Boreal Fire Emissions and Depth of Burn in Organic Soils," *Global Biogeochemical Cycles* 29 (2015).

45. Smith, Stephen, "Earthquakes and Volcanoes," Thunderbolts.info website, September 8, 2011; Chalmers, Keely, "Oregon Fires Burning Underground Pose New Threat," KGW8 News website, September 21, 2020.
46. Lee, Jamie (Aplanetruth3), "Paradise Lost #51: Fire Embers Are Programmable Swarming Nanobots," YouTube website, July 28, 2019.

5. GRAVITY, ÆTHER, AND TESLA'S ATMOSPHERIC ENGINE

1. De Aquino, Fran, "High-power ELF radiation generated by modulated HF heating of the ionosphere can cause Earthquakes, Cyclones and localized heating," Maranhao State University, Physics Department, S.Luis/MA, Brazil, 2011, available on the HAL Open Science website.
2. Coats, Callum, "Victor Schauberger—Living Energies," Argos Vu website, December 4, 2017.
3. See The Thunderbolts Project, YouTube website.
4. Seifer, Marc J., *Transcending the Speed of Light: Consciousness, Quantum Physics, and the Fifth Dimension* (Rochester, Vt.: Inner Traditions, 2008).

6. THE WEAPONIZATION OF MAGNETISM

1. Maher, Barbara, et al., "Magnetite Pollution Nanoparticles in the Human Brain," *PNAS* 113, no. 39 (September 6, 2016): 10797–801, noting, "We identify the abundant presence in the human brain of magnetite nanoparticles [less than 200 nm in diameter] that match precisely the high-temperature magnetite nanospheres formed by combustion and/or friction-derived heating, which are prolific in urban, airborne particulate matter (PM)."
2. Meskhidze, Helen and James Owen Weatherall, "Torsion in the Classical Spacetime Context." *Cambridge University Press*, 27 October 2023.
3. Becker, Robert O., *Cross Currents: The Perils of Electropollution, the Promise of Electromedicine* (New York: Penguin, 1990); Becker, Robert O., and Gary Selden, *The Body Electric* (New York: Morrow, 1985), chapter 15, "Maxwell's Silver Hammer." For papers by Dr. Pall, search "Martin L. Pall" on ResearchGate website.
4. Presman, Aleksandr S., *Electromagnetic Fields and Life* (New York: Springer, 1970). Full text available online at Archive.org website.

5. For a magnetism-based "solution" to the UFO mystery see these books by Bruce Cathie, *The Harmonic Conquest of Space* (Kempton, IL: Adventures Unlimited Press, 1998); *The Energy Grid: Harmonic 695: The Pulse of the Universe* (Kempton, IL: Adventures Unlimited Press, 1997); *Harmonic 695: The UFO and Antigravity* (self-pub., 1981). Cathie, a New Zealand pilot, turned down working for the CIA twice, for which he too has been wiped from science history.
6. Fanthorpe, Lionel, and Patricia Fanthorpe, "The Philadelphia Experiment," chapter 7 in *Unsolved Mysteries of the Sea* (Toronto: Dundurn Press, 2004).
7. Freeland, Elana, *Under an Ionized Sky* (Port Townsend, WA: Feral House, 2018).
8. Eckert, Tiffany, "Alarming Suicide Data Released by Lane County," Jefferson Public Radio website, February 9, 2023; van Wijngaarden, Edwin, et al., "Exposure to Electromagnetic Fields and Suicide among Electric Utility Workers," *Western Journal of Medicine* 173, no. 2 (2000): 94–100.
9. Harris, Shane, and John Hudson, "'Havana Syndrome' Not Caused by Energy Weapon or Foreign Adversary, Intelligence Review Finds," *Washington Post* website, March 1, 2023.
10. Miller, Korin, "What Is Havana Syndrome? Symptoms and Theories," Health website, November 10, 2023.
11. Mothershead, J. L., Z. F. Dembek, T. A. Hann, C. G. Owens, and A. Wu, "Havana Syndrome: Directed Attack or Cricket Noise?," *Joint Force Quarterly* 108 (National Defense University Press), January 2023.
12. National Academies of Sciences, Engineering, Medicine, "An Assessment of Illness in U.S. Government Employees and Their Families at Overseas Embassies" (Washington, DC: National Academies Press, 2020).
13. Ozimek, Tom, "Havana Syndrome 'Absolutely' the Result of Deliberate Attacks," Epoch Times website, May 10, 2021.
14. Weinberger, Sharon, "The Secret History of Diplomats and Invisible Weapons," *Foreign Policy*, August 25, 2017. This is a disinformation piece, *Foreign Policy* being the American counterpart to the British flagship *The Economist.*
15. Cathie, *The Harmonic Conquest of Space.*
16. Brookhaven National Laboratory, "Accelerators at Brookhaven Lab," BNL website, accessed July 8, 2024; Valerian, Valdamar, *Matrix III, Volume One* (Yelm, WA: Leading Edge Research, 1991), 310–11.
17. Mason, Harry, "Bright Skies: Top Secret Weapons Testing? Part 5," *Nexus* 5, no. 1 (Dec. 1997–Jan. 1998), series available at Biblioteca Pleyades website.

18. Truthstream Media, “The Real Secrets Hidden in Antarctica . . . Revealed,” YouTube website, April 8, 2017.
19. Reese, Greg, “Whistleblower Claims Advanced Technology in Antarctica Can Cause Earthquakes,” Reese Report website, June 14, 2023.
20. “Business Collaboration with Nazi Germany,” Wikipedia website; Sutton, Anthony, *Wall Street and the Rise of Hitler* (Seal Beach, CA: ’76 Press, 1976).
21. Picknett, Lynn, and Clive Prince, *The Stargate Conspiracy: The Truth about Extraterrestrial Life and the Mysteries of Ancient Egypt* (Berkeley, 2001).
22. Freeland, “Boots on the Ground,” chapter 8 in *Under an Ionized Sky.*
23. Becker, Robert O., *Cross Currents: The Perils of Electropollution, The Promise of Electromedicine* (New York: Penguin, 1990), 234–39.
24. McAuliffe, Kathleen, “Magnetic Fields That Control Our Behavior,” *Omni* 7, no.5 (Feb. 1985) 40.
25. Gruss, Mike, “AMOS Conference/JSpOC Upgrade on Track for 2016, although Parts of the Overhaul Face Delays,” Space News website, September 11, 2014.
26. For a graphic representation of this VLF “ring,” see NASA Goddard, “NASA’s Van Allen Probes Find Human-Made Bubble Shrouding Earth,” YouTube website, May 17, 2017.
27. Yirka, Bob, “CERN Council Endorses Building Larger Supercollider,” Phys.org website, June 22, 2020.
28. Steiner, Rudolf, “Origins of Natural Science, Lecture V,” Rudolf Steiner Archives website, December 28, 1923.
29. Xin, Ling, “China’s Premier Particle Collider Set for Major Upgrade,” Physics World website, May 2, 2022.
30. Energy Sciences Network, “A Nationwide Platform for Science Discovery,” ESnet website, accessed July 8, 2024.
31. Marc J. Seifer in his book *Wizard: The Life and Times of Nikola Tesla* (Citadel, 2016).
32. Defense Advanced Research Projects Agency, “Materials for Transduction (MATRIX) (Archived),” DARPA.mil website, accessed July 8, 2024.
33. Space Weather Prediction Center, “G1 and G2 Watches for 30 NOV–1 DEC, 2023,” SWPC.NOAA website, December 1, 2023.
34. Paron-Carrasco, F. J., and A. De Santis, “The South Atlantic Anomaly: The Key for a Possible Geomagnetic Reversal,” *Frontiers in Earth Science* 4 (2016).
35. Liberatore, Stacy, “Mysterious Shock Wave CRACKS Earth’s Magnetosphere,” *Daily Mail* website, December 19, 2022.

36. Dinmore, Naomi, "Large Hadron Collider reaches its first stable beams in 2024," CERN website, 5 April 2024. "Although the solar eclipse on 8 April will not affect the beams in the LHC, the gravitational pull of the moon, like the tides, changes the shape of the LHC because the machine is so big."
37. GOD-TV, "CERN, Transhumanism, Science & Prophecy—Apocalypse and the End Times," interview of Anthony Patch by Paul McGuire, YouTube website, November 9, 2018.
38. Ghosh, Pallab, "Scientists Get Closer to Solving the Mystery of Antimatter," BBC News website, September 27, 2023; Tedx Talks, "Matter and Antimatter | Prof. Jeffrey S. Hangst | TEDxDanubia," YouTube website, September 27, 2023.
39. Laurence Kotlikoff, "Is Our Government Intentionally Hiding $21 Trillion in Spending?" *Forbes* website, July 21, 2018.
40. Bergun, Norman R., *Ringmakers of Saturn* (Edinburgh: Pentland Press, 1986).
41. Becker, *Cross Currents*, 71.
42. Cathie, *The Energy Grid: Harmonic 695, the Pulse of the Universe*, 147.
43. Cathie, *The Energy Grid: Harmonic 695, the Pulse of the Universe*, 146–47.
44. Lakhovsky, Georges, *The Secret of Life* (1939; repr., Whitefish, MT: Kessinger, 2003), Lakhovsky recommended using spiral loops of copper wire to set up electromagnetic fields harmonically tuned to the life forces. He invented the multiple wave oscillator (wavelengths 10 cm to 400 meters).
45. Hodsden, Suzanne, "MIT Researchers Develop Wireless, Noninvasive Deep Brain Stimulation Approach," Med Device Online website, March 17, 2015.
46. Ahmed, I. A. M., V. Karloukovski, and L. Calderon-Garciduenas, "Magnetite Pollution Nanoparticles in the Human Brain," *PNAS* 113, no. 39 (2016): 10797–801.
47. Carrington, Damian, "Alzheimer's-Linked Nanoparticles, Found in Pollution, Are Showing up in People's Brains," *Business Insider* website, September 6, 2016.
48. Crenshaw, Jim, "More Evidence That PCR Nasal Swab Test Vaccinates You and Implants the Mark of the Beast!," interview of Dr. Lorraine Day, Bitchute website, October 6, 2021.
49. Rense "Incredible Photos of Hydrogel-Lithium Releasing Hollow Nylon Fibers in Nasal Swabs in Slovakia. See What Is Released in Those PCR 'Tests'—Photos," Rense website, May 5, 2021.
50. Main, Douglas, "Potentially Toxic Magnetic Nanoparticle Pollution Found in Human Brains," *Newsweek* website, September 5, 2016.

51. Al-Deen, F. N., C. Selomulya, C. Ma, and R. L. Coppel, "Superparamagnetic Nanoparticle Delivery of DNA Vaccine," *Methods in Molecular Biology* 1142 (2014): 181–94.
52. Plank, C., U. Schillinger, F. Scherer, C. Bergemann, J. S. Rémy, F. Krötz, M. Anton, J. Lausier, and J. Rosenecker, "The Magnetofection Method: Using Magnetic Force to Enhance Gene Delivery," *Biological Chemistry* 384, no. 5 (2003): 737–47.
53. Bertram, J., "MATra—Magnet Assisted Transfection: Combining Nanotechnology and Magnetic Forces to Improve Intracellular Delivery of Nucleic Acids," *Current Pharmaceutical Biotechnology* 7, no 4 (2006): 277–85.
54. See two sources: J. Bertram, "Magnet Assisted Transfection: combining nanotechnology and magnetic forces to improve intracellular delivery of nucleic acids," *Current Pharmaceutical Biotechnology*, August 2006, available at the PubMed website; and Petra Dames et al., "Targeted delivery of magnetic aerosol droplets to the lungs," *Nature Nanotechnology*, 22 July 2007, available at the *Nature* website.
55. Costandi, Mo, "Genetically Engineered 'Magneto' Protein Remotely Controls Brain and Behaviour," The Guardian website, March 24, 2016.
56. Chang, D. C., P. Q. Gao, and B. L. Maxwell, "High Efficiency Gene Transfection by Electroporation Using a Radio-Frequency Electric Field," *Biochimica et Biophysica Acta* 1092, no. 2 (1991): 153–60.
57. Dormehl, Luke, "Scientists Use Magnetism 'Mind Control' to Make a Mouse Move Around," Digital Trends website, Aug. 18, 2017.
58. Kirschvink, J. L., and J. L. Gould, "Biogenic Magnetite as a Basis for Magnetic Field Detection in Animals," *Bio Systems* 13, no. 3 (1981): 181–201.
59. Koch J. W., and W. M. Beery, "Observations of Radio Wave Phase Characteristics on a High-Frequency Auroral Path," *Journal of Research of the National Bureau of Standards, Section D: Radio Propagation* 66D, no. 3 (1962): 291–96.
60. Freeland, Elana, *Sub Rosa America: A Deep State History,* 4 vols (self-pub., CreateSpace, 2012).
61. David Geffen School of Medicine, "The Helical Heart," YouTube website, July 10, 2015; Torrent-Guasp, F., G. D. Buckberg, C. Clemente, J. L. Cox, H. C. Coghlan, and M. Gharib, "The Structure and Function of the Helical Heart and Its Buttress Wrapping. I. The Normal Macroscopic Structure of the Heart," *Seminars in Thoracic and Cardiovascular Surgery* 13, no. 1 (2001): 301–19.

62. Chu, Jennifer, "The Growth of an Organism Rides on a Pattern of Waves," MIT News website, March 23, 2020.
63. Pollack, Gerald H., *The Fourth Phase of Water: Beyond Solid, Liquid, and Vapor* (Seattle, WA: Ebner & Sons, 2013).
64. NASA, "Cassini: Saturn's Perplexing Hexagon," NASA Science website, accessed July 8, 2024; David Icke in Lotus Sun, "Saturn's Cymatic Hexagram/Hexagon Frequency Bombardment," YouTube website, November 10, 2017, tying in cymatics and noting that sound creates frequencies which lead to symbols that impact us.
65. Ali, Haider, "Young Pakistani Student Shines Internationally by Photographing Electric Honeycomb Phenomena," Daily Pakistan website, October 5, 2017.
66. Mount, Andrew, "8 Minutes of Enlightenment (38)—Torsion Fields," YouTube website, September 28, 2017.
67. Æther Force, "The Prime Mover—Primary Physics of Karl Schappeller," Æther Force website, accessed July 8, 2024.
68. "New Physics Study Finds the Real 'God' Particle," YogaEsoteric website, October 19, 2018; Tewari, Paramahamsa, et al., "Structural Relation between the Vacuum Space and the Electron," *Physics Essays* 31, no. 1 (2018): 108–29.
69. Perkins, Robert, "Engineers Create Stable Plasma Ring in Open Air," Phys.org website, November 15, 201. When Tewari was making the plasma torus, nearby cell phones went ballistic because the plasma ring was emitting distinct radio frequencies.
70. "Earth Lights: Spooklights and Ghost Lights," Inamidst website, accessed July 8, 2024.
71. Steiner, Rudolf, Lecture 2, "The Mystery of the Double. Geographic Medicine," in *Secret Brotherhoods and the Mystery of the Human Double* (Hillside, Forest Row, UK: Rudolf Steiner Press, 2004), regarding the relationship between evil and electricity.
72. Powell, Robert, "The Second Coming and the New Age," chapter 9 in *Hermetic Astrology, Vol. II: Astrological Biography* (San Rafael, CA: Sophia Foundation Press, 2006).
73. Cleland, C. E., and S. D. Copley, "The Possibility of Alternative Microbial Life on Earth," *International Journal of Astrobiology* 4, no. 3–4 (2006): 165–73; McKie, Robin, "'Shadow Biosphere' Theory Gaining Scientific Support," Raw Story website, April 13, 2013.

74. Toomey, David, *Weird Life: The Search for Life That Is Very, Very Different from Our Own* (New York: W.W. Norton, 2014).
75. McKie, "'Shadow Biosphere' Theory Gaining Scientific Support."
76. Gessner, David, "How Big Oil Seduced and Dumped This Utah Town," Mother Jones website, March 19, 2013.
77. Burns, Ryan Patrick, *Skinwalker & Beyond* (self-pub., Lulu, 2011).
78. Kehe, Jason, "The Biggest Threat to Humanity? Black Goo," Wired website, August 24, 2022.
79. Gessner, "How Big Oil Seduced and Dumped This Utah Town."
80. Arad Branding, "Introducing Gilsonite Magnetic + the Best Purchase Price," Arad Branding website, accessed July 8, 2024.
81. Tripp, Bryce T., "Gilsonite–An Unusual Utah Resource," Utah Geological Survey website, July 2004.

7. THE NUCLEAR BEHEMOTH

1. "Charter for Merged Bank," *St. Louis Post-Dispatch*, June 6, 1958.
2. NASA Johnson, "Orion: Trial by Fire," October 8, 2014, Winner of the 2015 Lone Star Emmy for Informational / Instructional Video.
3. See *Keesing's Record of World Events*, (Keesing's Historisch Archief), entries for June 29, 1962, May 11, 1962, and August 5, 1962.
4. Gillespie, Tom, "Chernobyl's Mutant Wolves Appear to Have Developed Resistance to Cancer, Study Finds," Sky News website, February 11, 2024.
5. Freeland, Elana, "Radiation and the Wigner Effect," in *Under an Ionized Sky* (Port Townsend, WA: Feral House, 2018), 82–86; Rabbit Hole Radio, "Rabbit Hole Radio—Fukushima and the Wigner Effect 13 Years of Entropy," interview with RadChick, one of the experts on the Wigner Effect, Rumble website, March 10, 2024.
6. Phillips, Tony, "Radiation Clouds at Aviation Altitudes," Spaceweather website, January 20, 2017; Tobiska, W. K., et al., "Global Real-Time Dose Measurements Using the Automated Radiation Measurements for Aerospace Safety (ARMAS) System," *Space Weather* 14, no. 11 (2016): 1053–80, stating: "We show five cases from different aircraft: the source particles are dominated by galactic cosmic rays but include cycle variation and their effect on aviation radiation. However, we report on small radiation 'clouds' in specific magnetic

latitude regions and note that active geomagnetic variable space weather conditions may sufficiently modify the magnetospheric magnetic field that can enhance the radiation environment, particularly at high altitudes and middle to high latitudes."

7. Prindle, Drew, "Boeing Just Patented a Laser-Powered Nuclear Fusion Engine," Digital Trends website, July 13, 2015.
8. Boeing Co., U.S. Patent 9068562-B1, "Laser-powered propulsion system," issued June 30, 2015.
9. "Why Europe's Wild Boars Are Radioactive," Sky News website, August 31, 2023.
10. Cook, Nick, *The Hunt for Zero Point: Inside the Classified World of Antigravity Technology* (New York: Crown, 2003). In the 1990s, Cook was the aviation editor of *Jane's Defence Weekly* and an aerospace consultant.
11. Shimatsu, Yoichi, "Fukushima Year 12—Radioactivity Powered Storms Blast the USA," Rense website, March 8, 2023.
12. Das, Saswato R., "Military Experiments Target the Van Allen Belts," IEEE Spectrum website, October 1, 2007.
13. Tomaswick, Andy, "The Navy Is Testing Beaming Solar Power in Space," Universe Today website, June 19, 2020.
14. Darren Orf, "The Dogs of Chernobyl Are Experiencing Rapid Evolution, Study Suggests," *Popular Mechanics* website, April 1, 2023.
15. Paleja, Ameya, "Starlink Launches V2 Mini-Satellites with 'Space Lasers,'" InterestingEngineering website, October 3, 2023; Murphy, Kendall, and Katherine Schauer, "NASA's First Two-way End-to-End Laser Communications Relay System," NASA website, October 25, 2023.
16. Wagner, Andrew, "Communicating via Long-Distance Lasers," NASA *Spinoff* website, December 15, 2020.
17. Ancillary. U.S. Air Force Scientific Advisory Board, "New World Vistas: Air and Space Power for the 21st Century," Defense Technical Information Center website, June 24, 1996.
18. Baard, Mark, "Sentient World: War Games on the Grandest Scale. Sim Strife," The Register website, June 23, 2007.
19. Keller, John, "The Promise of Laser Communications to Revolutionize Optical Links," Laser Focus World website, September 19, 2023.
20. Velasco, Emily, "Recording Brain Activity with Laser Light," CalTech website, June 4, 2021.

21. Laakso, A., A. I. Partanen, H. Kokkola, A. Laaksonen, K. E. J. Lehtinen, and H. Korhonen. "Stratospheric Passenger Flights Are Likely an Inefficient Geoengineering Strategy," *Environmental Research Letters* 7, no. 3 (2012).
22. Baghal, Christian, "This Plane 'Doesn't Exist'—SR-75 Penetrator," Medium website, December 10, 2023.
23. Stilwell, Blake, "These Air Force 'Rods from God' Could Hit with the Force of a Nuclear Weapon," Military website, accessed July 9, 2024.
24. Graphics King, "Space X UFO Explosion—Slow Motion!," YouTube website, September 1, 2016.
25. Kincaid, Paul W., "USAF Redeploys X-37B Climate Chaos, Earthquake and Tsunami Inducing Orbital HAARP Weapon," Press Core website, May 23, 2015.
26. Cathie, Bruce L., *The Energy Grid: Harmonic 695, the Pulse of the Universe* (Kempton, IL: Adventures Unlimited Press, 1997).
27. Kuiken, Todd, "DARPA's Synthetic Biology Initiatives Could Militarize the Environment," Slate website, May 3, 2017.
28. Orf, "The Dogs of Chernobyl Are Experiencing Rapid Evolution, Study Suggests"; Spatola, Gabriella J., et al., "The Dogs of Chernobyl: Demographic Insights into Populations Inhabiting the Nuclear Exclusion Zone," *Science Advances* 9, no. 9 (2023).
29. Grossman, Karl, "Applause for Perseverance Ignores Plutonium Bullet We Dodged," Fairness and Accuracy in Reporting website, February 25, 2021.
30. Grossman, Karl, "Insane U.S. Plan to Spend Billions on Weaponizing Space Makes Defense Contractors Jump for Joy—But Rest of World Cowers in Horror at Prospect of New Arms Race Leading to World War III," CovertAction website, August 25, 2021.
31. Strange Sounds, "Interplanetary Shock Wave Hits Earth's Magnetic Field Creating Red, Yellow, Green Southern Lights and Blue Northern Lights," Strange Sounds website, April 22, 2018.
32. Lauri, Arto, "Arto Lauri 251: Latest Chemtrail," YouTube website, November 7, 2019.
33. Bendix, Aria, "A Group of Scientists Called the 'Ring of 5' Found Evidence of a Major Nuclear Accident [in 2017] That Went Undeclared in Russia," Business Insider website, August 8, 2019; regarding Russia's Mayak plutonium separation facility, this was the site of the September 29, 1957, Kyshtym explosion, the third worst "accident" after Chernobyl and Fukushima, see Saunier, O.,

D. Didier, A. Mathieu, and J. D. Le Brazidec, "Atmospheric Modeling and Source Reconstruction of Radioactive Ruthenium from an Undeclared Major Release in 2017," *PNAS* 116, no. 50 (2019): 24991–25000.

34. Reilly, John, "Too Much Wind and Solar Raises Power System Costs. Deep Decarbonization Requires Nuclear," Utility Dive website, December 2, 2019.
35. Rudgard, Olivia, "Take Me Out to the Wind Turbines," Bloomberg website, February 18, 2023.
36. "Jáchymov," Wikipedia website, accessed July 9, 2024.
37. International Atomic Energy Agency, "Safety Standards," IAEA website; also see Welsome, Eileen, *The Plutonium Files: America's Secret Medical Experiments in the Cold War* (New York: Dial Press, 1999). Welsome won a Pulitzer Prize in early 1993 for her three-part series, "The Plutonium Experiment," in the *Albuquerque Tribune*, which pressured President Clinton to set up an investigation of the Advisory Committee on Human Radiation Experiments, which turned out to be a whitewash.
38. Hsieh, Steven, "St. Louis Is Burning," *Rolling Stone* website, May 10, 2013.
39. "Talvivaara: Environmental Disaster in Finland," Nuclear Heritage website, May 6, 2016.
40. "Finland's Path to Final Disposal of Nuclear Waste," Nuclear Engineering International Magazine website, March 31, 2022.
41. Moret, Leuren, "Depleted Uranium: The Trojan Horse of Nuclear War," *Journal of International Issues* 8, no. 2 (2004): 101–18; Lewis, William, "Beyond Treason UPDATE: Leuren Moret Exposes the Military DU," YouTube website, April 3, 2008. Moret is an independent scientist with a BS in geology from UC Davis (1968) and an MA in Near Eastern Studies from UC Berkeley (1978); she has completed all but her dissertation for a PhD in the geosciences at UC Davis. Former Manhattan Project scientist and retired insider at the Livermore Lab Marion Fulk trained her on radiation issues. She was an Expert Witness at the International Criminal Tribunal for Afghanistan at Tokyo and contributed to a scientific report on depleted uranium for the United Nations subcommission investigating the illegality of depleted uranium munitions.
42. Wellerstein, Alex, "Fears of a German Dirty Bomb," The Nuclear Secrecy Blog website, September 6, 2013.
43. Brooks, James, "US and Israel Targeting DNA in Gaza?" Palestine Chronicle website, December 12, 2006.

44. Agorist, Matt, "The U.S. Government Has 'Lost' Enough Radioactive Material to Bomb Nagasaki 800 Times," The Last American Vagabond website, July 16, 2018.
45. Institute of Medicine (US) Committee on Thyroid Screening Related to I-131 Exposure; National Research Council (US) Committee on Exposure of the American People to I-131 from the Nevada Atomic Bomb Tests, *Exposure of the American People to Iodine-131 from Nevada Nuclear-Bomb Tests: Review of the National Cancer Institute Report and Public Health Implications* (Washington, DC: National Academies Press, 1999).
46. Stone, Richard, "'It's Like the Embers in a Barbecue Pit.' Nuclear Reactions Are Smoldering again at Chernobyl," Science website, May 5, 2021.
47. Bob Nichols, "Gamma Radiation in America RADIATION DANGER," Your Radiation This Week, private website, April 18, 2020.
48. Stern, David P., "Trapped Radiation," Phy6.org website, accessed July 9, 2024.
49. Howe, Bridget S., *Silent Holocaust: The Global Covert Control and Assassination of Private Citizens* (self-pub., Lulu Press, 2017).

8. *BIOS*

1. Bordoloi, Pritam, "Quantum Computers vs Supercomputers: How Do They Differ?," Analytics India Magazine website, January 11, 2024.
2. Humphries, Suzanne, and Roman Bystrianyk, *Dissolving Illusions: Disease, Vaccines, and The Forgotten History* (self-pub, CreateSpace, 2013).
3. Corbley, Andy, "Swedish Firm to Unlock the Electricity of the Sea with Largest Wave Power Station in the World," Good News Network website, December 17, 2022.
4. McCurry, Justin, "Fukushima Water to Be Released into Ocean in the Next Few Months, Says Japan," The Guardian website, January 13, 2023.
5. Cousins, Farron, "Six Years After Deepwater Horizon: Time for Serious Action," De Smog International website, April 20, 2016; Sneath, Sara, and Oliver Laughland, "They Cleaned up BP's Massive Oil Spill. Now They're Sick—and Want Justice," The Guardian website, April 20, 2023.
6. Krajnak, Kristine, et al., "Acute Effects of COREXIT EC9500A on Cardiovascular Functions in Rats," *Journal of Toxicology and Environmental Health* 74, no. 21 (2011): 1397–404; Sriram, Krishnan, et al., "Neurotoxicity

Following Acute Inhalation Exposure to the Oil Dispersant COREXIT EC9500A," *Journal of Toxicology and Environmental Health* 74, no. 21 (2011): 1405–18.

7. Elejalde-Ruizand, Alexia, Michael Hawthorne, and RedEye, "Putting Bottled Water to the Test," *Chicago Tribune* website, April 17, 2008; Landa, Jennifer, "More Than 24,500 Chemicals Found in Bottled Water," Fox News website, January 13, 2014.
8. WTHR-13, "AP probe finds drugs in drinking water," WTHR-13 website, March 15, 2008.
9. Fluoride Action Network, "Fluoride on Trial: Federal Trial on the Neurotoxicity of Fluoridation Wraps Up," FluorideAlert.org website, January 25, 2024; "Exposing Fluoride with Attorney Michael Connett," Children's Health Defense.org website, January 13, 2023.
10. Pollack, Gerald H., *The Fourth Phase of Water: Beyond Solid, Liquid, and Vapor* (Seattle, WA: Ebner & Sons, 2013).
11. Patten, Terry, and Michael Hutchison, "Interview with Lt. Col. Thomas E. Bearden (ret.)," *Megabrain Report*, February 4, 1991.
12. "WiGig Solution: Enabling Wire-Free and Cable-Free User Experiences," Electronic Products website, April 15, 2016.
13. Ou, L., B. Song, H. Liang, J. Liu, X. Feng, B. Deng, T. Sun, and L. Shao, "Toxicity of Graphene-Family Nanoparticles: A General Review of the Origins and Mechanisms," *Particle and Fibre Toxicology* 13, no. 57 (2016).
14. Love, John, "Monitoring Microorganisms: Science in Space January 2024," NASA website, January 18, 2024.
15. Singh, Rajni, "Fungi as Builders for Nanomaterials," Biotech Articles website, August 1, 2012.
16. Berger, Michael, "Viruses as Nanotechnology Building Blocks for Materials and Devices," Nanowerk website, accessed July 9, 2024.
17. Salalha, W., et al., "Encapsulation of Bacteria and Viruses in Electrospun Nanofibers," *Nanotechnology* 17: 4675–4681, September 28, 2006.
18. Bloomquist, Lisa, "Genetically Modifying Humans via Antibiotics? Something You Need to Know," *Activist Post* website, October 23, 2013.
19. Warren, Stephanie, "Bacteria Live At 33,000 Feet," Popular Science website, June 20, 2013.
20. Scott, Donald, "Mycoplasm: The Linking Pathogen in Neurosystemic Diseases," *Nexus Magazine* 8, no. 5 (August–September 2001).

21. Lo, Shyh-Ching, U.S. Patent 5242820A, "Pathogenic Mycoplasma," issued September 7, 1993. "Novel pathogenic mycoplasma" refers to the fact that this *Mycoplasma* has been gain of functioned.
22. Scott, "Mycoplasm: The Linking Pathogen in Neurosystemic Diseases."
23. Hammer, Joshua, "A Ghost in the War Machine," review of *Endpapers: Family Story of Books, War, Escape, and Home*, by Alexander Wolff, New York Review of Books website, April 8, 2021.
24. Hilborne, Lee H., and Beatrice Alexandra Golomb, *A Review of the Scientific Literature as It Pertains to the Gulf War Illnesses, Volume 1: Infectious Diseases* (Santa Monica, CA: RAND Corporation, 2000).
25. Romero, Miguel, et al., 2016 "Horizontal Gene Transfers from Bacteria to Entamoeba Complex: A Strategy for Dating Events along Species Divergence," *Journal of Parasitology Research* 2016, no.1 (January 2016): 3241027.
26. See Garcia-Alvarez, Rafaela, and Maria Vallet-Regi, "Bacteria and Cells as Alternative Nano-Carriers for Biomedical Applications," *Expert Opinion on Drug Delivery* 19, no. 1 (2022): 103–18, for a thorough overview of bacteria properties, except for electromagnetic capabilities.
27. Shaw, Jonathan, "Engineering Life: Synthetic Biology and the Frontiers of Technology." *Harvard Magazine* website, January–February 2020. *Synbio* engineers use horizontal gene transfers into single-celled organisms like the bacterium *E. coli* for building in vivo "circuits" and "switches."
28. Slavo, Mac, "DARPA Seeks 'Militarized Microbes' So They Can Spread Genetically Modified Bacteria," Activist Post website, November 14, 2019.
29. Brandon Keim, "Bacteria on the Radio: DNA Could Act as Antenna," Wired website, April 25, 2011; Emerging Technology from the arXiv, "Storing Data in DNA Is a Lot Easier Than Getting It Back Out," MIT Technology Review website, January 26, 2018.
30. Shipman, Matt, "Researchers Find Gold Nanoparticles Capable of 'Unzipping' DNA," Advanced Science News website, June 26, 2012.
31. Centers for Disease Control and Prevention, "Outpatient Antibiotic Prescribing—United States 2020," CDC Archive website, October 18, 2021.
32. Bloomquist, "Genetically Modifying Humans Via Antibiotics?"
33. Dockrill, Peter, "New Light-Activated Nanoparticles Kill Over 90% of Antibiotic-Resistant Bacteria," ScienceAlert website, January 19, 2016; Ragendiran, Keerthiga, et al., "Antimicrobial Activity and Mechanism of Functionalized Quantum Dots," *Polymers* (*Basel*) 11, no.10 (2019): 1670.

34. Moskowitz, Joel, "5G Wireless Technology: Millimeter Wave Health Effects," Electromagnetic Radiation Safety website, September 6, 2023, republished by the Center for Electrosmog Prevention.
35. Chew, S. C., and L. Yang, "Biofilms," in *The Encyclopedia of Food and Health*, ed. Benjamin Caballero, Paul M. Finglas, and Fidel Toldrá (Amsterdam: Academic Press, 2016), 407–15.
36. Halper, Evan, "Uninvited Guest Leaves Behind a Mess," *Los Angeles Times*, Enewspaper website, Oct. 30, 2019.
37. Vestby, L. K., T. Grønseth, R. Simm, and L. L. Nesse, "Bacterial Biofilm and Its Role in the Pathogenesis of Disease," *Antibiotics* 9, no. 2 (2020): 59.
38. Adnan, Mohd, et al., "Agrobacterium: A Potent Human Pathogen," *Reviews in Medical Microbiology* 24, no. 4 (2013): 94–97.
39. Ho, Mae-Wan, and Joe Cummins, "Agrobacterium and Morgellons Disease, A GM Connection?" Science in Society website, April 28, 2008.
40. Ho and Cummins, "Agrobacterium and Morgellons Disease."
41. Wang, Xiuping, et al., "Synergistic Antifungal Activity of Graphene Oxide and Fungicides Against Fusarium Head Blight In Vitro and In Vivo," *Nanomaterials (Basel)* 11, no. 9 (2021): 2393.
42. Lookoutfa Charlie, "Electronic Harassment—Delivery Method Explained in Detail," YouTube website, August 21, 2019. This targeted individual has been invaluable in that he has been describing (with plentiful 4-letter words, due to his suffering) what covert targeting is utilizing, such as fungus, graphene oxide, and biofilm.
43. Krotz, Dan, "Berkeley Lab Scientists Generate Electricity from Viruses," Berkeley Lab News Center website, May 13, 2012.
44. Cunlot-Ponsard, Mirelle, "Strontium Barium Niobate Thin Films for Dielectric and Electro-Optic Applications," in *Ferroelectrics—Material Aspects* ed. Mickael Lallart (IntechOpen, 2011). The electro-optic use of the Human 1.0 body covered in aerosol-delivered biofilm is now a reality.
45. Rodney M. Donlan, "Biofilms: Microbial Life on Surfaces," *Emerging Infectious Diseases* (CDC), September 2002, available at the National Library of Medicine online.
46. DARPA Public Affairs, "DARPA Selects Teams to Build Beneficial Biofilms," DARPA.mil website, March 21, 2023.
47. University of Glasgow, "Scientists Take First Step towards Creating 'Inorganic Life,'" Science Daily website, September 15, 2011.

48. Rappoport, Jon, "Rejecting Rockefeller Germ Theory Once and for All," No More Fake News (blog), July 27, 2022. The McKinlay essay was formerly required reading in U.S. medical schools.
49. Wernick, Adam, "It's Raining Viruses, but Don't Panic," The World website, March 9, 2018.
50. Montalk, "Chemtrails: Suppressing Human Evolution," Biblioteca Pleyades website, September 29, 2000.
51. Roth, Yehuda, et al., "Feasibility of Aerosol Vaccination in Humans," *Annals of Otology, Rhinology, and Laryngology* 112, no. 3 (2003): 264–70.53.
52. Garcia-Contreras, Lucila, et al., "Immunization by a bacterial aerosol," *PNAS* 105, no. 12 (2008): 4656–60.
53. Business Wire, "World-First Inhaled COVID-19 Vaccine, Developed in Partnership Between Aerogen® and CanSinoBIO, First Public Booster Immunization in China," Bio Space website, November 14, 2022.
54. ACRObiosystems, "Can 'No-needle' COVID-19 Protection Be Achieved with Inhaled Aerosol Vaccines?" News Medical Life Sciences website, January 27, 2023.
55. Pearson, Helen, "Virus Built from Scratch in Two Weeks," *Nature* website, November 14, 2003.
56. Central Intelligence Agency, "The Darker Bioweapons Future," CIA.gov website, November 3, 2003.
57. Couzin-Frankel, Jennifer, "Poliovirus Baked from Scratch," Science Advisor website, July 11, 2002.
58. Young, L. S., P. F. Searle, D. Onion, and V. Mautner, "Viral Gene Therapy Strategies: From Basic Science to Clinical Application," *Journal of Pathology* 208, no. 2 (2006): 299–318.
59. Willis, Adam, "Next-Generation Nonsurgical Neurotechnology," DARPA.mil website, accessed July 10, 2024.
60. Gent, Edd, "The government Is Serious about Creating Mind-Controlled Weapons," Live Science website, May 23, 2019.
61. Cowan, Thomas S., and Sally Fallon Morell, *The Contagion Myth: Why Viruses (Including Coronavirus) Are Not the Cause of Disease* (New York: Skyhorse, 2020), 6–7.
62. La Leva di Archimedes, "Louis Pasteur vs Antoine Béchamp and the Germ Theory of Disease Causation—1," La Leva website, May 14, 2004.
63. Lanka, Stefan, "The Misconception called Virus," *WissenschafftPlus* magazine, Archive.org website, January 2020.

64. Cowan and Morell, *The Contagion Myth*, 67, 75, 77. In Cowan's chapter on exosomes, he writes: "Exosomes are completely indistinguishable from what the virologists have been calling 'viruses'" (page 69).
65. Steiner, Rudolf, *The Manifestations of Karma* (Hillside, Forest Row, UK: Rudolf Steiner Press, 1968), 112–14.
66. Berger, "Viruses as Nanotechnology Building Blocks for Materials and Devices."
67. Crow, Diana, "6 Amazing Things to Watch in Synthetic Biology," Medium website, October 12, 2017.
68. Shukla, R. K., A. Badiye, K. Vajpayee, and N. Kapoor, "Genotoxic Potential of Nanoparticles: Structural and Functional Modifications in DNA," *Frontiers in Genetics* 12 (2021): 728250.

9. *SYNBIO*

1. Kautz-Vella, Harald, and Kristin Hauksdottir, "The Chemistry in Contrails: Assessing the Impact of Aerosols from Jet Fuel Impurities, Additives and Classified Military Operations on Nature," (lecture at the Open Mind Conference, Oslo, October 27, 2012), Academia.edu website.
2. Hambling, David, *Swarm Troopers: How Small Drones Will Conquer the World* (self-pub., Archangel Ink, 2015); Modern War Institute, "Dr. Charles Morgan on Psycho-Neurobiology and War," (video) West Point website, April 17, 2017.
3. Carnicom, Clifford, "Potassium Interference Is Expected," Carnicom Institute website, May 15, 2005.
4. Freeland, Elana, "Deconstructing Eastlund's 1987 Patent for HAARP," chapter 2 in *Chemtrails, HAARP, and the Full Spectrum Dominance of Planet Earth* (Port Townsend, WA: Feral House, 2014), 67.
5. Inspired, "You Won't Believe What is Really Happening in Acapulco"—Jeff Berwick Interview, Rumble website, October 28, 2023.
6. Centers for Disease Control and Prevention, "Epigenetics, Health, and Disease," CDC website, accessed July 10, 2024.
7. Ross, Sherwood, "U.S. Biowarfare Programs Have 13,000 Death Scientists Hard at Work," *Scoop* website, February 26, 2020.
8. Steven Salzberg, "Scientists Have Re-Created the Deadly 1918 Flu Virus. Why?" *Forbes* website, August 15, 2022. Recall what Arthur Firstenberg

wrote about the Spanish flu in his 2017 history of electricity, *The Invisible Rainbow* (AGB Press).

9. Hirai, Toshiro, et al., "Amorphous Silica Nanoparticles Size-Dependently Aggravate Atopic Dermatitis-Like Skin Lesions Following an Intradermal Injection," *Particle and Fibre Toxicology* 9, no. 3 (2012).
10. Falling Through the Cracks Radio Show, "Episode 93: Lyme Disease with Dr. Dietrich Klinghardt," interview by Rebecca Risk, Ananta Health website, October 21, 2017.
11. Horrock, Nicholas M., "Colby Describes C.I.A. Poison Work," *New York Times* website, September 17, 1975.
12. Albarelli, Hank P., Jr., and Zoe Martell, "Morgellons Victims across the U.S. and Europe," Voltaire Network website, June 12, 2010.
13. Middelveen, M. J., D. Burugu, A. Poruri, J. Burke, P. J. Mayne, E. Sapi, D. G. Kahn, and R. B. Stricker, "Association of Spirochetal Infection with Morgellons Disease," *F1000Research* 2, no. 25 (2013).
14. Albarelli Jr. and Martell, "Morgellons Victims across the U.S. and Europe."
15. Shreekant, Tiwari, and Siba Shanker Beriha, "Primary Bacteremia Caused by Rhizobium radiobacter in Neonate: A Rare Case Report," *Journal of Clinical & Diagnostic Research* 9, no.10 (2015): DD01–2.
16. Sylla, Zenebou, and Elizabeth Wolfe, "Bayer Ordered to Pay $2.25 Billion after Jury Concludes Roundup Weed Killer Caused a Man's Cancer, Attorneys Say," CNN website, January 30, 2024.
17. University of British Columbia, "Monsanto and Terminator Seeds," Open Case Studies, UBC website, accessed July 10, 2024.
18. Johnson, Lance D., "Chlormequat Crop Chemical Is Hormonally Castrating Males and Denaturing Females, Subtle Depopulating the Earth," Natural News website, March 4, 2024; Temkin, Alexis M., et al., "A Pilot Study of Chlormequat in Food and Urine from Adults in the United States from 2017 to 2023," *Journal of Exposure Science & Environmental Epidemiology* 34, no. 2 (2024): 317–21.
19. Ho, Mae-Wan, and Joe Cummins, "Agrobacterium and Morgellons Disease, A GM Connection?" Science in Society website, April 28, 2008.
20. Institute for Collaborative Biotechnologies, ICB website, accessed July 10, 2024.
21. Centers for Disease Control Newsroom, "CDC to Launch Study on Unexplained Illness," CDC Archive website, January 16, 2008.

22. CBS, "Federal Study of Morgellons Yields No Answers," CBS News website, January 25, 2012.
23. Carnicom quotes are primarily taken from "Synthetic Biological Life Forms—CDB, Morgellons, Live Blood Findings in Post C19 Injection Era," interview of Clifford Carnicom by Dr. Ana Maria Mihalcea, Rumble website, December 26, 2022; also see Bandyopadhyay, Anirban, *Nanobrain: The Making of an Artificial Brain from a Time Crystal* (Boca Raton, FL: CRC Press, 2020).
24. McMahon, Mary, "What is the Three-Domain System?" All the Science website, May 21, 2024.
25. Freeland, "Morgellons: The Fibers We Breathe and Eat," chapter 8 in *Chemtrails, HAARP, and the Full Spectrum Dominance of Planet Earth.*
26. Whatsgoingon9, "Carnicom Institute Disclosure Project—Overview with Clifford Carnicom," Bitchute website, March 4, 2022, Parts 2 (1:24:52 / November 4, 2020) and 3 (1:17:38 / November 17, 2020) are at "BREAKING: Carnicom Disclosure Project," transparentmediatruth website, Part 1 (March 4, 2022) having been removed by YouTube.
27. Prodanov, M. F., V. V. Vashchenko, and A. K. Srivastava, "Progress Toward Blue-Emitting (460–475 nm) Nanomaterials in Display Applications," *Nanophotonics* 10, no. 7 (2021): 1801–36.
28. Carnicom, Clifford, "Morgellons: A New Classification," Carnicom Institute website, February 3, 2010, last updated November 27, 2015; Carnicom, Clifford, "The New Biology," Carnicom Institute website, January 18, 2014, last updated November 28, 2015.
29. Carnicom, Clifford, "Blood Alterations I: Coagulations," Carnicom Institute website, August 27, 2022.
30. Carnicom, Clifford, "Blood Alterations III: Transformation," Carnicom Institute website, August 28, 2022.
31. Carnicom, Clifford, "Blood Alterations VI: Implications and Consequences," Carnicom Institute website, October 9, 2022.
32. Carnicom, Clifford, "The Transformation of a Species?" Carnicom Institute website, November 28, 2019.
33. Carnicom, Clifford, "Blood Testing: Lasers, Morgellons & Fungus (?)" Carnicom Institute website, November 21, 2007.
34. Carnicom, Clifford, "'Morgellons': The Wine-Peroxide Test," Carnicom Institute website, March 9, 2008.

35. Book of Ours, "The New Constitution: Living War Crimes," YouTube website, January 1, 2023.
36. Fitts, Catherine Austin, "$21 Trillion Dollars Is Missing from the US Government. That Is $65,000 Per Person—As Much as the National Debt!," Solari Report website, accessed July 10, 2024.
37. Global Biodefense, "Operation Warp Speed's Big Vaccine Contracts Could Stay Secret," Global Biodefense website, September 29, 2020.
38. Chaturvedi, Rashmi, et al., "Got a Problem? Agent-Based Modeling becomes Mainstream," *Global Economics and Management Review* 18, no. 2 (2013): 33–39, note the public-private partnerships of universities, the U.S. Navy, and private enterprise. More at Chaturvedi, Alok, "Computational Challenges for a Sentient World Simulation" Purdue Homeland Security Institute, Archive.org website, March 10, 2006.
39. Adams, Mike, "Now 20 US States Liquefy Vaccine-Murdered People and Spread Their Flesh Goo on Food Crops as 'Fertilizer,'" Natural News website, May 17, 2021.
40. Carnicom, Clifford, "Morgellons: First Observation," Carnicom Institute website, August 12, 2006; Carnicom, Clifford, "Morgellons: Morphology Confirmed," Carnicom Institute website, November 15, 2007.
41. New China TV, "Chinese Scientists Develop Shape-Shifting Material," YouTube website, January 9, 2016.
42. University of Toronto, "'Person-on-a-Chip': Engineers Grow 3-D Heart, Liver Tissues for Better Drug Testing," Phys.org website, March 7, 2016.
43. Heller, M. J., and R. H. Tullis, "Self-Organizing Molecular Photonic Structures Based on Functionalized Synthetic Nucleic Acid (DNA) Polymers," *Nanotechnology* 2, no. 4 (1991): 165–71.
44. Albarelli, Hank P., Jr., and Zoe Martell, "National Security Secrecy: Morgellons Victims Across the US and Europe," Voltaire Network website, June 12, 2010.
45. Barbillon, G., "Plasmonics and Its Applications," *Materials* (*Basel*) 12, no. 9 (2019): 1502.
46. Maier, S. A., M. L. Brongersma, P. G. Kik, S. Meltzer, A. A. G. Requicha, and H. A. Atwater, "Plasmonics: A Route to Nanoscale Optical Devices," *Advanced Materials* 13, no. 19 (2001): 1501–05.
47. Hern, Alex, "Elon Musk Says Neuralink Has Implanted Its First Brain Chip in Human," The Guardian website, January 30, 2024.

48. Valdastri, P., A. Menciassi, A. Arena, C. Caccamo, and P. Dario, "An Implantable Telemetry Platform System for In Vivo Monitoring of Physiological Parameters," *IEEE Transactions on Information Technology in Biomedicine* 8, no. 3 (2004): 271–78.
49. Newitz, Annalee, "Scientists Just Invented the Neural Lace," Gizmodo website, June 15, 2015.
50. Kautz-Vella and Hauksdottir, "The Chemistry in Contrails." In my synthetic biology book, I will discuss Kautz-Vella, Harald, "Environmental Medicine's Approach to Geoengineering-Induced Disease," Archive.org website, accessed July 11, 2024.
51. Kehe, Jason, "The Biggest Threat to Humanity? Black Goo," Wired website, August 24, 2022.
52. Srinivasan, T. M., "Biophotons as Subtle Energy Carriers," *International Journal of Yoga* 10, no. 2 (2017): 57–58.
53. Reference in caption on page 19 of Kautz, Harald, "Environmental Medicines Approach to Geoengineering-Induced Disease: 2. Fiber Disease, Intestinal Pseudo-Parasites, Delusional Parasitosis & the Self-Assembly of Nano-Bots. The Multiple Facets of the Morgellon Condition Explained," Aquarius Technologies website, accessed July 11, 2024.
54. Kostyuk, N., P. Cole, N. Meghanathan, R. D. Isokpehi, and H. H. P. Cohly, "Gas Discharge Visualization: An Imaging and Modeling Tool for Medical Biometrics," *International Journal of Biomedical Imaging* 2011 (2011): 196460. See Professor Konstantin Korotkov's patented *bioelectrography device* similar to Kirlian photography in the nine-minute video from Become Superhuman, "What Is Gas Discharge Visualization (GDV)?," YouTube website, August 23, 2012.
55. Gillin, Murray, Loris Gillin, and Deva Paul, "Mind Control Using Holography and Dissociation: A Process Model," Angel-Fire website, March 2000.
56. Kautz-Vella and Hauksdottir, "The Chemistry in Contrails," 28.
57. Kautz, "Environmental Medicines Approach to Geoengineering-Induced Disease." Note Kautz's reference to James Clerk Maxwell's "original equations" regarding the variety of fields and wave forms existing at right-angle (orthogonal) rotations in hyperspace outside our electromagnetic fields—equations changed in 1881 so that materialistic Western science would not make public the "spiritual" dimensions of the world. See Freeland, "Æther, Plasma, and Scalar Waves," chapter 2 in *Under an Ionized Sky* (2018).

58. Purdey, M., "Elevated Silver, Barium and Strontium in Antlers, Vegetation and Soils Sourced from CWD [Chronic Wasting Disease] Cluster Areas: Do Ag/Ba/Sr Piezoelectric Crystals Represent the Transmissible Pathogenic Agent in Tses [Transmissible Spongiform Encephalopathies]?," *Medical Hypotheses* 63, no. 2 (2004): 211–25. Kautz-Vella, "The Chemistry of Contrails," 54, is dedicated to M. Purdey "who lost his life due to his love for the truth."
59. Kautz-Vella and Hauksdottir, "The Chemistry of Contrails," 54.
60. Aarhus University, "Researchers Create Synthetic Nanopores Made from DNA," Phys.org website, December 13, 2019.
61. Rice University, "Nanotubes Assemble! Rice Introduces Teslaphoresis," YouTube website, April 14, 2016.
62. Steiner, Rudolf, "The Occult Movement in the Nineteenth Century, (GA 254)," lecture 5, Dornach, Rudolf Steiner Archives website, October 18, 1915.
63. Tohoku University, "Graphene Becomes Superconductive—Electrons with 'No Mass' Flow with 'No Resistance,'" Phys.org, website, February 16, 2016.
64. Roeters, S. J., et al., "Ice-Nucleating Proteins Are Activated by Low Temperatures to Control the Structure of Interfacial Water," *Nature Communications* 12, no. 1183 (2021).
65. Peplow, M., "Graphene Sandwich Makes New Form of Ice," *Nature* (March 25, 2015): 17175; Häusler, T., P. Gebhardt, D. Iglesias, C. Rameshan, S. Marchesan, D. Eder, and H. Grothe, "Ice Nucleation Activity of Graphene and Graphene Oxide," *Journal of Physical Chemistry* 122, no. 15 (2018): 8182–90.
66. Whale, T. F., M. Rosillo-Lopez, B. J. Murray, and C. G. Salzmann, "Ice Nucleation Properties of Oxidized Carbon Nanomaterials," *Journal of Physical Chemistry Letters* 6, no. 15 (2015): 3012–16.
67. Schutt, F., et al., "Electrically Powered Repeatable Air Explosions Using Microtubular Graphene Assemblies," *Materials Today* 48 (2021): 7–17.
68. Singh, N., G. J. S. Jenkins, R. Asadi, and S. H. Doak, "Potential Toxicity of Superparamagnetic Iron Oxide Nanoparticles (SPION)," *Nano Reviews* 1 (2010).
69. Aich, N., J. Plazas-Tuttle, and N. B. Saleh, "A Critical Review of Nanohybrids: Synthesis, Applications and Environmental Implications," *Environmental Chemistry* 11, no. 6 (2014): 609–23.
70. CORDIS (Community Research and Development Information Service, European Commission), "Graphene Boosts GHz Signals into Terahertz Territory," CORDIS website, September 7, 2023.

71. Zhang, Marina, "Psychosis, Panic Attacks, Hallucinations: Bizarre Psychiatric Cases Among the COVID Vaxxed," Epoch Times website, October 3, 2023.
72. Scalia, Tanya, et al., "From Protosolar Space to Space Exploration: The Role of Graphene in Space Technology and Economy," *Nanomaterials* (*Basel*) 13, no. 4 (2023): 680.
73. Jiang, Y., et al., "Synergistically Chemical and Thermal Coupling between Graphene Oxide and Graphene Fluoride for Enhancing Aluminum Combustion," *ACS Applied Material Interfaces* 12, no. 6 (2020): 7451–58.
74. Betsaida B. Laguipo, Angela, "Brain Implants Made of Graphene—What Is Possible?," AZO Materials website, July 10, 2019.
75. Wang, M., U. Joshi, and J. H. Pikul, "Powering Electronics by Scavenging Energy from External Metals," *ACS Energy Letters* 5, no. 3 (2020): 758–65.
76. Hewitt, John, "Transparent Graphene-Based Display Could Enable Contact Lens Computers," Extreme Tech website, June 11, 2013.
77. World Orders Review, "26 GHZ [5G Spectrum] The 'Graphene' Sweet Spot (a Serious 'Health Risk,' Especially for the Jabbed," Brighteon website, September 14, 2022.
78. The Exposé, "Scientists Prove Graphene Nanobots Are in the Covid Vaccines, Shedding from the Vaccinated to the Unvaccinated; But There Is a Way to Remove Them," The Exposé website, May 18, 2023. The assumption that "shedding" of graphene oxide is responsible for finding it in the blood of the uninoculated is challenged by those who understand that superconductive graphene is capable of transmitting via 5G/6G, including Bluetooth.
79. The Exposé, "Scientists Prove Graphene Nanobots Are in the Covid Vaccines."
80. Freeman, Makia, "Hydrogel Biosensor: Implantable Nanotech to be Used in COVID Vaccines?," The Freedom Articles website, September 2, 2020.
81. Tommy Truthful, "Black GOO: The Fallen Angel Tech Revolutionizing Consciousness Transfer," Truth Mafia website, October 21, 2023.
82. Grolltex, "The Future of Graphene and 5G," Grolltex website, April 2, 2018.
83. Lv, X. S., Y. Qiu, Z. Y. Wang, G. M. Jiang, Y. T. Chen, X. H. Xu, and R. H. Hurt, "Aerosol Synthesis of Phase-Controlled Iron-Graphene Nanohybrids through Feooh Nanorod Intermediates," *Environmental Science: Nano* 3 (2016): 1215–21.

84. Xiang, C., Y. Zhang, W. Guo, and X. J. Liang, "Biomimetic Carbon Nanotubes for Neurological Disease Therapeutics as Inherent Medication," *Acta Pharmaceutica Sinica B* 10, no. 2 (2020): 239–48. Synapses connect neural circuits for computation and transmission of information.
85. Tommy Truthful, "Black GOO."
86. Rauti, R., M. Medelin, L. Newman, S. Vranic, G. Reina, A. Bianco, M. Prato, K. Kostarelos, and L. Ballerini, "Graphene Oxide Flakes Tune Excitatory Neurotransmission In Vivo by Targeting Hippocampal Synapses," *Nano Letters* 19, no. 5 (2019): 2858–70.
87. Ma, Baojin, et al., "Reaction between Graphene Oxide and Intracellular Glutathione Affects Cell Viability and Proliferation," *ACS Appl Mater Interfaces* 13, no. 3, (2021): 3528–35. Regarding NAC, "NAC adheres to the rGO surface as demonstrated by several spectroscopy techniques and avoids GO-mediated oxidation of glutathione." Palmieri, Valentina, et al., "Biocompatible N-acetyl cysteine Reduces Graphene Oxide and Persists at the Surface as a Green Radical Scavenger," *Chem Commun* (*Cambridge*) 55, no. 29 (2019): 4186–89.
88. See Karl. C, "The Blood Reaches New Levels of Tragic Alteration," Man Against the Microbes - Substack website, April 15, 2024.
89. Science Magazine, "This Long-Lasting Hydrogel Could Be Used to Replace Damaged Human Tissues," YouTube website, November 2021; Kim, Junsoo, et al., "Fracture, Fatigue, and Friction of Polymers in which Entanglements Greatly Outnumber Cross-Links," *Science* 374, no. 6564 (2021): 212–16.
90. Huff, Ethan, "Bombshell: Disposable Blue Face Masks Found to Contain Toxic, Asbestos-Like Substance That Destroys Lungs," Natural News website, March 30, 2021, noting, "The masks contain microscopic graphene particles that, when inhaled, could cause severe lung damage."
91. Solum, Celeste, "Talking Points for Graphene Hydrogel Quantum Dot Application and Mechanisms," Lost Art Radio website, accessed July 11, 2024.
92. Manchester University, "Manchester Scientists Develop Graphene Sensors That Could Revolutionise the Internet of Things," Manchester University website, January 8, 2018.
93. Orwell City, "Dr. José Luis Sevillano: 'Something Is Radiating Already with Qualities That Aren't Non-Ionizing but Ionizing,'" Orwell City website, August 10, 2021.

94. Kimgary, "Streets of Philadelphia, What's Going on Today. It Gets Much Worse Than Other Day," YouTube website, July 18, 2021.
95. McCloskey, Tyrone, "A Spoonful of Reality May NOT Help the (Nano) Medicine Go Down," (blog), Piece of Mindful website, accessed July 11, 2024.
96. Health Ranger, "Decarbonization: Terraforming of Planet Earth Is Now Underway . . . Giant Machines to Be Installed in Iowa to Suck 'Life Molecules' Out of the Atmosphere and Cause Global Crops to FAIL," Citizen News, A Final Warning website, November 10, 2021; also see the Heartland Greenway website.

10. 5G/6G, THE INTERNET OF THINGS, AND THE WIRELESS BODY AREA NETWORK (WBAN)

1. Rajkotia, Purva, "The 6G Future: How 6G Will Transform Our Lives," In Compliance magazine website, September 30, 2022.
2. Pu, Lena, "5G: Health Risks, Surveillance and Bioweaponry," Silicon Valley Health Institute, Archive.org website, September 19, 2019.
3. Nightflight, "Lena Pu—5G & The Wuhan Debacle," Bitchute website, February 19, 2020.
4. Hardell, L., and M. Carlberg, "Health Risks from Radiofrequency Radiation, Including 5G, Should Be Assessed by Experts with No Conflicts of Interest," *Oncology Letters* 20, no 4 (2020): 15.
5. Becker, Robert O., *Cross Currents: The Perils of Electropollution, the Promise of Electromedicine* (New York: Penguin, 1990), 235.
6. Pall, Martin L., "The Dangers of EMF and 5G, July 10, 2018," interview by Patrick Timpone, Soundcloud website, July 10, 2018.
7. Hambling, David, "Moscow's Remote-Controlled Heart Attacks," Military website, February 14, 2006, discussing a paper presented at the 2nd European Working Group on Non-Lethal Weapons 2003 by A. F. Korolev et al., "Bioelectrodynamic Criterion of the NLW Effectiveness Estimation and the Interaction Mechanisms of the Multilayer Skin Tissues with Electromagnetic Radiation," explaining how radio frequency weapons like the Active Denial System (ADS) affect the skin.
8. Fauteux, André, "Why Manmade Electromagnetic Fields Are the Most Damaging, according to Dimitris Panagopoulos," *La Maison du 21e siècle* magazine website, October 6, 2023.

9. Sheetz, Michael, "FCC Approves SpaceX to Deploy up to 1 Million Small Antennas for Starlink Internet Network," CNBC website, March 20, 2020.
10. Webb, Whitney, "Meet the IDF-Linked Cybersecurity Group 'Protecting' U.S. Hospitals 'Pro Bono,'" Unlimited Hangout website, August 27, 2020.
11. Betzalei, N., P. B. Ishai, and Y. Feldman, "The Human Skin as a Sub-THz receiver—Does 5G Pose a Danger to It or Not?," *Environmental Research* 163 (2018): 208–16.
12. Rubik, B., and R. R. Brown, "Evidence for a Connection between Coronavirus Disease-19 and Exposure to Radiofrequency Radiation from Wireless Communications Including 5G," *Journal of Clinical and Translational Research* 7, no. 5 (2021): 666–81; Fullerton Informer, "Joe Imbriano Warned Us in 2018—60 GHz Blocks Oxygen Uptake," YouTube website, February 10, 2018.
13. Belton, Padraig, "Why Does 5G Only Pose a Problem for US Airplanes?" Light Reading website, January 17, 2022.
14. Patriots for Truth, "How They Plan to Control Everything in Your Life," Patriots4Truth website, January 12, 2018.
15. Henderson, Dean, "Human Wetware and the 5G Computer Weapon," Left Hook Archives October 25, 2019, Dean Henderson—Substack website, April 9, 2024.
16. Smith, Carole, "Intrusive Brain Reading Surveillance Technology: Hacking the Mind," Global Research, website, December 13, 2007.
17. Firstenberg, Arthur, "5G Deep Penetration," *New Dawn* magazine, issue 167, March-April 2018. Firstenberg is the author *The Invisible Rainbow: A History of Electricity and Life* (2017).
18. RT, "Pentagon to Dish out $600mn in Contracts for '5G Dual-Use Experimentation' at 5 U.S. Military Sites, Including to 'Aid Lethality,'" RT website, October 9, 2020.
19. U.S. Deptartment of Defense, "DOD Names Seven Installations as Sites for Second Round of 5G Technology Testing, Experimentation," news release, Department of Defense website, June 3, 2020.
20. Nelson, Joyce, "5G Corporate Grail. Microwave Radiation," Global Research website, November 9, 2018, quoting Jody McCutcheon.
21. Firstenberg, Arthur, "5G—From Blankets to Bullets," Cellular Phone Task Force website, January 22, 2018.

22. Freeman, Makia, "5G and IoT: Total Technological Control Grid Being Rolled Out Fast," The Freedom Articles website, March 22, 2017.
23. Thomas, John P., "Can New 5G Technology and Smart Meters be Used as Weapons?," Health Impact News website, September 4, 2018.
24. Bathgate, Bill, "Electrical Engineer Explains How 'Smart' Meters Can Lead to Higher Bills," interview by Jill McManus, West View News website, April 9, 2023.
25. Williams, Jamie, "Win! Landmark Seventh Circuit Decision Says Fourth Amendment Applies to Smart Meter Data," Electronic Frontier Foundation website, August 21, 2018.
26. Clarke, Chris, "Are California Smart Meters Causing Fires?" PBSSoCal website, October 16, 2012; "Lawsuits Claim Faulty PG&E Smart Meters Started House Fires," ABC30 website, November 17, 2017; "The Evidence Mounts—Were Smart Meters and/or Other Directed Energy Weapons Involved in the California Fires? Is the 'Deep State' Murdering American People?" Radiation Dangers website, accessed July 11, 2024.
27. Riley, Michael, "U.S. Agencies Said to Swap Data with Thousands of Firms," Bloomberg website, July 15, 2013.
28. Bates, Albert, "Airport Androids Attack Human Gene Pool," Medium website, January 8, 2010.
29. Garcia-Martinez, Javier, "Here's What Will Happen When 30 Billion Devices Are Connected to the Internet," World Economic Forum website, June 23, 2016.
30. Rajkotia, "The 6G Future: How 6G Will Transform Our Lives."
31. Rajkotia, "The 6G Future: How 6G Will Transform Our Lives."
32. Rajkotia, "The 6G Future: How 6G Will Transform Our Lives."
33. Rajkotia, "The 6G Future: How 6G Will Transform Our Lives."
34. Wellers, Daniel, "Is This the Future of the Internet of Things?" World Economic Forum website, November 27, 2015.
35. Regis, Ed, *Nano: The Emerging Science of Nanotechnology: Remaking the World—Molecule By Molecule* (New York: Little, Brown and Company, 1995).
36. Regis, *Nano*, 92.
37. Sakamoto, Rex, "This Is the World's Smallest Computer," CBS News website, April 6, 2015.
38. Solon, Olivia, "Transfer a Secret Audio Message by Poking Someone with Your Finger," Wired website, September 13, 2013.

39. Truthstream Media, "Smart Phones Will Send Data Through Your Bones Next," YouTube video, June 26, 2017, discussing bone conduction.
40. Liu, Xiao, "Tracking How Our Bodies Work Could Change Our Lives," World Economic Forum website, June 4, 2020.
41. Soni, Jitendra, "Fast Evolving Botnet Targets Millions of IoT Devices," Tech Radar website, April 13, 2020, noting, "This botnet, known as dark nexus, is capable of launching a range of various DDoS [distributed denial of services] attacks, spreading multiple malware strains, and can infect devices running on 12 different CPU architectures."
42. Newcomb, Tim, "Scientists Want to Use People as Antennas to Power 6G," *Popular Mechanics* website, January 6, 2023.
43. Newcomb, "Scientists Want to Use People as Antennas to Power 6G." Remember Dr. Martin Pall's reference to sweat ducts in the skin acting as helical antennas.
44. Goovaerts, Diana, "U.S. Taps Apple, Google, Nokia, Qualcomm for 6G Boost," Fierce Network website, April 28, 2021.
45. Horowitz, Jeremy, "Researchers Say 6G Will Stream Human Brain-Caliber AI to Wireless Devices," Venture Beat website, June 14, 2019; Samsung Newsroom, "Samsung's 6G White Paper Lays Out the Company's Vision for the Next Generation of Communications Technology," Samsung website, July 14, 2020.
46. Constable, Trevor James, *The Cosmic Pulse of Life: The Revolutionary Biological Power Behind UFOs* (Bayside, CA: Borderland Sciences Research Foundation, 1976), IR photos on pages 397–448.
47. United States Space Force, "Space Based Infrared System (SBIRS)," Space Force website, accessed July 11, 2014.
48. United States Space Force, "Space Based Infrared System (SBIRS)."
49. Cheong, Elaine, "Actor-Oriented Programming for Wireless Sensor Networks," (PhD dissertation, UC Berkeley, 2007). (My emphasis.)
50. United States Space Force, "Space Based Infrared System (SBIRS)."
51. See, Hexa-X.eu website.
52. Hexa-X, "About," Hexa-X.eu website, accessed July 12, 2024.
53. Hawkins, Joshua, "6G May Be Fast Enough for Us to Communicate with Futuristic Holograms," BGR website, May 6, 2024.
54. Tohoku University, "Graphene Becomes Superconductive—Electrons with 'No Mass' Flow with 'No Resistance,'" Phys.org website, February 16, 2016.

55. Vetter, Peter, and Magnus Frodigh, "Hexa-X—The Joint European Initiative to Shape 6G," Ericsson website (blog), January 27, 2021.
56. Pollack, Gerald H., *The Fourth Phase of Water: Beyond Solid, Liquid, and Vapor* (Seattle, WA: Ebner & Sons, 2013).
57. Rugeland, Patrik, "Hexa-X: 6G Technology and Its Evolution So Far," Ericsson website (blog), July 15, 2021.
58. Tachover, Dafna, "Washington Post Is Right That 5G Is a Lie, But Wrong About the Reason," Le Vaud Sans Antennes website, September 15, 2020.
59. Tachover, Dafna, "5G AirGig: What Is It and Should You Be Worried?" Le Vaud Sans Antennes website, March 17, 2020.
60. Marti, Ernst, *The Four Ethers: Contributions to Rudolf Steiner's Science of the Ethers Elements-Ethers-Formative Forces,* translated by Eva Lauterbach and James Langbecker. (Roselle, IL: Schaumburg Publications, 1984).
61. Steiner, Rudolf, *The Electronic Doppelganger: The Mystery of the Double in the Age of the Internet,* edited by Andreas Neider, translated from the German by Simon Luke Breslaw (Hillside, Forest Row, UK: Rudolf Steiner Press, 2016), 132.
62. Firstenberg, Arthur, *The Invisible Rainbow: A History of Electricity and Life* (AGB Press, 2017), 7.

EPILOGUE. THE GREAT CHAIN OF BEING

1. Steiner, Rudolf, "The Overcoming of Evil, GA273," Rudolf Steiner Archives website, November 4, 1917.
2. See the Aztec timeline from the sixth century to 1525 at "Aztec Timeline," Aztec-History website, accessed July 11, 2024.
3. Steiner, Rudolf, "Body, Soul and Spirit," chapter 1 in *Theosophy,* Rudolf Steiner Archives website, 1904.
4. Davies, Paul, *The Demon in the Machine: How Hidden Webs of Information Are Solving the Mystery of Life* (Chicago: University of Chicago Press, 2018), 63, 73.
5. Fight4Freedom, "What's Really Going on? Dietrich Klinghardt," YouTube website, January 26, 2020.
6. Horsley, Jasun, "Rudolph Steiner in 1913 on a Vaccine to Close Down the Spiritual Senses," YouTube website, January 6, 2021.

APPENDIX 1. INVISIBLE MINDSETS

1. Miller, Edward, "Behind the Phoenix Program," *New York Times* website, December 29, 2017.
2. Valley, Paul E. and Michael A. Aquino, "From PSYOP to MindWar: The Psychology of Victory," 7th Psychological Operations Group, U.S. Army Reserve, Presidio of San Francisco, California, 1980, Archive.org website.
3. Valley and Aquino, "From PSYOP to MindWar," n20.
4. "Julianne McKinney, Director, Electronic Surveillance Project, January 1995: 'Homemaker's Recipe' for a Successful Revolution in Military Affairs," Everyday Concerned Citizen website, accessed July 11, 2024.
5. Thomas, Timothy, "The Mind Has No Firewall," *U.S. Army War College Quarterly: Parameters* 28, no. 1 (1998): 84–92.
6. Thomas, "The Mind Has No Firewall."
7. Forwood, Anthony, "The 'Conspiracy' against Michael Aquino—Satanic Pedophile," Exposing the Truth website (blog), February 11, 2014.
8. Retta, Mary, "How the Pentagon Influences US Colleges: A History of Military Influence in Higher Education," TeenVogue website, September 27, 2023.
9. Farrell, Bryan, and Emily Schwartz Greco, "Penn State's Frightening Defense," Foreign Policy in Focus website, April 25, 2008. Note that during the Covid-19 lockdown between 2021 and 2022, the rate for death by synthetic opioids (like fentanyl) increased 4.1 percent from 21.8 to 22.7, while rates for heroin, natural and semisynthetic opioids, and methadone declined: Spencer, Merianne, et al., "Drug Overdose Deaths in the United States, 2002-2022," NCHS Data Brief No. 491, CDC website, March 2024. Also see Gold, Mark S., "The Fentanyl Death Crisis in America," Psychology Today website, June 4, 2024.
10. Webb, Whitney, "Lifting of US Propaganda Ban Gives New Meaning to Old Song," Mint Press News website, February 12, 2018.
11. Office of the Under Secretary of Defense for Acquisition, Technology and Logistics, "Report of the Defense Science Board Task Force on the Creation and Dissemination of All Forms of Information in Support of Psychological Operations (PSYOP) in Time of Military Conflict," Defense Technical Information Center website, May 2000.
12. "Active-Duty Army Whistleblower Lt. Col. Daniel Davis: U.S. Deceiving Public on Afghan War," interview, Democracy Now! website, April 11, 2012,

in which he states: "Senior ranking US military leaders have so distorted the truth when communicating with the US Congress and American people in regards to conditions on the ground in Afghanistan that the truth has become unrecognizable."

13. See RT, "NDAA 2013: Congress Approves Domestic Deceptive Propaganda," RT website, May 22, 2012.
14. Joint Chiefs of Staff, "Joint Publication 3-13," Defense Technical Information Center website, November 27, 2012.
15. Weinberger, Sharon, "You Can't Handle the Truth: Psy-ops Propaganda Goes Mainstream," Slate website, September 19, 2005.
16. Duncan, Robert, *The Matrix Deciphered: Psychic Warfare Top-Secret Mind Interfacing Technology* (self-pub., Higher Order Thinkers Publishing, 2006), available at Archive.org website.
17. Grad, Peter, "Researchers Say Chatbot Exhibits Self-Awareness," Tech Xplore website, September 12, 2023.
18. Duncan, Robert, *The Matrix Deciphered*, 2007. Available at the Internet Archive website.
19. Chaturvedi, Rashmi, et al., "Got a Problem? Agent-Based Modeling becomes Mainstream," *Global Economics and Management Review* 18, no. 2 (2013): 33–39.
20. Cerri, T., J. Jfcom, and A. Chaturvedi, "Sentient World Simulation (SWS): A Continuously Running Model of the Real World—a Concept Paper for Comments," Semantic Scholar website, 2006.
21. Deplazes, A., and M. Huppenbauer, "Synthetic Organisms and Living Machines," *Systems and Synthetic Biology* 3 (2009): 55–63.
22. University of Glasgow, "Scientists Take First Step Towards Creating 'Inorganic Life,'" Phys.org website, September 12, 2011. Nanotechnology is not mentioned in the article despite the fact that "inorganic chemical compounds" that can self-replicate depend on it. See also Cooper, G. J. T., P. J. Kitson, R. Winter, M. Zagnoni, D. Long, and L. Cronin, "Modular Redox-Active Inorganic Chemical Cells: iCHELLS," *Angewandte Chemie* 50, no. 44 (2011): 10373–76.
23. Reyburn, Matthew, "Synthetic Biology Will Change Warfare," *U.S. Naval Institute: Proceedings* 147, no. 11 (November 2021), U.S. Naval Institute website.
24. Reyburn, "Synthetic Biology Will Change Warfare."

25. Finkbeiner, Ann, "Jason—A Secretive Group of Cold War Science Advisors—Is Fighting to Survive in the 21st Century," Science website, June 27, 2019.
26. Lash, John, *The Seeker's Handbook: The Complete Guide to Spiritual Pathfinding* (New York: Harmony Books, 1990).
27. Bondarev, G. A., *The World and Humanity at the Crossroads of the Occult-Political Movements of Our Time,* chapter 6, "God and World Evil" (in manuscript only).
28. Dalberg, John Emerich Edward (Lord Acton), *Lectures on the French Revolution*, ed. John Neville Figgis and Reginald Vere Laurence (Indianapolis: Liberty Fund, 2000).
29. Thoresen, Are, *Experiences from the Threshold and Beyond: Understood through Anthroposophy* (Forest Row, E. Sussex, UK: Temple Lodge Publishing, 2019). Thoreson is a veterinarian and anthroposophist.
30. Broad, William J., "The Core of the Earth May Be a Gigantic Crystal Made of Iron," *New York Times* website, April 4, 1995.
31. Balaguru, S., R. Uppal, R. Pal Vaid, and B. P. Kumar, "Investigation of the Spinal Cord as a Natural Receptor Antenna for Incident Electromagnetic Waves and Possible Impact on the Central Nervous System," *Electromagnetic Biology and Medicine* 31, no. 2 (2012): 101–11.
32. Freeman, Makia, "Exposing the Occult Corona-Initiation Ritual," The Freedom Articles website, July 14, 2020.
33. Lash, *The Seeker's Handbook*, 325.
34. Delaforge, Gaetan, "The Templar Tradition Yesterday and Today," Masonic World website, accessed July 12, 2024.

APPENDIX 2. QUANTUM, SCALAR, AND HYPERSPACE

1. Snyder, Michael, "Pre-Crime and Mind Control Technologies Are Already Here," Activist Post website, August 6, 2013.
2. Val Valerian, "The Ground-Wave Emergency Network (GWEN) System," Educate Yourself website, accessed July 12, 2024.
3. Whittaker, E. T., "On the Partial Differential Equations of Mathematical Physics," *Mathematische Annalen* 57 (1903): 333–55 (available on KRW Jones website).
4. Farrell, Joseph P., *Covert Wars and Breakaway Civilizations: The Secret Space*

Program, Celestial Psyops, and Hidden Conflicts (Kempton, IL: Adventures Unlimited Press, 2012).

5. Bearden, T. E., "Solutions to Tesla's Secrets and the Soviet Tesla Weapons" (Chula Vista, CA: Tesla Book, 1981), 18. Ridiculed his entire professional career for his insights, Bearden kindly referred to modern physics as "inconsistent" (general relativity, electromagnetics, quantum mechanics, what a photon is, what force fields are, etc.).
6. Bearden, "Solutions to Tesla's Secrets and the Soviet Tesla Weapons," 3; also Farrell, *Covert Wars and Breakaway Civilizations*, 264–68. Note 4-dimensional should be 5-dimensional (with gravity).
7. Bearden, T. E., "Background for Pursuing Scalar Electromagnetics," New Physics website, February 1992.
8. "Thomas E. Bearden," Rational Wiki website, accessed July 12, 2024.
9. Verismo, Christi, "Twelve Things You Should Know About Scalar Weapons," Millennium Report, February 21, 2019 (originally published 2012 in *Commander X's Guide to Incredible Conspiracies* [Inner Light - Global Communications, 2012]).

APPENDIX 3. TRINITY

1. Jungk, Robert, *Brighter Than a Thousand Suns: A Personal History of the Atomic Scientists* (Mariner Books, 1970).
2. Dorothy Day, "We Go on Record: The CW Response to Hiroshima," Catholic Worker website, September 1, 1945.

APPENDIX 6. A MEETING

1. Carnicom, Clifford, "A Meeting," Carnicom Institute website, July 26, 2003, updated August 17, 2003.

APPENDIX 7. VISITORS TO WWW.CARNICOM.COM, AUG. 26, 1999

1. Carnicom, Clifford, "Visitors to www.Carnicom.com," Carnicom Institute website, accessed July 12, 2024.

APPENDIX 8. MOVERS AND SHAKERS

1. Reallygraceful, "Who Controls the Gates Family?" YouTube website, May 21, 2020.
2. Engdahl, F. William, "Colossal Financial Pyramid: BlackRock and The WEF 'Great Reset,'" Global Research website (June 20, 2021), December 9, 2023.
3. "Executive Order 13528—Establishing Council of Governors," press release, Obama White House Archives website, January 11, 2010.
4. Hartmann, Thom, "Time for the U.S. To Dump the Word 'Homeland,'" Truthout.org website, September 23, 2014.
5. U.S. Department of Homeland Security, "Agreement Between the Government of the United States of America and the Government of the Kingdom of Sweden on Cooperation in Science and Technology for Homeland Security Matters," DHS.gov website, April 12, 2007.
6. U.S. Department of Homeland Security, "Readout of Secretary Napolitano's Meeting with Swedish Deputy Foreign Minister Frank Belfrage," Office of the Press Secretary, DHS.gov website, December 16, 2011.
7. For an honest appraisal of this travesty, see Skidmore, Mark, "FASAB Standard and the Authority of the Director of National Intelligence to Waive SEC Financial Reporting," Solari Report website, November 2020.
8. Taibbi, Matt, "Has the Government Legalized Secret Defense Spending?" *Rolling Stone* website, January 16, 2019.
9. Fitts, Catherine Austin, "1st Quarter 2019 Wrap Up. Will ESG Turn the Red Button Green?," Solari Report website, December 10, 2019.
10. Assange, Julian, "Google Is Not What It Seems," Wikileaks website, accessed July 12, 2024.
11. Levine, Yasha, "Oakland Emails Give Another Glimpse into the Google-Military-Surveillance Complex," *Pando Daily*, Archive Today website, March 7, 2014.
12. Levine, Yasha, "Google's Earth: How the Tech Giant Is Helping the State Spy on Us," The Guardian website, December 20, 2018.
13. Ahmed, Nafeez, "How the CIA Made Google," Medium website, January 22, 2015.
14. Ahmed, "How the CIA Made Google."

15. One article among many, Liebelson, Dana, and Chris Mooney, "Climate Intelligence Agency: The CIA Is Now Funding Research into Manipulating the Climate," Slate website, July 17, 2013.
16. Ahmed, "How the CIA Made Google."
17. Hall, James, "CIA Funding of Tech Companies," Activist Post website, December 16, 2015.
18. Langone, Michael D., "History of the American Family Foundation," *Cultic Studies Review* 1, no. 1 (2002): 3–50.
19. Lifeboat Foundation, "Launch of NanoShield Fund," press release, Lifeboat Foundation website, January 7, 2008.
20. Van Nedervelde, Philippe, "SecurityPreserver," Lifeboat Foundation website, accessed July 12, 2024.
21. Lifeboat Foundation, "Edward Snowden Named 2013 Lifeboat Foundation Guardian Award winner," press release, Lifeboat Foundation website, December 23, 2013.
22. Gangstalking Surfers, "COINTELPRO & the Truth about Organized Stalking & 21st Century Torture," Tag Archives: Directed Energy Weapons (DEWs), Gangstalking Surfers website, August 26, 2015.
23. Livingston, Vic, "U.S. Silently Tortures Americans with Cell Tower Electromagnetic Neuroweapon," VicLivingston.blogspot website, December 29, 2011.
24. SAIC, "Who We Serve—Homeland Security," SAIC website, accessed July 12, 2014.
25. Cave, Damien, "It's Time for ICANN to Go," Salon website, July 2, 2002. The Internet Corporation for Assigned Names and Numbers of California (ICANN) was formed in 1998 to administrate for the Internet Assigned Numbers Authority (IANA).
26. Wood, Patrick, "Technocracy: The Real Reason Why the UN Wants Control Over the Internet," Technocracy News website, October 27, 2016.
27. U.S. Office of Personnel Management, "Senior Executive Service FAQ—What Does the SES Insignia Symbolize?," OPM website, accessed July 12, 2014.
28. Administration of Barack Obama, "2015 Executive Order 13714—Strengthening the Senior Executive Service," GovInfo website, December 15, 2015.

29. For more on the SES, see American Intelligence Media, "Deep State—Shadow Government Revealed: Senior Executive Service," Aim4truth website, January 3, 2018.
30. Von Reitz, Anna, "Serco, SES, Comments," New Human New Earth Communities website, August 11, 2018.
31. Sardi, Bill, "Who Runs the World? BlackRock and Vanguard," LewRockwell website, April 21, 2021.

Index

About the Author

Elana Freeland is a writer, ghostwriter, teacher, storyteller, and lecturer who since the late 1960s has researched and written prolifically on the Deep State and tandem issues like geoengineering, MK-Ultra mind control, ritual abuse, targeting, and directed energy weapons. Elana came of age in the eye of the sixties maelstrom. She graduated with a bachelor of arts in creative writing and a few hours short of a bachelor of science in biology. Years later, she obtained a master of arts degree from St. John's College in Santa Fe, New Mexico, concentrating on historiography.

Elana's first book in her geoengineering trilogy is *Chemtrails, HAARP, and the Full Spectrum Dominance of Planet Earth* (Feral House, June 2014); the second book is *Under an Ionized Sky: From Chemtrails to Space Fence Lockdown* (Feral House, February 2018); and the third book is the 650-page self-published *Geoengineered Transhumanism: How the Environment Has Been Weaponized by Chemicals, Electromagnetism, & Nanotechnology for Synthetic Biology* (2021). The present book, *The Geoengineered Transhuman: The Hidden Technologies of HAARP, Chemtrails, 5G/6G, Nanotechnology, Synthetic Biology, and the Scientific Effort to Transform Humanity* (Inner Traditions, 2024), goes even deeper down the rabbit hole into the esoterics of Transhumanism.

Her four-book fictional series *Sub Rosa America: A Deep State History* (2012) exposes the hidden history of America since President John F. Kennedy's televised assassination. (An extensive bibliography of what she read in order to write *Sub Rosa America* is available at https://www.elanafreeland.com under "Sub Rosa America.")

Elana posts regularly on three social media sites: on Facebook under EMF Planetary Engineering and gab.com under "Geoengineered Transhumanism" as well as on X (formerly Twitter), with her videos on Odysee and Bitchute.